Recht und Verfassung in Afrika – Law and Constitution in Africa

Band/Volume 35

Sanita van Wyk

The Impact of Climate Change Law on the Principle of State Sovereignty over Natural Resources

Nomos

UNIVERSITEIT
STELLENBOSCH
UNIVERSITY

Konrad
Adenauer
Stiftung

Stellenbosch University (South Africa) grants permission for this publication and the copyright in the dissertation rests with Stellenbosch University. The research conducted in order to complete this research project was conducted in fulfilment of the requirements for the degree of Doctor of Laws in the Faculty of Law at Stellenbosch University.

This publication was produced with the support and in cooperation with the Climate Policy and Energy Security Programme for Sub-Saharan Africa of the Konrad Adenauer Stiftung.
For more information on the Programme see http://www.kas.de/climate-energy-africa/en/

The Deutsche Nationalbibliothek lists this publication in the Deutsche Nationalbibliografie; detailed bibliographic data are available on the Internet at http://dnb.d-nb.de

a.t.: Stellenbosch, South Africa, University, Diss., 2017

ISBN 978-3-8487-4291-2 (Print)
978-3-8452-8554-2 (ePDF)

British Library Cataloguing-in-Publication Data
A catalogue record for this book is available from the British Library.

ISBN 978-3-8487-4291-2 (Print)
978-3-8452-8554-2 (ePDF)

Library of Congress Cataloging-in-Publication Data
van Wyk, Sanita
The Impact of Climate Change Law on the Principle of State Sovereignty over Natural Resources
Sanita van Wyk
373 p.
Includes bibliographic references.

ISBN 978-3-8487-4291-2 (Print)
978-3-8452-8554-2 (ePDF)

1. Edition 2017

Preface

Stellenbosch University (South Africa) grants permission for this publication and the copyright in the dissertation rests with Stellenbosch University. The research conducted in order to complete this research project was conducted in fulfilment of the requirements for the degree of Doctor of Laws in the Faculty of Law at Stellenbosch University.

I hereby express my gratitude to my research project supervisor, Prof Dr Oliver C. Ruppel of the Law Faculty of Stellenbosch University, for giving me the opportunity to conduct my research under his guidance. Without his insight and support the research project could not have been completed. I also wish to extend my gratitude to my teachers at the faculty, namely Prof Dr Lourens M. du Plessis and Prof Dr Annika Rudman, to whom I am indebted for igniting my passion for international law as a young law student.

The adventure of researching and writing this project led me to spend two memorable years in Germany. In this regard, I wish to express my gratitude to the European Commission, and the Erasmus Mundus scholarship team at the Katholieke Universiteit Leuven in Belgium and at Freie Universität Berlin in Germany, which administered and coordinated my scholarship at Freie Universität Berlin. I also wish to convey my appreciation to Prof Dr Sigrid Boysen for her generous guidance during the years I spent at Freie Universität Berlin.

This research project is dedicated to my family, my own invaluable natural resource.

Inhaltsverzeichnis

List of Abbreviations

AAPL	Asian Agricultural Products Ltd.
Aarhus Convention	Convention on Access to Information, Public Participation in Decision-Making and Access to Justice in Environmental Matters 1998
AAU	Assigned Amount Unit
ACHPR	African Commission on Human and Peoples' Rights
ACP	African Caribbean and Pacific Countries
AF	Adaptation Fund
African Convention on the Conservation of Nature and Natural Resources	African Convention on the Conservation of Nature and Natural Resources 1968 (revised in 2003)
Agenda 21	Agenda 21 1992
Agreement for the Promotion and Protection of Investments	Agreement for the Promotion and Protection of Investments 1987
Agreement on Promotion, Protection and Guarantee of Investments among Member States of the Organisation of the Islamic Conference	Agreement on Promotion, Protection and Guarantee of Investments among Member States of the Organisation of the Islamic Conference 1981
Agreement on the Conservation of Nature and Natural Resources	ASEAN Agreement on the Conservation of Nature and Natural Resources 1985
American Convention on Human Rights	American Convention on Human Rights 1969
Antarctic Treaty	Antarctic Treaty 1959
AOSIS	Alliance of Small Island States
arts.	articles
ASEAN	Association of Southeast Asian Nations
Atlantic Charter	Atlantic Charter 1941
Banjul Charter	African Charter on Human and Peoples' Rights 1981
Bergen Conference	Bergen Ministerial Conference on Sustainable Development 1990

Bern Convention	Convention on the Conservation of European Wildlife and Natural Habitats 1979
Biodiversity Convention	United Nations Convention on Biological Diversity 1992
BIT	Bilateral Investment Treaty
Bonn Convention	Convention on the Conservation of Migratory Species of Wild Animals 1979
Brundtland Commission	World Commission on Environment and Development
Canberra Convention	Convention for the Conservation of Antarctic Marine Living Resources 1980
CDM	Clean Development Mechanism
CER	Certified Emission Reduction
CERDS	Charter of Economic Rights and Duties of States 1974
CESCR	United Nations Committee on Economic, Social and Cultural Rights
ch	chapter
Constitution of the World Health Organization	Constitution of the World Health Organization 1946
Convention Concerning Indigenous and Tribal Peoples in Independent Countries	Convention Concerning Indigenous and Tribal Peoples in Independent Countries 1989
Convention Concerning the Protection of the World Cultural and Natural Heritage	Convention Concerning the Protection of the World Cultural and Natural Heritage 1972
Convention for the Protection of the Marine Environment of the North-East Atlantic	Convention for the Protection of the Marine Environment of the North-East Atlantic 1992
Convention on Long-Range Transboundary Air Pollution	Convention on Long-Range Transboundary Air Pollution 1979
Convention on the Continental Shelf	Convention on the Continental Shelf 1958
Convention on the Law of Non-navigational Uses of International Watercourses	Convention on the Law of Non-navigational Uses of International Watercourses 1997
Convention on the Law of the Sea	United Nations Convention on the Law of the Sea 1982

COP	Conference of the Parties
CPF	Collaborative Partnership on Forests
Declaration on the Granting of Independence to Colonial Countries and Peoples	Declaration on the Granting of Independence to Colonial Countries and Peoples 1960
Declaration on the Right to Development	Declaration on the Right to Development 1986
doc	document
Draft OECD Convention on the Protection of Foreign Property	Draft OECD Convention on the Protection of Foreign Property 1967
Dutch Constitution	Constitution of the Kingdom of the Netherlands 2002
Earth Charter	Earth Charter 2000
EC	European Commission
ECHR	European Convention on Human Rights 1950
ECOSOC	United Nations Economic and Social Council
ed	editor or edition
eds	editors
EIT	Economies in Transition
Energy Charter Protocol	Energy Charter Protocol on Energy Efficiency and related Environmental Aspects of the Energy Charter Treaty 1994
Energy Charter Treaty	Energy Charter Treaty 1994
ERU	Emission Reduction Unit
EU	European Union
FAO	Food and Agriculture Organization of the United Nations
FWCC	First World Climate Conference
GCF	Green Climate Fund
GEF	Global Environment Facility
GHG	Greenhouse gas
Hague Convention (II)	Convention (II) with Respect to the Laws and Customs of War on Land 1899 (and its annex of Regulations Concerning the Laws and Customs of War on Land)
Havana Charter	Final Act of the United Nations Conference on Trade and Employment 1948

Helsinki Rules	Helsinki Rules on the Uses of the Waters of International Rivers 1966
IACHR	Inter-American Commission on Human Rights
ICC	International Chamber of Commerce
ICCPR	International Covenant on Civil and Political Rights 1966
ICESCR	International Covenant on Economic, Social and Cultural Rights 1966
ICJ	International Court of Justice
ICSID	International Centre for Settlement of Investment Disputes
ICSID Convention	Convention on the Settlement of Investment Disputes between States and Nationals of Other States 1965
ILA	International Law Association
INC/FCCC	Intergovernmental Negotiating Committee for a Framework Convention on Climate Change
INDC	Intended Nationally Determined Contribution
International Convention for the Regulation of Whaling	International Convention for the Regulation of Whaling 1946
International Covenants on Human Rights	International Covenant on Economic, Social and Cultural Rights 1966 & International Covenant on Civil and Political Rights 1966
International Tropical Timber Agreement	International Tropical Timber Agreement 2006
IPCC	Intergovernmental Panel on Climate Change
JISC	Joint Implementation Supervisory Committee
Johannesburg Declaration on Sustainable Development	Johannesburg Declaration on Sustainable Development 2002
Johannesburg Principles on the Role of Law and Sustainable Development	Johannesburg Principles on the Role of Law and Sustainable Development 2002
Kyoto Protocol	Kyoto Protocol 1997
LDC	Least Developed Countries
LDCF	Least Developed Countries Fund
Lomé Convention	Lomé IV Convention 1995
LULUCF	Land use, land-use change and forestry

Madrid Protocol	Protocol on Environmental Protection to the Antarctic Treaty 1991
Montevideo Convention	Montevideo Convention on Rights and Duties of States 1933
Montreal Protocol	Montreal Protocol on Substances that Deplete the Ozone Layer 1985
MPRDA	Mineral and Petroleum Resources Development Act 2002
NAFTA	North American Free Trade Agreement 1991
NAMA	Nationally Appropriate Mitigation Action
NDC	Nationally Determined Contribution
NIEO	New International Economic Order
NIEO Declaration	Declaration on the Establishment of a New International Economic Order 1974
Noordwijk Conference	Noordwijk Ministerial Conference on Atmospheric Pollution and Climatic Change 1989
Noordwijk Declaration	Noordwijk Declaration Atmospheric Pollution and Climatic Change 1989
OECD	Organisation for Economic Co-operation and Development
Oslo Principles	Oslo Principles on Global Climate Change 2015
par	paragraph
Paris Agreement	Paris Agreement 2015
pg	page
Plan of Implementation of the World Summit on Sustainable Development	Plan of Implementation of the World Summit on Sustainable Development 2002
PSNR	principle of sovereignty over natural resources
QELRO	Quantified Emission Limitation and Reduction Objective
REDD	Reducing Emissions from Deforestation and Forest Degradation
Rio Declaration	Declaration of the United Nations Conference on Environment and Development 1992
RMU	Removal Unit
SCCF	Special Climate Change Fund
SDM	Sustainable Development Mechanism

Seoul Declaration	Declaration on the Progressive Development of Principles of Public International Law relating to a New International Economic Order 1986
SERAC	Social and Economic Rights Action Center
Stockholm Declaration	Declaration of the United Nations Conference on the Human Environment 1972
SWCC	Second World Climate Conference
TNC	Transnational Corporation
Tokyo Convention	Tokyo Convention for the High Seas Fisheries of the North Pacific Ocean 1952
Treaty for Amazonian Cooperation	Treaty for Amazonian Cooperation 1978
TWCC	Third World Climate Conference
UK	United Kingdom
UN	United Nations
UNCCD	United Nations Convention to Combat Desertification 1994
UNCITRAL	United Nations Commission on International Trade Law
UNCTAD	United Nations Conference on Trade and Development
UNDRIP	United Nations Declaration on the Rights of Indigenous Peoples 2007
UNECE	United Nations Economic Commission for Europe
UNEP	United Nations Environment Programme
UNESCO	United Nations Educational, Scientific and Cultural Organization
UNFCCC	United Nations Framework Convention on Climate Change 1992
UNFF	United Nations Forum on Forests
UNGA	United Nations General Assembly
Unified Agreement for the Investment of Arab Capital in the Arab States	Unified Agreement for the Investment of Arab Capital in the Arab States 1980
UNITAR	United Nations Institute for Training and Research
United Nations Charter	Charter of the United Nations 1945
UNWC	United Nations Water Conference

USA	United States of America
Vienna Convention for the Protection of the Ozone Layer	Vienna Convention for the Protection of the Ozone Layer 1985
Vienna Convention on Succession of States in Respect of State Property, Archives and Debts	Vienna Convention on Succession of States in Respect of State Property, Archives and Debts 1983
Vienna Convention on Succession of States in Respect of Treaties	Vienna Convention on Succession of States in Respect of Treaties 1978
Vienna Conventions on Succession of States	Vienna Convention on Succession of States in Respect of Treaties 1978 & Vienna Convention on Succession of States in respect of State Property, Archives and Debts 1983
vol	volume
WHO	World Health Organization
WMO	World Meteorological Organization
World Charter for Nature	World Charter for Nature 1982
WTO	World Trade Organization

CHAPTER 1: Introduction

1 1 *Background to research question*

This research project explores the impact of climate change on the international law principle of sovereignty over natural resources (PSNR). On the one hand, climate change is arguably the defining environmental, political and legal crisis of the 21st century.[1] On the other hand, the concept of PSNR already came into existence and to the forefront in the 1950s as it developed from the efforts of developing states during this time to ensure economic freedom and development which was not dominated by the interests of developed states.[2] PSNR grew from two main United Nations concerns: state sovereignty (understood to include the right to development) and human rights (in particular the right to self-determination).[3] Legally, the principle originated from United Nations General Assembly (UNGA) resolutions in 1952.[4] The UNGA resolutions, which are specifi-

---

1 It must be noted that the reference to the term 'climate change' throughout the research project can include the environmental, political and the legal component of climate change. The research project is entitled "the impact of climate change law on the principle of state sovereignty over natural resources", as it highlights the legal component of climate change.

2 L Barrera-Hernández "Sovereignty over Natural Resources under Examination: The Inter-American System for Human Rights and Natural Resource Allocation" (2006) 12 *Annual Survey of International & Comparative Law* 43 45; N Schrijver *Sovereignty Over Natural Resources, Balancing Rights and Duties* (1997) 20.

3 NJ Schrijver "Permanent Sovereignty over natural resources versus the common heritage of mankind: complementary or contradictory principles of international economic law?" in P De Waart et al (eds) *International Law and Development* (1988) 87 87; L Barrera-Hernández "Sovereignty over Natural Resources under Examination: The Inter-American System for Human Rights and Natural Resource Allocation" (2006) 12 *Annual Survey of International & Comparative Law* 43 45; P Sands *Principles of International Environmental Law* 3ed (2003) 233-235.

4 For early development of the principle, see the following UNGA resolutions: 523 (VI) of 12 January 1952, 626 (VII) of 21 December 1952, 1314 (XIII) of 12 December 1958, 837 (IX) of 14 December 1954, 1515 (XV) of 15 December 1960, 1803 (XVII) of 14 December 1962, 2158 (XXI) of 25 November 1966, 2386 (XXIII) of 19 December 1968, 2692 (XXV) of 11 December 1970, 3016 (XXVII) of 18 December 1972, 3037 (XXVII) of 19 December 1972, 88 (XII) of 19 October 1973,

cally considered to mark the genesis of PSNR, are resolutions 523 (VI) of 1952 and 626 (VII) of 1952.[5] Since 1952, a multitude of UNGA resolutions have addressed PSNR and have been instrumental in developing an international legal understanding of the principle.[6] In 1954, the Commission on Human Rights proposed that a commission be established with the task of conducting a study into the right of states and people to permanent sovereignty over natural resources.[7] Accordingly, in 1958, a Commission on Permanent Sovereignty over Natural Resources was established, whose work led to the adoption of UNGA resolution 1803 (XVII) of 14 December 1962. This resolution is still today considered as the landmark resolution on PSNR.[8]

PSNR, like climate change, encompasses environmental, political and legal considerations. An examination of the historical development of PSNR by means of a consideration of the various UNGA resolutions that shaped the principle illustrates that it is a principle capable of evolving in accordance with major political and legal occurrences. This makes PSNR a pertinent principle to consider in the context of climate change as an occurrence with environmental, political and legal implications. As submitted by Schrijver, in predicting the future application of PSNR:

> "It would be incorrect to assert that, as a result of the trend towards globalization, fragmentation and privatization, the principle of permanent sovereignty is dead or no longer serves any function in international law and in interna-

---

3082 (XXVIII) of 6 December 1973, 3171 (XXVIII) of 17 December 1973, 3021 (S-VI) of 1 May 1974, 3201 (S-VI) of 1 May 1974 and 3281 (XXIX) of 12 December 1974.

5 UNGA resolution 523 (VI) of 12 January 1952 and UNGA resolution 626 (VII) of 21 December 1952.

6 These resolutions include the following UNGA resolutions: 523 (VI) of 12 January 1952, 626 (VII) of 21 December 1952, 1314 (XIII) of 12 December 1958, 837 (IX) of 14 December 1954, 1515 (XV) of 15 December 1960, 1803 (XVII) of 14 December 1962, 2158 (XXI) of 25 November 1966, 2386 (XXIII) of 19 December 1968, 2692 (XXV) of 11 December 1970, 3016 (XXVII) of 18 December 1972, 3037 (XXVII) of 19 December 1972, 88 (XII) of 19 October 1973, 3082 (XXVIII) of 6 December 1973, 3171 (XXVIII) of 17 December 1973, 3021 (S-VI) of 1 May 1974, 3201 (S-VI) of 1 May 1974 and 3281 (XXIX) of 12 December 1974.

7 N Schrijver Sovereignty Over Natural Resources, Balancing Rights and Duties (1997) 57.

8 G Abi-Saab "Permanent Sovereignty over Natural Resources and Economic Activities" in M Bedjaoui (ed) International Law: Achievements and Prospects (1991) 597 601.

> tional relations. [...] The challenge in the next two or three decades will be how to balance permanent sovereignty over natural resources with other basic principles and emerging norms of international law – including the duty to observe international agreements, grant fair treatment of foreign investors, pursue sustainable development at national and international levels and to respect human [...] rights – and in this way to serve the best interests of present and future generations."[9]

Climate change is considered an emerging area of international environmental law, as described above by Schrijver, which ought to be balanced with PSNR. Accordingly, it is possible and beneficial to consider whether the gravity of climate change warrants the reinterpretation and reconfiguration of PSNR, in order to contribute to the international legal climate change regime. Such a reinterpretation and reconfiguration of PSNR will pertain to the rights and duties of states and people, and will aim, in particular, to promote a principle focused on people rather than on states. The phrase 'reinterpretation and reconfiguration' should be clarified: firstly, by 'reinterpreting' the principle, the principle is interpreted or rather construed in a different light, meaning in this case that the principle is construed in light of the grave effects of climate change and in the context of climate change; and, secondly, by 'reconfiguring' the principle, constitutive elements of the principle are organised in a different form, meaning in this case that people's rights as a constitutive element of PSNR are organised in such a way as to promote or focus on people's rights, rather than state rights.

PSNR emerged from two United Nations concerns, namely economic development of underdeveloped countries and human rights.[10] Accordingly, PSNR rests on pillars of state sovereignty and self-determination of people.[11] However, it is still commonly accepted that PSNR mainly serves to safeguard state sovereignty in the use, exploitation and disposal of natural resources. By reinterpreting and reconfiguring PSNR to focus on peo-

---

9 N Schrijver Sovereignty Over Natural Resources, Balancing Rights and Duties (1997) 379-380.

10 NJ Schrijver "Permanent Sovereignty over natural resources versus the common heritage of mankind: complementary or contradictory principles of international economic law?" in P De Waart et al (eds) International Law and Development (1988) 87 87.

11 L Barrera-Hernández "Sovereignty over Natural Resources under Examination: The Inter-American System for Human Rights and Natural Resource Allocation" 2006 (12) *Annual Survey of International & Comparative Law* 43 45; P Sands *Principles of International Environmental Law* 3ed (2003) 233-235.

ple's rights in the context of natural resources, a noteworthy movement back to one of the original pillars of PSNR, namely economic self-determination of people, is observed. An interpretation of PSNR that has a stronger focus on the rights of people than on the rights and duties of states, can be a valuable legal tool in the global fight against climate change. One of the greatest obstacles to the international legal system, in particular the international legal climate change regime, is state compliance to climate commitments. If people were to exercise their rights in terms of PSNR, it could present a way in which to hold states accountable for their climate commitments and improve state compliance within the international legal climate change regime. This reinterpretation and reconfiguration of existing constituents of PSNR could represent a new frontier in the fight against climate change and the promotion of state compliance with the international legal climate change regime.

1 2 Research question

The primary research question of this research project is the following:

- Does climate change[12] have an impact on the principle of sovereignty over natural resources?

In order to investigate the primary research question efficiently, the scope of investigation is delineated by the secondary research questions, which are the following:

- What does the principle of sovereignty over natural resources entail?
- What does the international legal climate change regime entail?
- Does the gravity of climate change warrant a reinterpretation and reconfiguration of the principle of sovereignty over natural resources that could increase the effectiveness of the international legal climate change regime?
- What is the outcome of the reinterpretation and reconfiguration of the principle of sovereignty over natural resources as a principle that is

12 Note that the reference to 'climate change' throughout the research project can include the environmental, political and the legal component of climate change. The research project is entitled 'the impact of climate change law on the principle of state sovereignty over natural resources', as it highlights the legal component of climate change.

able to increase the effectiveness of the international legal climate change regime?

1 3 Hypotheses

The starting point for the abovementioned research questions is the hypotheses mentioned below. It is hypothesised that PSNR is a principle that is capable of evolving and capable of being affected by occurrences with political and legal implications, such as climate change. In this regard, the answer to the primary research question would most likely be that climate change does not necessarily automatically have an impact on PSNR, but the scale and gravity of the effects of climate change and the magnitude of the legal response required to address climate change, certainly warrant an impact on PSNR. It is hypothesised that this impact on PSNR warrants and necessitates the reinterpretation and reconfiguration of PSNR whereby state rights are constricted, and state duties and people's rights gain momentum. Furthermore, people's rights in terms of PSNR could form a basis upon which people can institute climate litigation against states, and thereby increases state compliance to climate commitments in terms of the international legal climate change regime.

1 4 Research objectives and assumptions

This research project sets out to determine whether the gravity of climate change warrants an impact on PSNR in view of its global environmental, legal and political significance. The research project further aims to reinterpret and propose further reinterpretation of PSNR as a principle focused on people, rather than a principle focused on states, that is capable of being employed as a valuable tool in increasing the effectiveness of the international legal climate change regime. It is an objective of the research project to show that people's rights in terms of PSNR, when enforced, can present a valuable addition to ensuring state compliance with climate commitments made by states in terms of the international legal climate change regime. In this sense the implementation of people's rights in terms of PSNR is considered as a negative incentive for state compliance with climate commitments. This consideration requires the following initial steps or assumptions: firstly, that the state which wishes to ensure compliance has actually made a climate commitment in terms of the Kyoto Protocol 1997 (Kyoto Protocol) or the Paris Agreement 2015 (Paris Agreement); secondly, that the state implements a national legal system whereby people are able to hold the state accountable; and thirdly, it is possible that only the nationals of a specific state might be able to hold their state accountable. These assumptions are discussed in greater detail in Section 4.4.1.

The overarching objective of the research project is to make a positive contribution to the global efforts to address the problem of climate change through the international law principle of PSNR.

1 5 Research design and methodology

This research project predominantly applies doctrinal research methods.[13] The international legal sources considered in order to investigate the primary and secondary research questions include: international and regional instruments such as treaties and conventions; United Nations publications and documentation, such as UNGA resolutions; decisions of international and regional courts and tribunals; literature produced by governmental and non-governmental or other independent agencies; and academic literature in the form of books, chapters in books, and journal articles and reports.

The inherent interdisciplinary nature of the research regarding international environmental law means that the abovementioned sources of research emanate not only from the field of law, but also from the fields of politics, economics and science.

The research conducted for this research project was partly reform-oriented, since the purpose was to examine the current understanding of PSNR and (owing to the gravity of climate change) the adequacy or suitability of the principle as a tool to combat climate change. Furthermore, this research project makes recommendations concerning how PSNR may be reconfigured and implemented to ensure that it is a suitable tool by which to ensure states' compliance with climate commitments made in

13 The doctrinal research methodology is understood as explained by Hutchinson. Hutchinson compares the doctrinal research method to a social science literature review, but submits that it extends further than a literature review as literature is not only located, but also analysed. [T Hutchinson Researching and Writing in Law 3ed (2010) 37-38.].

terms of the international legal climate change regime. In this context, principles that guide the discussion of the reform of PNSR include the principles of equity, common but differentiated responsibility and sustainable development as well as the precautionary principle and the polluter-pays principle, which are embedded within the international legal climate change regime as per section 3.7.2.

1 6 Chapter overview

Besides the introductory chapter, the research project contains four additional chapters.

Chapter 2 sets out to determine what PSNR is and whether the nature of PSNR allows it to be impacted by the gravity of climate change. The chapter achieves this objective by examining the origin of the principle, the legal nature of the principle, and the terminology that the principle employs. The examination of the origin of the principle illustrates the evolutionary character of the principle, meaning that it is possible for the effects of climate change to have an impact on the principle, specifically by affecting how the principle can be reinterpreted (such as, in this case, interpreting the principle as a principle focused on people instead of a principle focused on states). By considering the terminology employed for PSNR, the object and subjects of PSNR are also determined. The object of the principle is determined to be natural resources, and it is illustrated in this context how the nexus between PSNR and climate change is strengthened by also considering the object of the international legal climate change regime. Chapter 2 investigates states and people as the subjects of PSNR. The examination of states as the subject of the principle highlights state sovereignty as a root of the principle and a contributing factor to the evolutionary character of the principle. In the investigation of the subjects of the principle, it is found that the focus has shifted throughout history between states and people as the main subject of PSNR, which contributes to establishing PSNR as a traditionally state-focused principle and once again as a principle which is able to evolve. Chapter 2 sets the scene for Chapter 3 of the research project that will examine whether the gravity of climate change has an impact on PSNR or if the gravity of climate change warrants an impact on PSNR. Chapter 2 also sets the scene to reconfigure PSNR as a principle focused on people that will be capable of improving the effectiveness of the international legal climate change regime, in Chapter 4.

Chapter 3 establishes the nexus between climate change and the principle of sovereignty over natural resources, which forms the basis on which the impact of the gravity of climate change on the principle is considered. The chapter sets out the scientific basis of climate change and outlines the environmental, social and economic consequences of climate change, and the resulting effects on the well-being of people, all in an effort to illustrate the scale and gravity of the problem of climate change. As was submitted above, the term 'climate change' within the phrase 'the impact of the gravity of climate change on PSNR', should be read to incorporate all aspects (environmental, political and legal) related to climate change, including the international legal climate change regime. This chapter examines the international legal climate change regime, by defining the term, outlining the development of the regime, and examining the most important constituents of the regime, namely the United Nations Framework Convention on Climate Change 1992 (UNFCCC), the Kyoto Protocol and the latest addition, the Paris Agreement. During the process of examining the international legal climate change regime, principles on which the regime is founded, namely equity, the principle of common but differentiated responsibility, the precautionary principle, the polluter-pays principle, and the principle of sustainable development, are also discussed. Throughout the chapter, the embedding of PSNR within the international legal climate change regime is also considered, adding to the establishment of the nexus between PSNR and climate change. This chapter therefore addresses the primary research objective of the research project, which is to determine whether the gravity of climate change has an impact or warrants an impact on the principle of sovereignty over natural resources. As previously submitted, it is accepted that this impact might not be an automatic impact. It is suggested and illustrated in Chapter 3 that the scale and gravity of the effects of climate change are so significant, that a reinterpretation and reconfiguration of any principle of international environmental law that is able to contribute to making the international legal climate change regime more effective, is warranted. This first section of Chapter 3 therefore determines that the gravity of climate change does necessitate an impact on PSNR or, more accurately, does warrant an impact on PSNR. The secondary research objective of the research project is to enquire how the gravity of climate change impacts on PSNR. This question can be phrased more clearly by asking: How should PSNR be reinterpreted in order to enable the principle to contribute to increasing the effectiveness of the international legal climate change regime? Chapter 3 therefore also considers the international legal climate change regime and one specific source of ineffectiveness, namely the Kyoto Protocol. In addition to explaining the functionality of the Kyoto Protocol, the contribution of the Kyoto Protocol to the ineffectiveness of the regime is also considered, in particular in regard to the component of state compliance with climate commitments. This discussion shows that the international legal regime to combat climate change calls for a solution to the ineffectiveness of the present regime, and Chapter 4 shows that it is possible that a revision of PSNR can offer this solution. In order to increase state compliance with the international legal climate change regime, a viable solution would be to promote and implement the rights of people in terms of PSNR, which can in turn contribute to increasing state compliance with climate commitments. The solution is introduced in Chapter 3 and examined in Chapter 4.

Chapter 4 contains an examination of the consequences of the effects of climate change on the principle of sovereignty over natural resources and explores the expansion of the means to combat climate change. The chapter commences with an examination of the most relevant rights and duties of states in terms of PSNR, which serves to complete the examination of the principle commenced in Chapter 2, but also illustrates why the effects of climate change warrant the reinterpretation of the principle and a shift in focus from states to people as the foremost subject of the principle. The consequences of the impact of climate change, which form part of the general global movement in support of environmental conservation and sustainable development, affect the interpretation and application of the rights and duties of states, so as to bring about a restriction of the rights of states and an expansion of the duties of states. These consequences of the effects of climate change law on PSNR can be described as impacts of climate change law on PSNR. Other consequences of the effects of climate change law on PSNR can serve to increase the effectiveness of the international legal climate change regime by increasing state compliance. In this regard, Chapter 4 repositions PSNR as a principle focused on people, which means that it is reinterpreted to give effect to people's rights in terms of PSNR. This repositioning of PSNR is able to constitute a valuable addition to the international legal climate change regime and possibly increase its effectiveness. This is because, as is shown in Chapter 4, people have the right to dispose freely of natural resources in terms of PSNR, which if implemented can constitute a basis upon which people are able to hold states accountable for commitments made in terms of the Kyoto Protocol and the Paris Agreement, which can present a negative incentive for states to adhere to their climate commitments.

Chapter 5 concludes the research project. The final chapter reflects on the research objectives and hypotheses of the research project. Furthermore, the chapter summarises the chapters of the research project chapters and their pertinent conclusions in regard to the primary and secondary research questions and objectives. Chapter 5 also includes an outline of the research outcomes, and the primary and secondary research questions and objectives of the research project are holistically addressed, independent of the chapter structure. Chapter 5 ends by framing the research outcomes of the research project as recommendations.

CHAPTER 2: The principle of sovereignty over natural resources

2 1 Introduction

In order to answer the research question, which asks whether the gravity of climate change warrants an impact on PSNR, it is necessary to understand what the principle entails and whether the nature of the principle allows it to evolve in accordance with this impact. This chapter will examine what PSNR is by considering its origin and legal nature. Furthermore, this chapter will examine the terminology most frequently employed by PSNR, so as to set out the research project's understanding of the principle and to demarcate certain boundaries for this study. The chapter determines the main object of PSNR as natural resources, and the subjects of PSNR as the state and people of the state. In addition, this chapter will illustrate that the principle has been more focused on states rather than people, and will set the scene to reinterpret and reconfigure PSNR as a principle focused on people in succeeding chapters.

This chapter illustrates that PSNR is a principle capable of evolving according to large-scale occurrences with political and legal implications, such as climate change, which means that it is capable of being reinterpreted and reconfigured. Finally, the chapter introduces the idea of reinterpreting and reconfiguring PSNR as a means of increasing state compliance with the international legal climate change regime.

2 2 Foundation of the principle of sovereignty over natural resources

PSNR is considered to be a fundamental principle of international law and an international legal right.[14] PSNR can be described as a source of rules by which a state manages its natural resources or the rules by which a state

14 SR Chowdhury "Permanent Sovereignty Over Natural Resources: Substratum of the Seoul Declaration" in P De Waart et al (eds) *International Law and Development* (1988) 59 80; K Hossain "Introduction" in K Hossain & SR Chowdhury (eds) *Permanent Sovereignty Over Natural Resources in International Law, Prin-*

exercises control over its natural resources and receives benefit from them.[15] The dependence of states on natural resources was officially acknowledged in international law documents from as early as 1941.[16] Thus, although the principle has only formally existed since the early 1950s, the issues to which PSNR pertained, such as state sovereignty, natural resources and economic activities, were already familiar issues in international law.[17]

The concept of PSNR came into existence in the 1950s in reaction to unequal arrangements between developing and developed states, regarding natural resource management. In other words, PSNR developed from the efforts of developing states to ensure economic freedom and development not dominated by the interests of developed states, which were deemed economically and politically more powerful.[18] Schrijver accurately and concisely described PSNR as growing from two main United Nations concerns: firstly, economic development of underdeveloped countries and, secondly, human rights.[19] It is still possible today to describe PSNR as be-

---

ciple and Practice (1984) ix ix; UNGA resolution 1803 (XVII) of 14 December 1962; P Sands *Principles of International Environmental Law* 2ed (2003) 237.

15 SR Chowdhury "Permanent Sovereignty Over Natural Resources: Substratum of the Seoul Declaration" in P De Waart et al (eds) *International Law and Development* (1988) 59 80; K Hossain "Introduction" in K Hossain & SR Chowdhury (eds) *Permanent Sovereignty over Natural Resources in International Law, Principle and Practice* (1984) ix ix.

16 The Atlantic Charter reads as follows: "The President of the United States of America and the Prime Minister, Mr. Churchill, representing His Majesty's Government in the United Kingdom, being met together, deem it right to make known certain common principles in the national policies of their respective countries on which they base their hopes for a better future for the world. [...] [T]hey will endeavor, with due respect for their existing obligations, to further the enjoyment by all States, great or small, victor or vanquished, of access, on equal terms, to the trade and to the raw materials of the world which are needed for their economic prosperity".

17 G Abi-Saab "Permanent Sovereignty over Natural Resources and Economic Activities" in M Bedjaoui (ed) *International Law: Achievements and Prospects* (1991) 597 597.

18 L Barrera-Hernández "Sovereignty over Natural Resources under Examination: The Inter-American System for Human Rights and Natural Resource Allocation" (2006) 12 *Annual Survey of International & Comparative Law* 43 45; N Schrijver *Sovereignty Over Natural Resources, Balancing Rights and Duties* (1997) 20.

19 NJ Schrijver "Permanent Sovereignty over natural resources versus the common heritage of mankind: complementary or contradictory principles of international

ing concerned with state sovereignty (understood to include the right to development) and human rights (in particular the right to self-determination).[20]

Legally, the principle originated from UNGA resolutions in 1952.[21] The UNGA resolutions, which are specifically considered to mark the genesis of PSNR, are resolutions 523 (VI) of 1952 and 626 (VII) of 1952.[22] Resolution 523 (VI) of 1952 is particularly important for a number of reasons. Firstly, underdeveloped countries were distinctly highlighted as the right-bearers of PSNR.[23] Secondly, underdeveloped countries were also specifically obliged to utilise their resources, not only according to their own national interests, but also in the interests (and for the expansion) of the world economy.[24] Since 1952, many UNGA resolutions have

economic law?" in P De Waart et al (eds) *International Law and Development* (1988) 87 87.

20 L Barrera-Hernández "Sovereignty over Natural Resources under Examination: The Inter-American System for Human Rights and Natural Resource Allocation" (2006) 12 *Annual Survey of International & Comparative Law* 43 45; P Sands *Principles of International Environmental Law* 3ed (2003) 233-235.

21 For early development of the principle, see the following UNGA resolutions: 523 (VI) of 12 January 1952, 626 (VII) of 21 December 1952, 1314 (XIII) of 12 December 1958, 837 (IX) of 14 December 1954, 1515 (XV) of 15 December 1960, 1803 (XVII) of 14 December 1962, 2158 (XXI) of 25 November 1966, 2386 (XXIII) of 19 December 1968, 2692 (XXV) of 11 December 1970, 3016 (XXVII) of 18 December 1972, 3037 (XXVII) of 19 December 1972, 88 (XII) of 19 October 1973, 3082 (XXVIII) of 6 December 1973, 3171 (XXVIII) of 17 December 1973, 3021 (S-VI) of 1 May 1974, 3201 (S-VI) of 1 May 1974 and 3281 (XXIX) of 12 December 1974.

22 UNGA resolution 523 (VI) of 12 January 1952 and UNGA resolution 626 (VII) of 21 December 1952.

23 N Schrijver *Sovereignty Over Natural Resources, Balancing Rights and Duties* (1997) 41.

24 N Schrijver *Sovereignty Over Natural Resources, Balancing Rights and Duties* (1997) 41.
UNGA resolution 523 (VI) of 12 January 1952 reads as follows: "Considering that the under-developed countries have the right to determine freely the use of their natural resources and that they must utilize such resources in order to be in a better position to further the realization of their plans of economic development in accordance with their national interests, and to further the expansion of the world economy".

mentioned PSNR and have been instrumental in developing the principle.[25] In 1954, the Commission on Human Rights proposed that a commission be established with the task of conducting a study into the right of states and people to permanent sovereignty over natural resources.[26] Accordingly, in 1958, a Commission on Permanent Sovereignty over Natural Resources was established, whose work led to the adoption of UNGA resolution 1803 (XVII) of 14 December 1962. This resolution is still today considered as the landmark resolution on PSNR.[27]

It follows that PSNR historically had two main functions. Firstly, PSNR served to ensure that people living under colonial rule could receive the benefit arising from the exploitation of natural resources by their colonial ruler in their country.[28] Accordingly, PSNR also played an important role later during the decolonisation period.[29] Secondly, PSNR served to protect newly independent or developing states, after decolonisation, from the influence or exploitation of foreign states and foreign companies wishing to do business in that state which involved natural resources.[30] Later on, PSNR played a role in the establishment of the New International Economic Order (NIEO).[31] Over the years, in practice, the principle has been applied in situations where transnational companies exploit resources of

---

25 These resolutions include the following UNGA resolutions: 523 (VI) of 12 January 1952, 626 (VII) of 21 December 1952, 1314 (XIII) of 12 December 1958, 837 (IX) of 14 December 1954, 1515 (XV) of 15 December 1960, 1803 (XVII) of 14 December 1962, 2158 (XXI) of 25 November 1966, 2386 (XXIII) of 19 December 1968, 2692 (XXV) of 11 December 1970, 3016 (XXVII) of 18 December 1972, 3037 (XXVII) of 19 December 1972, 88 (XII) of 19 October 1973, 3082 (XXVIII) of 6 December 1973, 3171 (XXVIII) of 17 December 1973, 3021 (S-VI) of 1 May 1974, 3201 (S-VI) of 1 May 1974 and 3281 (XXIX) of 12 December 1974.

26 N Schrijver Sovereignty Over Natural Resources, Balancing Rights and Duties (1997) 57.

27 G Abi-Saab "Permanent Sovereignty over Natural Resources and Economic Activities" in M Bedjaoui (ed) International Law: Achievements and Prospects (1991) 597 601.

28 N Schrijver *Sovereignty Over Natural Resources, Balancing Rights and Duties* (1997) 24.

29 As seen, for instance, in the UNGA resolution 3201 (S-VI) of 1 May 1974, entitled 'Declaration on the Establishment of a New International Economic Order'.

30 N Schrijver *Sovereignty Over Natural Resources, Balancing Rights and Duties* (1997) 24.

31 UNGA resolution 3201 (S-VI) of 1 May 1974.

states beyond the borders of their own state,[32] which can also be referred to as the natural resource trade of transnational companies. In this context, prominent relevant issues in terms of PSNR include nationalisation, compensation and dispute settlement.[33] PSNR is also relevant in other contexts, both national and international, and including areas pertaining to the environment, foreign investment, state succession, war or human security, human rights, state sovereignty and sustainable development.[34] It is accepted that PSNR gives rise to both rights (or claims) and duties (or obligations) for states, as explained in a succeeding chapter of the research project.[35] Throughout history the focus of the principle has shifted between the rights and duties of states. The principle has been more successful in ensuring state rights, than it has been in ensuring that states fulfil their duties in terms of the principle.

This chapter will lay the groundwork for showing that it is possible to reinterpret and reconfigure PSNR so that it becomes more focused on people and serves the well-being of people, by presenting a means by which people can hold the state accountable for commitments made in terms of the international legal climate change regime. Such a reinterpretation and reconfiguration is necessary because of the increase in global natural resource exploitation and inequitable distribution of resources by states within their own borders, but – most pertinent in the context of the research project – because of the increasing risks posed to people by climate change.[36] Although economic self-determination of people is an original pillar of PSNR, it is considered a modern approach to interpret PSNR as a

---

32 K Hossain "Introduction" in K Hossain & SR Chowdhury (eds) *Permanent Sovereignty over Natural Resources in International Law, Principle and Practice* (1984) ix ix.

33 SR Chowdhury "Permanent sovereignty over natural resources" in K Hossain & SR Chowdhury (eds) *Permanent Sovereignty over Natural Resources in International Law, Principle and Practice* (1984) 1 9.

34 Partially listed by Schrijver in N Schrijver *Sovereignty Over Natural Resources, Balancing Rights and Duties* (1997) 4-7.

35 N Schrijver *Sovereignty Over Natural Resources, Balancing Rights and Duties* (1997) 306-364.

36 D Çambou & S Smis "Permanent sovereignty over natural resources from a human rights perspective: Natural resources exploitation and indigenous peoples' rights in the arctic" (2013) 22(1) Michigan State International Law Review 347 360.

principle concerned with the interests and well-being of people within a state.[37]

2 3 Legal nature of the principle of sovereignty over natural resources

This paragraph will illuminate the most important aspects of the legal nature of PSNR. The legal nature of the principle was not always clear, and the principle used to be considered as a political statement rather than a legal principle.[38] PSNR is today considered to be a fundamental principle of international law and an international legal right.[39] As discussed in Section 2.2, the principle was brought into formal existence during the 1950s by the UNGA.[40] The UNGA resolutions represent the general "world

---

37 SR Chowdhury "Permanent Sovereignty Over Natural Resources: Substratum of the Seoul Declaration" in P De Waart et al (eds) *International Law and Development* (1988) 59 62; K Hossain "Introduction" in K Hossain & SR Chowdhury (eds) *Permanent Sovereignty over Natural Resources in International Law, Principle and Practice* (1984) ix xiii.

38 KN Gess "Permanent sovereignty over natural resources: An analytical review of the United Nations declaration and its genesis" (1964) 13 *International and Comparative Law Quarterly* 398 414.

39 SR Chowdhury "Permanent Sovereignty Over Natural Resources: Substratum of the Seoul Declaration" in P De Waart et al (eds) *International Law and Development* (1988) 59 80; K Hossain "Introduction" in K Hossain & SR Chowdhury (eds) *Permanent Sovereignty Over Natural Resources in International Law, Principle and Practice* (1984) ix ix; UNGA resolution 1803 (XVII) of 14 December 1962; P Sands *Principles of International Environmental Law* 2ed (2003) 237.

40 Article 7 of the UN Charter establishes the General Assembly as an organ or body of the United Nations, and Article 9 and 10 of the UN Charter sets out the composition, functions and powers of the General Assembly.
Article 7(1)-(2) of the UN Charter reads as follows: (1) "There are established as the principal organs of the United Nations: a General Assembly, a Security Council, an Economic and Social Council, a Trusteeship Council, an International Court of Justice, and a Secretariat"; (2) "Such subsidiary organs as may be found necessary may be established in accordance with the present Charter".
Article 9(1)-(2) of the UN Charter reads as follows: (1) "The General Assembly shall consist of all the Members of the United Nations"; (2) "Each Member shall have not more than five representatives in the General Assembly".
Article 10 of the UN Charter reads as follows: "The General Assembly may discuss any questions or any matters within the scope of the present Charter or relating to the powers and functions of any organs provided for in the present Charter, and, except as provided in Article 12, may make recommendations to the Mem-

opinion" on a specific matter and accordingly are useful to ascertain the nature of PSNR, bearing in mind that resolutions only have recommendatory power.[41] This means that states are not bound to these resolutions, but the resolutions can be used to justify the conduct of a state.[42] The resolutions serve an important role in advancing the adoption of certain soft guidelines, standards or recommended practices for states in the international arena.[43] Therefore, these resolutions play a vital role in the formation of customary international law.[44]

In discussing the legal nature of PSNR, it should be noted that this principle is one of the "UN-principles".[45] Ideally, UN-principles such as PSNR represent an accumulation of ideas, reinforced by public opinion and a fair number of governments.[46] UN-principles can be recognised on the basis that they are principles affecting the United Nations as well as its

---

bers of the United Nations or to the Security Council or to both on any such questions or matters.".

41 P Birnie & A Boyle *International Law & the Environment* 2ed (2002) 23; NJ Schrijver "Permanent sovereignty over natural resources versus the common heritage of mankind: complementary or contradictory principles of international economic law?" in P de Waart et al (eds) *International law and development* (1988) 87 91.
Also see Article 10 of the UN Charter, which reads as follows: "The General Assembly may discuss any questions or any matters within the scope of the present Charter or relating to the powers and functions of any organs provided for in the present Charter, and, except as provided in Article 12, may make recommendations to the Members of the United Nations or to the Security Council or to both on any such questions or matters.".

42 P Birnie "International Environmental law: Its Adequacy for Present and Future Needs" in A Hurrell & B Kingsbury (eds) *The International Politics of the Environment* (1992) 51 59.

43 P Birnie "International Environmental law: Its Adequacy for Present and Future Needs" in A Hurrell & B Kingsbury (eds) *The International Politics of the Environment* (1992) 51 59.

44 NJ Schrijver "Permanent sovereignty over natural resources versus the common heritage of mankind: complementary or contradictory principles of international economic law?" in P de Waart et al (eds) *International Law and Development* (1988) 87 91.

45 W Lang "UN Principles and International Environmental Law" (1999) 3 *Max Planck Yearbook of United Nations Law* 157-172.

46 Lang W "UN Principles and International Environmental Law" (1999) 3 *Max Planck Yearbook of United Nations Law* 157 164.

member states and stemming from a United Nations body (in the case of PSNR, the General Assembly).[47]

In determining the contemporary nature of the principle, it is useful to consider Principle 21 of the Declaration of the United Nations Conference on the Human Environment 1972 (Stockholm Declaration), which reads as follows:

> "States have, in accordance with the Charter of the United Nations and the principles of international law, the sovereign right to exploit their own resources pursuant to their own environmental policies, and the responsibility to ensure that activities within their jurisdiction or control do not cause damage to the environment of other states or of areas beyond the limits of national jurisdiction."[48]

Twenty years later, the Declaration of the United Nations Conference on Environment and Development 1992 (Rio Declaration) re-stated Principle 21 of the Stockholm Declaration. However, the Rio Declaration added two words, which illustrate a shift in the international political focus at the time to include the issue of development:

> "States have, in accordance with the Charter of the United Nations and the principles of international law, the sovereign right to exploit their own resources pursuant to their own environmental *and developmental* policies, and the responsibility to ensure that activities within their jurisdiction or control do not cause damage to the environment of other states or of areas beyond the limits of national jurisdiction."[49] (emphasis added)

UNGA resolution 1803 (XVII) of 1962, which indicates that PSNR is to be considered an international legal right,[50] is accepted as forming part of customary international law.[51] In addition, it was confirmed by the International Court of Justice that Principle 21 of the Stockholm Declaration, encompassing PSNR, is considered to be part of customary international

---

47 Lang W "UN Principles and International Environmental Law" (1999) 3 *Max Planck Yearbook of United Nations Law* 157 159.

48 Principle 21, Stockholm Declaration.

49 Principle 2, Rio Declaration.

50 P Sands *Principles of International Environmental Law* 2ed (2003) 237.

51 ICJ Judgment on the Case Concerning Armed Activities on the Territory of the Congo (Democratic Republic of the Congo v Uganda) 19 December 2005 *ICJ Reports* 168-283 251; P Sands *Principles of International Environmental Law* 2ed (2003) 236.

law.[52] Therefore, it can be submitted that PSNR forms part of customary international law.

PSNR is also embedded in many international instruments. The most noteworthy instruments to mention PSNR or elements of the principle include the following: the International Covenant on Civil and Political Rights 1966 (ICCPR) and the International Covenant on Economic, Social and Cultural Rights 1966 (ICESCR) (International Covenants on Human Rights);[53] the Vienna Convention on Succession of States in Respect of Treaties 1978[54] (Vienna Convention on Succession of States in Respect of Treaties) and the Vienna Convention on Succession of States in Respect of State Property, Archives and Debts 1983[55] (Vienna Convention on Succession of States in Respect of State Property, Archives and Debts) (Vienna Conventions on Succession of States); the African Charter on Human and Peoples' Rights 1981 (Banjul Charter);[56] the United Nations Conven-

52 K Hossain "Introduction" in K Hossain & SR Chowdhury (eds) *Permanent Sovereignty Over Natural Resources in International Law, Principle and Practice* (1984) ix ix.

53 Article 1(2) of the ICCPR and the ICESCR read as follows: "All peoples may, for their own ends, freely dispose of their natural wealth and resources without prejudice to any obligations arising out of international economic co-operation, based upon the principle of mutual benefit, and international law. In no case may a people be deprived of its own means of subsistence."
Article 47 of the ICCPR and article 25 of the ICESCR read as follows: "Nothing in the present Covenant shall be interpreted as impairing the inherent right of all peoples to enjoy and utilize fully and freely their natural wealth and resources.".

54 Article 13 of the Vienna Convention on Succession of States in Respect of Treaties reads as follows: "Nothing in the present Convention shall affect the principles of international law affirming the permanent sovereignty of every people and every State over its natural wealth and resources.".

55 The Vienna Convention on Succession of States in respect of State Property, Archives and Debts contains two references to PSNR. Article 15(4) reads as follows: "Agreements concluded between the predecessor State and the newly independent State to determine succession to State property of the predecessor State otherwise than by the application of paragraphs 1 to 3 shall not infringe the principle of the permanent sovereignty of every people over its wealth and natural resources." Article 38(2) reads as follows: "The agreement referred to in paragraph 1 shall not infringe the principle of the permanent sovereignty of every people over its wealth and natural resources, nor shall its implementation endanger the fundamental economic equilibria of the newly independent State.".

56 Article 21(1)-(5) of the Banjul Charter reads as follows: (1) "All peoples shall freely dispose of their wealth and natural resources. This right shall be exercised in the exclusive interest of the people. In no case shall a people be deprived of it"; (2)

tion on the Law of the Sea 1982 (Convention on the Law of the Sea);[57] the United Nations Convention on Biological Diversity 1992 (Biodiversity Convention);[58] the UNFCCC;[59] the African Convention on the Conservation of Nature and Natural Resources 1968 (revised in 2003) (African

---

"In case of spoliation the dispossessed people shall have the right to the lawful recovery of its property as well as to an adequate compensation"; (3) "The free disposal of wealth and natural resources shall be exercised without prejudice to the obligation of promoting international economic cooperation based on mutual respect, equitable exchange and the principles of international law"; (4) "States parties to the present Charter shall individually and collectively exercise the right to free disposal of their wealth and natural resources with a view to strengthening African unity and solidarity"; (5) "States parties to the present Charter shall undertake to eliminate all forms of foreign economic exploitation particularly that practiced by international monopolies so as to enable their peoples to fully benefit from the advantages derived from their national resources.".

57 The Convention on the Law of the Sea mentions issues concerning sovereignty and/or natural resources throughout the convention. However, article 193 is specifically relevant in regard to PSNR as it is entitled "Sovereign right of States to exploit their natural resources". Article 193 reads as follows: "States have the sovereign right to exploit their natural resources pursuant to their environmental policies and in accordance with their duty to protect and preserve the marine environment.".

58 The Biodiversity Convention contains a few references relevant in regard to PSNR. The parties to the Convention reaffirm the following: "States have sovereign rights over their own biological resources." Article 3 reads as follows: "States have, in accordance with the Charter of the United Nations and the principles of international law, the sovereign right to exploit their own resources pursuant to their own environmental policies, and the responsibility to ensure that activities within their jurisdiction or control do not cause damage to the environment of other States or of areas beyond the limits of national jurisdiction." Article 15(1) reads as follows: "Recognizing the sovereign rights of States over their natural resources, the authority to determine access to genetic resources rests with the national governments and is subject to national legislation.".

59 In the UNFCCC the parties to the convention recall the following: "States have, in accordance with the Charter of the United Nations and the principles of international law, the sovereign right to exploit their own resources pursuant to their own environmental and developmental policies, and the responsibility to ensure that activities within their jurisdiction or control do not cause damage to the environment of other States or of areas beyond the limits of national jurisdiction".

Convention on the Conservation of Nature and Natural Resources);[60] and the Energy Charter Treaty 1994 (Energy Charter Treaty)[61].

The international law relationship between customary international law and treaty law is complex. Treaties usually encapsulate or replace custom, but custom can also replace treaty provisions.[62] The hierarchy, as it pertains especially to these two sources of international law, can be difficult to establish. Usually, the rule is that the source of law that falls later in time will be superior to the source of law that falls earlier in time – in other words, lex posterior derogat priori (a later law repeals an earlier law).[63]

---

60 The African Convention on the Conservation of Nature and Natural Resources contains the following re-affirming statement in its preamble: "States have, in accordance with the Charter of the United Nations and the principles of international law, a sovereign right to exploit their own resources pursuant to their environmental and developmental policies, and the responsibility to ensure that activities within their jurisdiction or control do not cause damage to the environment of other States or of areas beyond the limits of national jurisdiction".

61 The most relevant provision found in the Energy Charter Treaty is article 18 (entitled "Sovereignty over energy resources"). Article 18(1)-(4) reads as follows: (1) "The Contracting Parties recognize state sovereignty and sovereign rights over energy resources. They reaffirm that these must be exercised in accordance with and subject to the rules of international law"; (2) "Without affecting the objectives of promoting access to energy resources, and exploration and development thereof on a commercial basis, the Treaty shall in no way prejudice the rules in Contracting Parties governing the system of property ownership of energy resources"; (3) "Each state continues to hold in particular the rights to decide the geographical areas within its Area to be made available for exploration and development of its energy resources, the optimalization of their recovery and the rate at which they may be depleted or otherwise exploited, to specify and enjoy any taxes, royalties or other financial payments payable by virtue of such exploration and exploitation, and to regulate the environmental and safety aspects of such exploration, development and reclamation within its Area, and to participate in such exploration and exploitation, inter alia, through direct participation by the government or through state enterprises"; (4) "The Contracting Parties undertake to facilitate access to energy resources, inter alia, by allocating in a non-discriminatory manner on the basis of published criteria authorizations, licences, concessions and contracts to prospect and explore for or to exploit or extract energy resources.".

62 MN Shaw *International Law* 6ed (2008) 123.

63 P Malanczuk *Akehurst's Modern Introduction to International Law* 7ed (1997) 56; MN Shaw *International Law* 6ed (2008) 123.
According to Malanczuk two other principles must also be observed when there is a conflict between treaty law and customary international law, namely *lex posterior generalis non derogat priori speciali*, which means "a later law, general in nature, does not repeal an earlier law which is more special in nature", and *lex spe-*

However, according to Shaw, where the same rule is contained in treaty law and customary international law, the rules are allowed to exist concurrently.[64]

However, the situation differs where a rule of customary international law or treaty law is deemed to be an obligation *erga omnes* or a rule of *jus cogens*. An obligation *erga omnes* or a rule of *jus cogens* is a norm or obligation "deemed to be of a different or higher status than others, whether derived from custom or treaty".[65] An obligation erga omnes and a rule of *jus cogens* differ in nature – the former having a procedural focus and the latter a substantive focus.[66] The status of PSNR as a rule of *jus cogens* is often debated.

The Vienna Convention on the Law of Treaties 1969 defines a peremptory norm (or a rule of *jus cogens*) as follows:

> "[A] peremptory norm of general international law is a norm accepted and recognized by the international community of States as a whole as a norm from which no derogation is permitted and which can be modified only by a subsequent norm of general international law having the same character."[67]

It has been argued that PSNR cannot be afforded the status of a rule of *jus cogens*.[68] This opinion is mainly based on the fact that PSNR cannot satisfy the requirements for a norm of *jus cogens*, according to article 53 of the Vienna Convention on the Law of Treaties 1969.[69] The definition for a rule of *jus cogens*, according to article 53 of the Vienna Convention on the

---

cialis derogat legi generali, which means "a special law prevails over a general one". [P Malanczuk *Akehurst's Modern Introduction to International Law* 7ed (1997) 56.].

64 MN Shaw *International Law* 6ed (2008) 123-124.

65 MN Shaw *International Law* 6ed (2008) 124.

66 MN Shaw *International Law* 6ed (2008) 124.
For a further explanation of obligations *erga omnes* see: MN Shaw *International Law* 6ed (2008) 124-126; P Malanczuk *Akehurst's Modern Introduction to International Law* 7ed (1997) 58-60.

67 Article 53, Vienna Convention on the Law of Treaties.

68 N Schrijver *Sovereignty Over Natural Resources, Balancing Rights and Duties* (1997) 374-377; P Malanczuk *Akehurst's Modern Introduction to International Law* 7ed (1997) 60; R Pereira & O Gough "Permanent sovereignty over natural resources in the 21st century: Natural resource governance and the right to self-determination of indigenous peoples under international law" (2013) 14 *Melbourne Journal of International Law* 13.

69 N Schrijver *Sovereignty Over Natural Resources, Balancing Rights and Duties* (1997) 374-377.

Law of Treaties 1969, sets out certain requirements which must be met if the norm is to attain *jus cogens* status. These requirements are the following: the norm must be "accepted and recognised by the international community of states" and "no derogation is permitted" from the norm.[70] It is important that "the international community of States as a whole"[71] must have accepted the norm as a norm of *jus cogens*, which can also mean a "very large majority of states",[72] as long as the recognition of the norm is not subject to the "cultural and ideological differences"[73] of different states.

Another argument made in support of the notion that PSNR is not a norm of *jus cogens* includes that the purpose of a norm of *jus cogens* is to ensure the "common good of the international community" as opposed to the individual interests of states.[74] It is doubtful whether the original intention behind the inception of PSNR was to ensure the "common good of the international community",[75] although in an ideal world PSNR can be considered to have the intention of ensuring the common good of the international community.

On the other hand, it has been contended that PSNR does have *jus cogens* status[76] owing to the fairly regular use of certain adjectives, such as 'permanent', 'full' and 'inalienable', used in conjunction with PSNR.[77] This issue of terminology is discussed further below in the current chapter. Another argument in support of the *jus cogens* status of PSNR is the fact that PSNR developed from self-determination of people, which does have

70 Article 53, Vienna Convention on the Law of Treaties.

71 Article 53, Vienna Convention on the Law of Treaties.

72 N Schrijver *Sovereignty Over Natural Resources, Balancing Rights and Duties* (1997) 376.

73 P Malanczuk *Akehurst's Modern Introduction to International Law* 7ed (1997) 58.

74 A Orakhelashvili *Peremptory Norms in International Law* (2006) 46-47.

75 A Orakhelashvili *Peremptory Norms in International Law* (2006) 46.

76 SR Chowdhury "Permanent sovereignty over natural resources" in K Hossain & SR Chowdhury (eds) *Permanent Sovereignty over Natural Resources in International Law, Principle and Practice* (1984) 1 7-9; I Brownlie *Principles of Public International Law* 5ed (1998) 515.

77 NJ Schrijver "Permanent sovereignty over natural resources versus the common heritage of mankind: complementary or contradictory principles of international economic law?" in P De Waart et al (eds) *International Law and Development* (1988) 87 91.

the status of a rule of *jus cogens*.[78] Put differently, PSNR has been described as an instance of the principle of self-determination which forms a norm of *jus cogens*.[79] Although the contrasting opinions on this point should be noted, it does not form part of the overall objective of this research project to debate the *jus cogens* status of the principle more extensively than has been done to this point. It should be noted that the overriding and more contemporary academic opinions do not seem to award PSNR the status of a norm of *jus cogens*, which is also the opinion followed in this research project.

To conclude on determining the legal nature of PSNR, it can be stated that the principle is a fundamental principle and a right of international law. As a UN-principle, it forms part of treaty law and customary international law, but it does not attain the status of a norm of *jus cogens*.

2 4 Objects of the principle of sovereignty over natural resources

2 4 1 Identifying key terminology

Understanding the terminology employed by PSNR means understanding the objects and subjects to which the principle applies. This section of the research project is concerned with the objects of PSNR. The principle is usually only phrased to include the term 'natural resources', but this is not the only existing object to which the principle applies. The principle has also been documented to include the following terms as its objects by various UNGA resolutions, international documents and instruments, and academic literature: 'natural wealth and resources'; 'natural and human resources'; 'resources and wealth'; and 'all economic activities'. These terms are often employed irregularly and used interchangeably.

The UNGA resolutions are responsible for the official legal inception of the principle and function as a recording of the development of the princi-

78 N Schrijver *Sovereignty Over Natural Resources, Balancing Rights and Duties* (1997) 375; K Hossain "Introduction" in K Hossain & SR Chowdhury (eds) *Permanent Sovereignty over Natural Resources in International Law, Principle and Practice* (1984) ix xiii.

79 G Jaenicke "Zur Frage des Internationalen Ordre Public" (1967) 7 *Berichte der Deutschen Gesellschaft für Völkerrecht* 77 as quoted by A Orakhelashvili *Peremptory Norms in International Law* (2006) 52.

ple throughout history. Therefore, considering UNGA resolutions can contribute to defining or ascertaining the meaning of the terms 'natural resources', 'natural wealth' and 'all economic activities', which are all terms used to refer to the object of PSNR.

In the 1950s and 1960s the UNGA resolutions demarcated "natural wealth and resources" as the objects of PSNR.[80] The International Covenants on Human Rights also include references to "natural wealth and resources" as the objects of PSNR.[81] Towards the late 1960s, the UNGA resolutions took the step of referring only to "natural resources" as the objects of PSNR.[82]

In the 1970s a more confusing period in regard to terminology can be distinguished. The UNGA resolution 2692 (XXV) of 1970 refers to both "natural resources" and "natural wealth and resources" as the objects of PSNR.[83] The UNGA resolution 3005 (XXVII) of 1972[84] refers to "national wealth and resources", while the UNGA resolution 3295 (XXIX) of 1974[85] refers to "natural and human resources" as the objects of PSNR.[86]

---

80 For example: UNGA resolution 626 (VII) of 21 December 1952 allows "natural wealth and resources" to be the object of PSNR; UNGA resolution 1314 (XIII) of 12 December 1958 speaks of "natural wealth and resources" as the object of PSNR; UNGA resolution 1514 (XV) of 14 December 1960 affirms "natural wealth and resources" as the object of PSNR; UNGA resolution 1515 (XV) of 15 December 1960 restated the term as "wealth and [...] natural resources"; UNGA resolution 1803 (XVII) of 14 December 1962 cements the usage of the term "natural wealth and resources".

81 Article 1.2 of both the ICCPR and the ICESCR read as follows: "All peoples may, for their own ends, freely dispose of their natural wealth and resources without prejudice to any obligations arising out of international economic co-operation, based upon the principle of mutual benefit, and international law. In no case may a people be deprived of its own means of subsistence."
Article 47 of the ICCPR and Article 25 of the ICESCR read as follows: "Nothing in the present Covenant shall be interpreted as impairing the inherent right of all peoples to enjoy and utilize fully and freely their natural wealth and resources.".

82 For example: UNGA resolutions 2158 (XXI) of 25 November 1966; 2386 (XXIII) of 19 December 1968; 2574 (XXIV) of 15 December 1969.

83 UNGA resolution 2692 (XXV) of 11 December 1970.

84 UNGA resolution 3005 (XXVII) of 15 December 1972.
Also see UNGA resolutions 3175 (XXVIII) of 17 December 1973 and 3336 (XXIX) of 17 December 1974.

85 UNGA resolution 3295 (XXIX) of 13 December 1974.

86 These terms are not examined in the same way as the terms 'natural resources', 'natural wealth' and 'economic activities' because the resolutions referring to these terms specifically deal with territories occupied by Israel and concerns issues

In the same year, the UNGA resolution 3336 (XXIX) of 1974[87] refers to "resources and wealth" as the objects of PSNR.[88] The UNGA resolutions 3016 (XXVII) of 1972[89] and 3171 (XXVIII) of 1973[90] then reaffirm that states have the right to exercise permanent sovereignty "over all their natural resources", but go on to extend[91] this right to "land within their international boundaries as well as those found in the sea-bed and the sub-soil thereof within their national jurisdiction and in the superjacent waters". In 1974, the UNGA adopted the Declaration on the Establishment of a New International Economic Order (NIEO Declaration) in resolution 3201 (S-VI) of 1974[92], whereby PSNR was extended to "all economic activities" of a state. In the same year, the UNGA also adopted resolution 3281 (XXIX) of 1974,[93] establishing the Charter of Economic Rights and Duties of States 1974 (CERDS), which also extended the objects of PSNR to include "wealth, natural resources and economic activities".[94]

It is clear that a basic understanding of the objects to which PSNR applies requires an understanding of the terms most frequently mentioned in UNGA resolutions since the 1950s. The terms most frequently mentioned as the objects of PSNR include: 'natural resources', 'natural wealth', and 'all economic activities'. These three terms together are understood to constitute 'natural resources' as the objects of PSNR for the purposes of this research project, and their meaning will be ascertained below.

---

of self-determination in addition to the principle of PSNR, and accordingly such an examination would fall outside the scope of the research project.

87 UNGA resolution 3336 (XXIX) of 17 December 1974.

88 UNGA resolution 3336 (XXIX) of 17 December 1974 "reaffirms the right of the Arab States and peoples whose territories are under Israeli occupation to full and effective permanent sovereignty over all their resources and wealth.".

89 UNGA resolution 3016 (XXVII) of 18 December 1972.

90 UNGA resolution 3171 (XXVIII) of 17 December 1973.

91 NJ Schrijver "Permanent sovereignty over natural resources versus the common heritage of mankind: complementary or contradictory principles of international economic law?" in P de Waart et al (eds) *International Law and Development* (1988) 87 89.

92 UNGA resolution 3201 (S-VI) of 1 May 1974.

93 UNGA resolution 3281 (XXIX) of 12 December 1974.
Also see: UNGA resolution 3486 (XXX) of 12 December 1975.

94 Article 2, CERDS; NJ Schrijver "Permanent sovereignty over natural resources versus the common heritage of mankind: complementary or contradictory principles of international economic law?" in P De Waart et al (eds) *International Law and Development* (1988) 87 89.

2 4 2 Natural resources

2 4 2 1 Definition of natural resources

The first of three terms that needs to be defined is 'natural resources'. This term can be defined from a wide range of perspectives, which include legal, geographical, political and economic perspectives. It is the international legal definition of 'natural resources' that is relevant to the research project. A uniform, general definition for 'natural resources' does not exist in international law.[95] In the succeeding paragraphs, some of the more useful definitions of 'natural resources' will be considered.

Elian vaguely interprets the term 'natural resources' in its widest sense so that it pertains to "the whole range of a country's endowments".[96] A more concrete definition of 'natural resources', also useful in an international legal sense, is the following geographical definition:

> "All the freely given material phenomena of nature within the zone of men's activities, at present a zone extending about twelve miles above the surface of the earth and about four miles below it, plus the additional non-material quality of situation or location."[97]

Blanco and Razzaque submit broadly that natural resources include "all aspects of the environment – the forests, the oceans, the air we breathe, mineral deposits, soil and freshwater – virtually anything that is not human-made and is of value to us."[98]

Some international law instruments provide their own definitions for the term 'natural resources'. These include, most notably, the African Convention on the Conservation of Nature and Natural Resources 1968 (revised in 2003) and the United Nations Convention on Biological Diversity 1992. The African Convention on the Conservation of Nature and Natural Resources 1968 (revised in 2003) defines 'natural resources' as

---

95 N Schrijver *Sovereignty Over Natural Resources, Balancing Rights and Duties* (1997) 15.

96 G Elian *The Principle of Sovereignty over Natural Resources* (1979) 11.

97 N Ginsburg "Natural Resources and Economic Development" (1957) 47 *Annals of the Association of American Geographers* 197 204.

98 RA Devlin & RQ Grafton *Economic Rights and Environmental Wrongs, Property Rights for the Common Good* (1998) 68.

follows: "renewable resources, tangible and non tangible, including soil, water, flora and fauna and non renewable resources."[99]

The United Nations Convention on Biological Diversity 1992, although not specifically defining the term 'natural resources', does define the term 'biological resources' as "genetic resources, organisms or parts thereof, populations, or any other biotic component of ecosystems with actual or potential use or value for humanity."[100]

The most useful definition of 'natural resources' in this context is found in a report to the Food and Agriculture Organization of the United Nations, where Cano provides a structured definition of natural resources in the context of PSNR.[101] Cano holds that natural resources pertain to "physical natural goods" rather than man-made goods, which he calls "cultural resources".[102] Cano includes the following categories of resources under the term 'natural resources': space;[103] energy;[104] land;[105]

---

99 Article V, African Convention on the Conservation of Nature and Natural Resources.

100 Article 2, Biodiversity Convention.

101 GJ Cano *A legal and institutional framework for natural resources management* 2ed (1983).

102 GJ Cano *A legal and institutional framework for natural resources management* 2ed (1983) 30.
Cano explains cultural resources as "created or transformed by human action" and it can be abstract (such as institutions, credit, or money) or concrete (such as manufactured goods). [GJ Cano *A legal and institutional framework for natural resources management* 2ed (1983) 31).].

103 Space refers to "[s]pace and its content: air, atmosphere, gases, mineral particles, different kinds of waves, winds". [GJ Cano *A legal and institutional framework for natural resources management* 2ed (1983) 30.].

104 Energy refers to "[e]nergy: from various sources such as solar, nuclear, Aeolian, hydraulic, geothermal, tidal, thermal (mineral or plant)". [GJ Cano *A legal and institutional framework for natural resources management* 2ed (1983) 30.].

105 Land refers to "[l]and and topography, including slopes capable of generating energy. Some consider the latter as an independent natural resource distinct from land itself. Land is referred to here as a non-agricultural resource distinct from soil." [GJ Cano *A legal and institutional framework for natural resources management* 2ed (1983) 30.].

scenic resources;[106] soil;[107] mineral deposits;[108] plant and animal wildlife;[109] non-maritime water resources;[110] the sea and seabed;[111] and geothermal resources.[112] The atmosphere (also relevant to Chapter 3 of the research project, which discusses climate change and the international legal climate change regime) is specifically mentioned by Cano as a 'natural resource' under the listing of 'space'.[113]

After careful consideration of the most prominent definitions offered for the term 'natural resources' in the context of the research project, Cano's definition, although older than some, still holds and is the most useful definition for the term 'natural resources' in the context of PSNR. Therefore, 'natural resources', as referred to in this research project, should be understood according to the definition formulated by Cano.

---

106 Scenic resources refer to "[p]anoramic or scenic resources: i.e., places the particular beauty of which makes them suited for recreational uses and aesthetic enjoyment." [GJ Cano *A legal and institutional framework for natural resources management* 2ed (1983) 30.].

107 Soil refers to "[s]oil (for agriculture or livestock use)." [GJ Cano *A legal and institutional framework for natural resources management* 2ed (1983) 30.].

108 Mineral deposits refer to "[m]ineral deposits: solid, liquid and gaseous." [GJ Cano *A legal and institutional framework for natural resources management* 2ed (1983) 30.].

109 Plant wildlife refer to "[p]lant wildlife (both land and aquatic)"; "Animal wildlife (land, aquatic, amphibious and aerial, including bacteria, insects and other primary forms of animal life)." [GJ Cano *A legal and institutional framework for natural resources management* 2ed (1983) 30.].

110 Non-maritime water resources refer to "[n]on-maritime water resources: i.e., water in its different forms – liquid, solid (ice, snow), or gaseous (steam, clouds) – wherever found: on the surface (rivers, lakes, etc.), underground, in the atmosphere." [GJ Cano *A legal and institutional framework for natural resources management* 2ed (1983) 30.].

111 The sea and seabed refer to "[t]he sea and seabed: including sea water and its content (both biological and mineral), the seabed and its subsoil with its mineral content." [GJ Cano *A legal and institutional framework for natural resources management* 2ed (1983) 30.].

112 Geothermal resources refer to "[g]eothermal resources: endogenous steam and hot water produced naturally inside the earth as a result of the passage thereof through natural thermic sources." [GJ Cano *A legal and institutional framework for natural resources management* 2ed (1983) 30.].

113 Space refers to "[s]pace and its content: air, atmosphere, gases, mineral particles, different kinds of waves, winds". [GJ Cano *A legal and institutional framework for natural resources management* 2ed (1983) 30.].

It should be noted that Ginsburg distinguishes resource "potentialities" from resources that are "available and understood".[114] This means that any definition for the term 'natural resources' would always need to be qualified, since there cannot be certainty about the ways in which the environment can be utilised.[115] Furthermore, the amount of non-renewable material that can be used from the yield of mining may improve as technology develops.[116] The present definition or understanding of the term 'natural resources' is accordingly incomplete, but should be clear for the purposes of the research project.

Though a suitable definition for the term 'natural resources' has been found, it should nevertheless be noted that natural resources can also be classified or categorised according to various criteria. The following paragraphs explain how natural resources can be understood according to classification by renewability or location.

2 4 2 2 Possibility of climate as a natural resource

In order to establish whether the climate can be considered as a natural resource and an object of PSNR, it should first be established whether climate can be considered to be a natural resource in general.

Hare defines climate in general as the "sequence of weather" that can be expected at a "given locality".[117] More specifically, Easterling defines climate as a "composite of all states of the atmosphere".[118] In Cano's definition of natural resources submitted above, the atmosphere is included as a natural resource.[119] Furthermore, the atmosphere is also considered to be

---

114 N Ginsburg "Natural Resources and Economic Development" (1957) 47 *Annals of the Association of American Geographers* 197 205.

115 N Ginsburg "Natural Resources and Economic Development" (1957) 47 *Annals of the Association of American Geographers* 197 204.

116 RA Devlin & RQ Grafton *Economic Rights and Environmental Wrongs, Property Rights for the Common Good* (1998) 69.

117 FK Hare "Climatic Variability and Change" in RW Kates et al (eds) *Climate Impact Assessment: Studies of the Interaction of Climate and Society* (1985) 87 88.

118 WE Easterling "Climate Trends and Prospects" in RN Sampson & D Hair (eds) *Natural Resources for the 21st Century* (1990) 32 32.

119 According to Cano, 'space' refers to "[s]pace and its content: air, atmosphere, gases, mineral particles, different kinds of waves, winds". [GJ Cano *A legal and institutional framework for natural resources management* 2ed (1983) 30.].

"a global environmental resource".[120] By this understanding, it is possible to consider climate as a natural resource.

The idea of classifying climate as a natural resource has existed since the 1940s.[121] In his article entitled "Climate as a Natural Resource", Landsberg designates climate as a natural resource.[122] Riebsame, too, contends that climate is a natural resource, because it can be managed or mismanaged.[123] It becomes even easier to envisage climate as a natural resource when considering climate in the economic context of renewable energy, such as wind or solar energy. It is also possible to apply economic concepts to climate such as scarcity, exhaustibility and renewability, which would generally be applied to natural resources.[124]

This research project is accordingly of the view that climate can be considered a natural resource and an object of PSNR. This strengthens the contention that a change in climate can have an impact on the evolution of PSNR, since this would constitute a change to the object of PSNR.

---

120 W Scholtz "Greening Permanent Sovereignty through the Common Concern in the Climate Change Regime: Awake Custodial Sovereignty!" in OC Ruppel et al (eds) Climate Change: International Law and Global Governance. Volume II: Policy, Diplomacy and Governance in a Changing Environment (2013) 201 206.

121 For example: H Landsberg "Climate as a Natural Resource" (1946) 3(4) *The Scientific Monthly* 293.

122 H Landsberg "Climate as a Natural Resource" (1946) 3(4) *The Scientific Monthly* 293 293.

123 WE Riebsame "Research in Climate-Society Interaction" in RW Kates et al (eds) *Climate Impact Assessment: Studies of the Interaction of Climate and Society* (1985) 1.

124 WE Easterling "Climate Trends and Prospects" in RN Sampson & D Hair (eds) *Natural Resources for the 21st Century* (1990) 32 32.
D'Arge considers the impacts of the climate on economic sectors extensively in his well-known paper delivered at the World Climate Conference in 1979: RC D'Arge "Climate and economic activity" (1979) *Proceedings of the World Climate Conference, A Conference of Experts on Climate and Mankind* 652-681.

2 4 2 3 Classification of natural resources

2 4 2 3 1 Classifying natural resources according to renewability and non-renewability

The first classification of natural resources for the purposes of the research project involves classifying natural resources into renewable and non-renewable resources. Renewable resources will not necessarily be permanently depleted by the use thereof as they are self-perpetuating,[125] while non-renewable resources inevitably will be permanently depleted by the use thereof. However, it should be noted that it is possible for a self-perpetuating, renewable resource to be permanently depleted, which can technically render it a non-renewable natural resource.[126] The African Development Bank Group simply refers to renewable resources as including, mainly "water, forestry and land resources".[127] A more comprehensive list of renewable natural resources includes "solar energy, wind, falling water, the heat of the earth (geothermal), plant materials (biomass), waves, ocean currents, temperature differences in the oceans and the energy of the tides".[128]

A list of non-renewable natural resources include: fossil fuels (coal, crude oil and natural gas); fertilisers (guano, nitrates, phosphates and sul-

125 RA Devlin & RQ Grafton *Economic Rights and Environmental Wrongs, Property Rights for the Common Good* (1998) 68.

126 RA Devlin & RQ Grafton *Economic Rights and Environmental Wrongs, Property Rights for the Common Good* (1998) 68.
Examples can include an old-growth forest, which would take one hundred years to be restored or an animal species that becomes extinct. [RA Devlin & RQ Grafton *Economic Rights and Environmental Wrongs, Property Rights for the Common Good* (1998) 68.].

127 African Development Bank Group "Natural Resources for Sustainable Development in Africa" (2007) (12) African Development Report <http://www.afdb.org/fileadmin/uploads/afdb/Documents/Publications/African%20Development%20Report%202007.PDF> (accessed 14-09-2016).

128 AJ Armstrong & J Hamrin "The Renewable Energy Policy Manual" (date of publication unavailable) (page numbers unavailable) United States Export Council for Renewable Energy <http://www.oas.org/dsd/publications/unit/oea79e/ch05.htm> (accessed 02-03-2015).

phur); precious metals and stones (diamonds, gems, gold and silver); radionuclides (radium and uranium); and salt.[129]

It is sufficient for the purposes of this research project to note that both renewable and non-renewable resources are categories of natural resources, thus forming part of the natural resources which are objects of PSNR.

2 4 2 3 2 Classifying natural resources according to location

The second classification of natural resources for the purposes of the research project can be done according to the location of the natural resources, or more accurately according to state territory. In this regard, three different categories of natural resources can be distinguished. The first category consists of natural resources that occur within the determined national territory of a state and fall under the national sovereignty of that state[130] (this includes the continental shelf or the exclusive economic zone of a state[131]). It is possible for most natural resources to occur

---

129 C Macaulay & PR Hensel "Natural Resources and Territorial Conflict" (March 2014) (14) <http://www.paulhensel.org/Research/isa14.pdf> (accessed 02-03-2015); African Development Bank Group "Natural Resources for Sustainable Development in Africa" (2007) (55) African Development Report <http://www.afdb.org/fileadmin/uploads/afdb/Documents/Publications/African%20Development%20Report%202007.PDF> (accessed 14-09-2016).

130 E Blanco & J Razzaque *Globalisation and Natural Resources Law: Challenges, Key Issues and Perspectives* (2011) 87.

131 K Odendahl "Conservation and Utilization of Natural Resources and Common Spaces" (2002) (page numbers unavailable) *Encyclopedia of Life Support Systems* <http://www.eolss.net/Sample-Chapters/C04/E4-21-06.pdf> (accessed 27-04-2012).
The Convention on the Law of the Sea defines the 'continental shelf' in article 76 as follows: "the seabed and subsoil of the submarine areas that extend beyond its territorial sea throughout the natural prolongation of its land territory to the outer edge of the continental margin, or to a distance of 200 nautical miles from the baselines from which the breadth of the territorial sea is measured where the outer edge of the continental margin does not extend up to that distance."
The Convention on the Law of the Sea defines the 'exclusive economic zone' in article 55 as follows: "an area beyond and adjacent to the territorial sea, subject to the specific legal regime established in this Part, under which the rights and jurisdiction of the coastal State and the rights and freedoms of other States are governed by the relevant provisions of this Convention.".

solely within the territory of one state. Examples of natural resources which may occur solely in the territory of one state include: oil, gas, minerals, forests, water and agricultural production.

The second category of natural resources consists of natural resources occurring within the determined national territory of more than one state. These natural resources are referred to as shared or transboundary natural resources.[132] Shared or transboundary natural resources are shared by states and are therefore subject to the regulations of more than one state.[133] This group of natural resources is the most challenging category to regulate, since the interest of more than one state applies.[134] Shared or transboundary natural resources are often the root of international environmental disputes, since all parties involved usually wish to promote their own interests. Examples of shared or transboundary natural resources most commonly include species which migrate, forests, international watercourses, enclosed or semi-enclosed seas and regional air masses.[135]

The third category of natural resources consists of those that lie outside the jurisdiction of any state. This category of natural resources can be fur-

132 K Odendahl "Conservation and Utilization of Natural Resources and Common Spaces" (2002) (page numbers unavailable) *Encyclopedia of Life Support Systems* <http://www.eolss.net/Sample-Chapters/C04/E4-21-06.pdf> (accessed 27-04-2012).

133 E Blanco & J Razzaque *Globalisation and Natural Resources Law: Challenges, Key Issues and Perspectives* (2011) 66, 87.

134 E Blanco & J Razzaque *Globalisation and Natural Resources Law: Challenges, Key Issues and Perspectives* (2011) 66, 12.

135 E Blanco & J Razzaque *Globalisation and Natural Resources Law: Challenges, Key Issues and Perspectives* (2011) 87.

ther divided into one of three categories: common areas,[136] common heritage of humankind,[137] and common concern of humankind.[138]

---

136 Common areas are also known as 'the global commons' and defined as natural resources (or areas) which are not situated in the territorial jurisdiction of any state and are not subject to the control of any one state. The natural resources forming a part of a common area are only subject to the no-harm rule (as contained in Principle 21 of the Stockholm Declaration) which has the status of an obligation *erga omnes*. Common areas may be situated outside the jurisdiction of a state for a number of reasons. These reasons include the "physical impossibility of extending control" over a resource or area, or the existence of an international agreement which prohibits the exercise of control over a resource or area. Examples of common areas or the global commons include the high seas (often referred to as the oceans) and their fisheries, Antarctica, the deep seabed and outer space. The atmosphere is differentiated from this group, especially from the deep seabed or outer space, on the basis that it partly falls within the jurisdiction of states. [J Vogler *The Global Commons, Environmental and Technological Governance* 2ed (2000) 6, 122; E Blanco & J Razzaque *Globalisation and Natural Resources Law: Challenges, Key Issues and Perspectives* (2011) 88-89.].

137 Although, the common heritage of mankind also technically forms a part of the global commons, it is grouped into a different category and distinguished on the basis that it is governed by a "special regime" because there are distinct factors which affect it. [E Blanco & J Razzaque *Globalisation and Natural Resources Law: Challenges, Key Issues and Perspectives* (2011) 89.] It should be noted at this point that the consequences for states resulting from the common heritage of mankind principle are not considered to be clear. [P Malanczuk *Akehurst's Modern Introduction to International Law* 7ed (1997) 208].

138 The term "common interest" has also made an appearance in earlier international treaties concerned with the exploitation of natural resources. These treaties include the International Convention for the Regulation of Whaling, the Tokyo Convention for the High Seas Fisheries of the North Pacific Ocean and Antarctic Treaty. [A Kiss & D Shelton *International Environmental Law* 3ed (2004) 32.]
The preamble of the International Convention for the Regulation of Whaling "[r]ecogniz[es] the interest of the nations of the world in safeguarding for future generations the great natural resources represented by the whale stocks" and "recogniz[es] that it is in the common interest to achieve the optimum level of whale stocks as rapidly as possible without causing wide-spread economic and nutritional distress."
The preamble of the Tokyo Convention for the High Seas Fisheries of the North Pacific Ocean reads as follows: "[the parties believe] that it will best serve the common interest of mankind, as well as the interests of the Contracting Parties, to ensure the maximum sustained productivity of the fishery resources of the North Pacific Ocean, and that each of the Parties should assume an obligation, on a free and equal footing, to encourage the conservation of such resources."

Natural resources which lie outside the jurisdiction of any state, known as the common concern of humankind, are particularly relevant to this research project.[139] Natural resources which are a common concern are distinguished by the fact that they are not subjected to any territorial boundaries, but their presence affects the all of humanity.[140] Natural resources which are a common concern are also sometimes referred to as global public goods.[141] Natural resources forming part of the common concern include:[142] biodiversity,[143] the ozone layer,[144] and world cultural[145] and natural heritage.[146] In the context of the research project, the atmosphere

---

The preamble of the Antarctic Treaty "[r]ecogniz[es] that it is in the interest of all mankind that Antarctica shall continue forever to be used exclusively for peaceful purposes and shall not become the scene or object of international discord".

139 In order to limit the scope of the research project, the common concern of humankind cannot be discussed in its entirety. However, the relevant aspects the common concern of humankind is included in this section of the research project as well as a corresponding brief explanation of the principle.

140 E Blanco & J Razzaque *Globalisation and Natural Resources Law: Challenges, Key Issues and Perspectives* (2011) 91.

141 E Blanco & J Razzaque *Globalisation and Natural Resources Law: Challenges, Key Issues and Perspectives* (2011) 91.

142 These are listed by E Blanco & J *Razzaque Globalisation and Natural Resources Law: Challenges, Key Issues and Perspectives* (2011) 91.

143 The Biodiversity Convention defines the term "biological resources" in article 2 as follows: "genetic resources, organisms or parts thereof, populations, or any other biotic component of ecosystems with actual or potential use or value for humanity".

144 See the Montreal Protocol.

145 Blanco and Razzaque consider cultural resources to form part of the common concern of humankind. [E Blanco & J Razzaque *Globalisation and Natural Resources Law: Challenges, Key Issues and Perspectives* (2011) 91.] However, Cano does not consider cultural resources to form a part of natural resources. Cano explains cultural resources as follows: "created or transformed by human action [and] […] may be abstract (institutions, credit, money etc.) or concrete. Among the latter are […] manufactured goods." [GJ Cano *A legal and institutional framework for natural resources management* (1975) 30-31.] However, take into account in this regard the context in which Cano defined natural resources - he defined the term for the FAO.

146 Article 1 of the Convention Concerning the Protection of the World Cultural and Natural Heritage considers "cultural heritage" to include the following: monuments ("architectural works, works of monumental sculpture and painting, elements or structures of an archaeological nature, inscriptions, cave dwellings and combinations of features, which are of outstanding universal value from the point of view of history, art or science"); groups of buildings ("groups of separate or

and climate change also form part of natural resources, as determined earlier in this chapter.

It is well-recognised that the atmosphere and climate change are considered to be a common concern of humankind.[147] The preamble of the UNFCCC specifically acknowledges that a "change in the Earth's climate and its adverse effects are a common concern of humankind". Therefore, not only does the UNFCCC consider climate change a common concern, but it also considers the environmental effects of climate change to be a common concern of humankind.[148] As will be noted in Chapter 3 when introducing the UNFCCC, PSNR is included in the preamble of the UNFCCC. It can thus be submitted that PSNR must be implemented in accordance with the common concern of humankind within the international legal climate change regime. It should be noted that this research project does not consider the common concern of humankind to run counter to PSNR, since the state duties in terms of PSNR include, *inter alia*, exercising

---

connected buildings which, because of their architecture, their homogeneity or their place in the landscape, are of outstanding universal value from the point of view of history, art or science"); and sites (works of man or the combined works of nature and man, and areas including archaeological sites which are of outstanding universal value from the historical, aesthetic, ethnological or anthropological point of view").
Article 21 of the Convention Concerning the Protection of the World Cultural and Natural Heritage considers "natural heritage" to include: "natural features consisting of physical and biological formations or groups of such formations, which are of outstanding universal value from the aesthetic or scientific point of view; geological and physiographical formations and precisely delineated areas which constitute the habitat of threatened species of animals and plants of outstanding universal value from the point of view of science or conservation; natural sites or precisely delineated natural areas of outstanding universal value from the point of view of science, conservation or natural beauty.".

147 FL Kirgis "Standing to Challenge Human Endeavors That Could Change the Climate" (1990) 84 *The American Journal of International Law* 525-530 530; E Blanco & J Razzaque *Globalisation and Natural Resources Law: Challenges, Key Issues and Perspectives* (2011) 91; J Vogler *The Global Commons, Environmental and Technological Governance* 2ed (2000) 122-123.
Furthermore, the UNGA passed resolution 43/53, entitled "Protection of Global Climate for Present and Future Generations of Mankind", which refers to climate change as a "common concern for mankind". [UNGA resolution 43/53 of 6 December 1988.].

148 D Shelton "Common Concern of Humanity" (2009) 1 *Iustum Aequum Salutare* 33 37.

PSNR for the well-being of people and in terms of the principle of sustainable development.[149]

2 4 3 Natural wealth

The second object of PSNR is natural wealth. The term 'natural resources' can be described as a sub-category of the term 'natural wealth' or as "supplies drawn from natural wealth".[150] Schrijver submits that the term 'natural wealth' includes: "land, soil, forests, wetlands, natural harbours, rivers, lakes, beaches, seas and oceans, flora and wildlife, rainfall and other beneficial climatic conditions, including the sun, the wind and natural sources of energy".[151]

In addition, Schrijver contends that 'natural wealth' includes activities based on natural wealth such as the tourist industry and electricity generation.[152] Upon examination it is found that the term 'natural wealth' is almost synonymous with broad terms such as 'environment' and 'ecosystem'.[153] Ultimately, the attempt to determine the exact meaning of the term 'natural wealth' is a futile exercise. It is concluded that the definition of the term 'natural wealth' is too wide to be practically useful in the context of this research project.

---

149 The state's rights and duties in terms of PSNR are discussed in Chapter 4 of the research project, and these two mentioned duties are included as state duties by N Schrijver Sovereignty Over Natural Resources, Balancing Rights and Duties (1997) 308, 324.

150 PSNR does not seem to extend to wealth in general or non-natural wealth. [N Schrijver Sovereignty Over Natural Resources, Balancing Rights and Duties (1997) 17, 19.].

151 N Schrijver *Sovereignty Over Natural Resources, Balancing Rights and Duties* (1997) 16.

152 NJ Schrijver "Permanent sovereignty over natural resources versus the common heritage of mankind: complementary or contradictory principles of international economic law?" in P De Waart et al (eds) *International Law and Development* (1988) 87 89.

153 N Schrijver *Sovereignty Over Natural Resources, Balancing Rights and Duties* (1997) 16-17.

2 4 4 Economic activities

The third term employed by United Nations documents in the 1970s as an object to PSNR is 'all economic activities'.[154] At that time, the term 'all economic activities' was understood to include: non-extractive industries, such as banking, insurance and technology.[155] More recently, the Energy Charter Treaty 1994[156] defined the term 'economic activities' as "the exploration, extraction, refining, production, storage, land transport, transmission, distribution, trade, marketing, or sale of Energy Materials and Products".[157]

It is found, for the purposes of this research project, that it is not necessary to understand the term 'all economic activities' as separate from the term 'natural resources' since "it is a rule that economic activities within the territorial jurisdiction of a State are subject to its sovereignty."[158] Accordingly, sovereignty over natural resources includes sovereignty over the economic value of natural resources.[159] Therefore, this research project includes all economic activities as part of natural resources, forming the object of PSNR.

154 In 1974 the UNGA resolution 3281 (XXIX) of 12 December 1974 adopted CERDS. Article 2 of CERDS expands the objects of PSNR even further so that it includes "wealth, natural resources and economic activities". [NJ Schrijver "Permanent sovereignty over natural resources versus the common heritage of mankind: complementary or contradictory principles of international economic law?" in P De Waart et al (eds) *International Law and Development* (1988) 87 89.].

155 NJ Schrijver "Permanent sovereignty over natural resources versus the common heritage of mankind: complementary or contradictory principles of international economic law?" in P De Waart, P Peters & E Denters (eds) *International Law and Development* (1988) 87 90; N Schrijver *Sovereignty Over Natural Resources, Balancing Rights and Duties* (1997) 18.

156 The Energy Charter Treaty and the Energy Charter Protocol on Energy Efficiency and Related Environmental Aspects were signed in December 1994 and entered into legal force in April 1998.

157 Article 1(5) of the Energy Charter Treaty 1994.

158 N Schrijver *Sovereignty Over Natural Resources, Balancing Rights and Duties* (1997) 19.

159 N Schrijver *Sovereignty Over Natural Resources, Balancing Rights and Duties* (1997) 19.

2 5 *Subjects of the principle of sovereignty over natural resources*

2 5 1 Ambiguity in UNGA resolutions and instruments of international law

2 5 1 1 Introduction

The subjects of international law include states, international organisations, regional organisations, non-governmental organisations, public companies, private companies and individuals.[160] However, in terms of PSNR, the state and people are usually construed as the subjects of PSNR.

Academic literature, UNGA resolutions and international instruments reveal the perceived subjects of PSNR at a given time, and illustrate how the focus has shifted between these subjects over time. The subjects of PSNR are states and people, but with the passage of time the focus has shifted between states and people as the subjects of PSNR and at times both groups were simultaneously designated to be the subjects of PSNR. For the most part, the ambiguous phrasing examined in determining the subjects of PSNR has supported the older, general interpretation that states are the only, or the main, subjects of PSNR. This inconsistency and ambiguity in referring to the subjects of PSNR will be illustrated below.

2 5 1 2 UNGA resolutions

According to Chowdhury, the period between 1952 until the adoption of the UNGA resolution 1803 (XVII) of 1962[161] can be distinguished as a time during which people specifically were emphasised as the subjects of PSNR.[162] However, Schrijver contends that people, nations and underdeveloped countries are highlighted as the subjects of PSNR during this time.[163] After inspection of the relevant UNGA resolutions of the time pe-

160 MN Shaw *International Law* 6 ed (2008) 196.

161 UNGA resolution 1803 (XVII) of 14 December 1962.

162 SR Chowdhury "Permanent sovereignty over natural resources" in K Hossain & SR Chowdhury (eds) *Permanent Sovereignty over Natural Resources in International Law, Principle and Practice* (1984) 1 3.

163 N Schrijver *Sovereignty Over Natural Resources, Balancing Rights and Duties* (1997) 8.

riod between 1952 until 1962, it does become clear that people alone were not emphasised as subjects of PSNR. For instance, the UNGA resolution 523 (VI) of 1952 considers "under-developed countries" as the subjects of PSNR, and the UNGA resolution 626 (VIII) of 1952 remembers "peoples" and "member states" as the subjects of PSNR, while the UNGA resolution 637 (VII) of 1952 acknowledges "peoples" as the subjects of the principle.

The 1960s are distinguished by the changes initiated by decolonisation.[164] The decolonisation period did bring about a definite shift in focus from people as the subjects of PSNR to developing countries or newly independent states as the subjects of PSNR.[165] In 1960, the UNGA passed resolution 1514 (XV) of 1960 containing the Declaration on the Granting of Independence to Colonial Countries and Peoples 1960 (Declaration on the Granting of Independence to Colonial Countries and Peoples).[166] Both countries and people form the subjects of PSNR according to this resolution. The UNGA resolution 1515 (XV) of 1960 considers "every state" as the subject of PSNR. UNGA resolution 1803 (XVII) of 1962 specifically underlines the right of people and nations to PSNR,[167] but at the same time confers to states the right to exercise sovereignty by stating that care should be taken "to ensure that there is no impairment, for any reason, of that State's sovereignty over its natural wealth and resources".[168] Some contend that it is in fact a prominent feature of resolution 1803 (XVII) that there is a lack of clarity in designating either people or states as the sub-

---

164 For a discussion on decolonisation, see J O'Brien *International Law* (2002) 42-46.

165 N Schrijver *Sovereignty Over Natural Resources, Balancing Rights and Duties* (1997) 8; P Malanczuk *Akehurst's Modern Introduction to International Law* 7 ed (1997) 28; UNGA resolution 3201 (S-VI) of 1 May 1974 entitled "Declaration on the Establishment of a New Economic Order".
The right to development, although relevant to many discussions involving PSNR, falls outside the scope of this research project. For a contemporary discussion on the right to development refer to: E Blanco & J Razzaque *Globalisation and Natural Resources Law: Challenges, Key Issues and Perspectives* (2011) 47-51.

166 UNGA resolution 1514 (XV) of 14 December 1960.

167 E Blanco & J Razzaque *Globalisation and Natural Resources Law: Challenges, Key Issues and Perspectives* (2011) 14.

168 UNGA resolution 1803 (XVII) of 14 December 1962.

jects of PSNR.[169] Others insist that resolution 1803 (XVII) does actually largely consist of statements affirming that states, in particular, are the subjects of PSNR.[170]

The findings above are not in line with Chowdhury's contention that people are to be considered the subjects of PSNR during this time, as the UNGA resolutions mentioned above prove that people alone cannot be considered as the only perceived subjects of PSNR during this time. Instead, the findings above are in agreement with the contention of Schrijver that people, states and underdeveloped countries were considered the main subjects of PSNR during the 1950s. This contention is in line with the historic functions of PSNR which were to ensure that benefit arises for people living under colonial rule, and to protect newly independent or developing states from the influence or exploitation of foreign states and foreign companies.[171] Furthermore, such a contention supports the fact that PSNR developed mainly from the efforts of developing states to ensure economic freedom and development, not dominated by the interests of developed states, as previously mentioned.[172]

The 1970s can be considered as the period when states gained further importance as the subjects of PSNR.[173] Moreover, the UNGA resolution 3016 (XXVII) of 1972[174] and the UNGA resolution 3171 (XXVIII) of 1973[175] reaffirm the rights of states as the subjects of PSNR. In addition, UNGA resolution 3281 (XXIX) of 1974, entitled 'Charter of Economic Rights and Duties of States', makes reference to states as the subjects of

169 E Duruigbo "Permanent Sovereignty and Peoples' Ownership of Natural Resources in International Law" (2006) 38 *The George Washington International Law Review* 33 43.

170 SM Schwebel "The Story of the U.N.'s Declaration on Permanent Sovereignty over Natural Resources" (1963) 49 *American Bar Association Journal* 463 464.

171 N Schrijver *Sovereignty Over Natural Resources, Balancing Rights and Duties* (1997) 24.

172 L Barrera-Hernández "Sovereignty over Natural Resources under Examination: The Inter-American System for Human Rights and Natural Resource Allocation" (2006) 12 *Annual Survey of International & Comparative Law* 43 45; N Schrijver *Sovereignty Over Natural Resources, Balancing Rights and Duties* (1997) 20.

173 N Schrijver *Sovereignty Over Natural Resources, Balancing Rights and Duties* (1997) 8.

174 UNGA resolution 3016 (XXVII) of 18 December 1972 entitled "Permanent sovereignty over natural resources of developing countries".

175 UNGA resolution 3171 (XXVIII) of 17 December 1973 entitled "Permanent sovereignty over natural resources".

PSNR in article 2.[176] The 1980s can also be earmarked as showing a predisposition to states as the subjects of PSNR.[177] Both the UNGA resolutions passed during the 1970s and 1980s that refer to people as the subjects of PSNR, are concerned only with those people still under colonial or alien control.[178] United Nations resolutions passed since the mid-1990s with regard to "permanent sovereignty over natural resources" all concern

176 UNGA resolution 3281 (XXIX) of 12 December 1974.

177 N Schrijver *Sovereignty Over Natural Resources, Balancing Rights and Duties* (1997) 9.

178 N Schrijver *Sovereignty Over Natural Resources, Balancing Rights and Duties* (1997) 8.
For examples of UNGA resolutions passed during the 1970's and the 1980's which do refer to people as the subjects of PSNR but which are only concerned with those people still under colonial or alien control, see inter alia UNGA resolutions: 3175 (XXVIII) of 17 December 1973 (entitled "Permanent sovereignty over national resources in the occupied Arab territories"); 3336 (XXIX) of 17 December 1974 (entitled "Permanent sovereignty over national resources in the occupied Arab territories"); 3295 (XXIX) of 13 December 1974 (entitled "Question of Namibia"); 3516 (XXX) of 15 December 1975 (entitled "permanent sovereignty over national resources in the occupied Arab territories"); 2145 (XXI) of 27 October 1966 (entitled "Question of South West Africa"); 2248 (S-V) of 19 May 1967 (entitled "Question of South West Africa"); 2372 (XXII) of 12 June 1968 (entitled "Question of South West Africa"); 33/40 of 13 December 1978 (entitled "Activities of foreign economic and other interests which are impending the implementation of the Declaration on the Granting of Independence to Colonial Countries and Peoples in Southern Rhodesia and Namibia and in all other Territories under colonial domination and efforts to eliminate colonialism, apartheid and racial discrimination in southern Africa"); 33/182 A of 21 December 1978 (entitled "Question of Namibia", "Situation in Namibia resulting from the illegal occupation of the territory of South Africa") and 33/182 C of 21 December 1978 (entitled "Question of Namibia", "Programme of work of the United nations Council for Namibia"); 3005 (XXVII) of 15 December 1972 (entitled "Report of the special committee to investigate Isreali practices affecting the human rights of the population of the occupied territories"); 32/161 of 19 December 1977 (entitled "Permanent sovereignty over national resources in the occupied Arab territories"); 34/136 of 14 December 1979 (entitled "Permanent sovereignty over national resources in the occupied Arab territories"); 35/110 of 5 December 1980 (entitled "Permanent sovereignty over national resources in the occupied Arab territories"); 36/173 of 17 December 1981 (entitled "Permanent sovereignty over national resources in the occupied Palestinian and other Arab territories"); 37/135 of 17 December 1982 (entitled "Permanent sovereignty over national resources in the occupied Palestinian and other Arab territories"); 38/144 of 19 December 1983 (entitled "Permanent sovereignty over national resources in the occupied Palestinian and other Arab territories").

the Palestinian people in the occupied Palestinian territory, including East Jerusalem, and the Arab population in the occupied Syrian Golan.[179]

To summarise the UNGA Resolutions' presentation of the subjects of PSNR, it can be said that the first resolutions were adopted to ensure economic development for underdeveloped countries at the time and therefore states were the subjects of PSNR.[180] Later, the human rights movement of the 1950s allowed people to become (and has since remained as) subjects of PSNR.[181] However, during the decolonisation period, people were removed as subjects to PSNR in the resolutions, allowing states and developing countries to become the main focus once again.[182]

2 5 1 3 Instruments of international law

In the quest to ascertain the true subjects of PSNR, international instruments may also prove helpful. The most noteworthy international instruments to mention PSNR have been listed in Section 2.3; however, these instruments are also relevant in the context of ascertaining the subjects of PSNR.

As instruments concerned with human rights, the International Covenants on Human Rights[183] emphasise people as the subjects of PSNR. Similarly, the Banjul Charter also focuses on people as the subjects

179 See UNGA resolutions: 53/196 of 22 February 1999; 54/230 of 22 February 2000; 55/209 of 15 February 2001; 56/204 of 21 February 2002; 58/229 of 25 February 2004; 57/269 of 5 March 2003; 59/251 of 7 March 2005; 60/183 of 31 January 2006; 61/184 of 25 January 2007; 62/181 of 31 January 2008; 63/201 of 28 January 2009; 64/185 of 29 January 2010; 65/179 of 22 March 2011; 66/225 of 25 March 2012.

180 N Schrijver Sovereignty Over Natural Resources, Balancing Rights and Duties (1997) 387.

181 N Schrijver Sovereignty Over Natural Resources, Balancing Rights and Duties (1997) 387.

182 N Schrijver Sovereignty Over Natural Resources, Balancing Rights and Duties (1997) 387.

183 Article 1.2 of both the ICCPR and the ICESCR read as follows: "All peoples may, for their own ends, freely dispose of their natural wealth and resources without prejudice to any obligations arising out of international economic co-operation, based upon the principle of mutual benefit, and international law. In no case may a people be deprived of its own means of subsistence."

of PSNR.[184] The Vienna Conventions on Succession of States,[185] change the focus to include both people and states as the subjects of PSNR. The Convention on the Law of the Sea focuses only on states as the subjects of PSNR.[186] UNFCCC also focuses on states as the subjects of PSNR.[187] The

---

Article 47 of the ICCPR and Article 25 of the ICESCR read as follows: "Nothing in the present Covenant shall be interpreted as impairing the inherent right of all peoples to enjoy and utilize fully and freely their natural wealth and resources." Also see UNGA resolution 1314 (XIII) of 12 December 1958.

184 Article 21(1)-(5) of the Banjul Charter reads as follows: (1) "All peoples shall freely dispose of their wealth and natural resources. This right shall be exercised in the exclusive interest of the people. In no case shall a people be deprived of it"; (2) "In case of spoliation the dispossessed people shall have the right to the lawful recovery of its property as well as to an adequate compensation"; (3) "The free disposal of wealth and natural resources shall be exercised without prejudice to the obligation of promoting international economic cooperation based on mutual respect, equitable exchange and the principles of international law"; (4) "States parties to the present Charter shall individually and collectively exercise the right to free disposal of their wealth and natural resources with a view to strengthening African unity and solidarity"; (5) "States parties to the present Charter shall undertake to eliminate all forms of foreign economic exploitation particularly that practiced by international monopolies so as to enable their peoples to fully benefit from the advantages derived from their national resources".

185 Article 13 of the Vienna Convention on Succession of States in Respect of Treaties reads as follows: "Nothing in the present Convention shall affect the principles of international law affirming the permanent sovereignty of every people and every State over its natural wealth and resources."
The Vienna Convention on Succession of States in respect of State Property, Archives and Debts contains two references to PSNR. Article 15(4) reads as follows: "Agreements concluded between the predecessor State and the newly independent State to determine succession to State property of the predecessor State otherwise than by the application of paragraphs 1 to 3 shall not infringe the principle of the permanent sovereignty of every people over its wealth and natural resources." Article 38(2) reads as follows: "The agreement referred to in paragraph 1 shall not infringe the principle of the permanent sovereignty of every people over its wealth and natural resources, nor shall its implementation endanger the fundamental economic equilibria of the newly independent State.".

186 The Convention on the Law of the Sea mentions issues concerning sovereignty or natural resources throughout the convention. However, Article 193 is specifically relevant to PSNR as it is entitled "Sovereign right of States to exploit their natural resources" and reads as follows: "States have the sovereign right to exploit their natural resources pursuant to their environmental policies and in accordance with their duty to protect and preserve the marine environment.".

187 In the UNFCCC, the parties to the convention recall that "States have, in accordance with the Charter of the United Nations and the principles of international

Biodiversity Convention again only refers to states as subject in the context of PSNR.[188] Other instruments also containing provisions relevant to PSNR and including states as the subjects of PSNR include the African Convention on the Conservation of Nature and Natural Resources[189] and the Energy Charter Treaty[190].

---

law, the sovereign right to exploit their own resources pursuant to their own environmental and developmental policies, and the responsibility to ensure that activities within their jurisdiction or control do not cause damage to the environment of other States or of areas beyond the limits of national jurisdiction".

188 The Biodiversity Convention contains a few references relevant to PSNR. The parties to the Convention reaffirm that "States have sovereign rights over their own biological resources." Article 3 reads as follows: "States have, in accordance with the Charter of the United Nations and the principles of international law, the sovereign right to exploit their own resources pursuant to their own environmental policies, and the responsibility to ensure that activities within their jurisdiction or control do not cause damage to the environment of other States or of areas beyond the limits of national jurisdiction." Article 15(1) reads as follows: "Recognizing the sovereign rights of States over their natural resources, the authority to determine access to genetic resources rests with the national governments and is subject to national legislation.".

189 The African Convention on the Conservation of Nature and Natural Resources contains the following re-affirming statement in its preamble: "States have, in accordance with the Charter of the United Nations and the principles of international law, a sovereign right to exploit their own resources pursuant to their environmental and developmental policies, and the responsibility to ensure that activities within their jurisdiction or control do not cause damage to the environment of other States or of areas beyond the limits of national jurisdiction".

190 The most relevant provision found in the Energy Charter Treaty is article 18, entitled "Sovereignty over energy resources". Article 18(1)-(4) reads as follows: (1) "The Contracting Parties recognize state sovereignty and sovereign rights over energy resources. They reaffirm that these must be exercised in accordance with and subject to the rules of international law"; (2) "Without affecting the objectives of promoting access to energy resources, and exploration and development thereof on a commercial basis, the Treaty shall in no way prejudice the rules in Contracting Parties governing the system of property ownership of energy resources"; (3) "Each state continues to hold in particular the rights to decide the geographical areas within its Area to be made available for exploration and development of its energy resources, the optimalization of their recovery and the rate at which they may be depleted or otherwise exploited, to specify and enjoy any taxes, royalties or other financial payments payable by virtue of such exploration and exploitation, and to regulate the environmental and safety aspects of such exploration, development and reclamation within its Area, and to participate in such exploration and exploitation, inter alia, through direct participation by the government or through state enterprises"; (4) "The Contracting Parties undertake

International instruments concerned with PSNR therefore clearly show a bias towards states, unless they are distinctly focused on human rights. In fact, human rights law is very useful in reapplying PSNR to people in a modern context under the broader banner of self-determination. Salomon and Sengupta also extend the application of human rights law to people and submit that "these rights can now be understood as being owed both to the state, and to people and groups within the state in different forms".[191]

However, it remains possible to argue that a prominent "state-centric atmosphere"[192] surrounding PSNR still exists. The most prominent three reasons for this continued state-centric approach may be, firstly, that the wording in early documents focused only on decolonisation of states;[193] secondly, that during the early years newly independent states focused mostly on keeping their independence and did not pay much regard to the needs of the people;[194] and, finally, that the United Nations, which is responsible for producing much of the relevant PSNR documentation, focused mainly on state actors.[195] The United Nations, in the drafting and enforcement of international instruments, is largely concerned with state action, even though the preamble of the Charter of the United Nations 1945 (UN Charter) begins by stating: "We the peoples of the United Nations …".

---

to facilitate access to energy resources, inter alia, by allocating in a non-discriminatory manner on the basis of published criteria authorizations, licences, concessions and contracts to prospect and explore for or to exploit or extract energy resources".

191 ME Salomon & A Sengupta The Right to Development: Obligations of States and the Rights of Minorities and Indigenous Peoples (2003) 35.

192 E Duruigbo "Permanent Sovereignty and Peoples' Ownership of Natural Resources in International Law" 2006 (38) The George Washington International Law Review 33 51.

193 N Schrijver Sovereignty Over Natural Resources, Balancing Rights and Duties (1997) 370.

194 N Schrijver Sovereignty Over Natural Resources, Balancing Rights and Duties (1997) 370.

195 N Schrijver Sovereignty Over Natural Resources, Balancing Rights and Duties (1997) 370.

2 5 2 States as the subjects of the principle of sovereignty over natural resources

2 5 2 1 Meaning of states

It has already been considered in Section 2.5.1 that, historically, there has been a shift in focus between states and people as the subjects of PSNR. The meaning of the term 'state' remains to be determined.

In terms of international law, the term 'state' should be understood to be synonymous with terms such as 'nation', 'country' or 'nation-state'.[196] Distinguishing these four terms is difficult owing to modern usage of the terms that allows for interchangeable use.[197] Furthermore, in the context of PSNR, terms such as 'underdeveloped country', 'developing country', or 'newly independent state' are also used to describe certain states.[198] These states are African states (not including South Africa until 1994), Asia (not including Japan), Latin America, and some countries in Europe such as Albania, Cyprus and Malta.[199]

The Treaty of Westphalia supported state interrelationships and territorial independence by allowing small states to enter into treaties among themselves, irrespective of the empire to which they belong.[200] The recognition of the concept of the state grew with time, especially with the creation of the League of Nations, but a continuing lack of clarity regarding the definition of the state remained.[201]

---

196 N Schrijver *Sovereignty Over Natural Resources, Balancing Rights and Duties* (1997) 10; W McClure *World Prosperity as sought through the economic work of the League of Nations* (1933) 3.

197 WC Opello & SJ Rosow *The Nation-State and Global Order: A Historical Introduction to Contemporary Politics* 2ed (2004) 3.
It seems that the term "nation" in particular is not as relevant to present day discussions of PSNR as it was to the discussions during 1950s. [N Schrijver *Sovereignty Over Natural Resources, Balancing Rights and Duties* (1997) 10.].

198 N Schrijver *Sovereignty Over Natural Resources, Balancing Rights and Duties* (1997) 10.

199 N *Schrijver Sovereignty Over Natural Resources, Balancing Rights and Duties* (1997) 10.

200 B Broms "States" in M Bedjaoui (ed) International Law: Achievements and Prospects (1991) 41 43.

201 B Broms "States" in M Bedjaoui (ed) International Law: Achievements and Prospects (1991) 41 43.

Even though there is no single meaning for the term 'state',[202] the term does possess some identifiable elements in terms of international law, as identified by the Montevideo Convention on Rights and Duties of States 1933 (Montevideo Convention). Article 1 of the Montevideo Convention sets out the qualifications for a state as follows: a permanent population, a defined territory, government, and the capacity to enter into relations with other states.

Although only binding for signatory states, the Montevideo Convention remains an important document representing the effort to define the term 'state'.[203] The first qualification, a permanent population, is easy to verify. The second qualification, a defined territory, is less straightforward to determine. Firstly, the territory need not be confined to a minimum area, but must consist of a basic area, even if boundaries are disputed.[204] The third qualification is to have a government, which is required to function at all times.[205] The government may need to leave the defined territory and it may change, but it will continue to be recognised as a government for the purposes of this qualification as long as it exists and functions.[206] The final qualification concerns the capacity to enter into treaties with other states. This qualification requires more than the will of the state to enter into a treaty with other states. It further requires the states with whom the treaty is entered into to judge the situation at hand and find that the state in

---

202 CW Morris *An Essay on the Modern State* (1998) 20.

203 For a more detailed exposition of statehood, see: T Jacobsen, C Sampford & R Thakur (eds) Re-envisioning sovereignty, The end of Westphalia? (2008) 5-12; Grant TD The Recognition of States, Law and Practice in Debate and Evolution (1999).

204 B Broms "States" in M Bedjaoui (ed) International Law: Achievements and Prospects (1991) 41 44.
Natural resources are owned mostly by the state, but can also be owned privately or by a community. According to Blanco and Razzaque, even when natural resources are owned privately or by a community the state retains the right, under territorial sovereignty, to regulate this ownership. In other words, the state regulates the way in which private or community ownership is exercised and people or communities will have a right to compensation upon expropriation of their property (in the form of natural resources). [E Blanco & J Razzaque *Globalisation and Natural Resources Law: Challenges, Key Issues and Perspectives* (2011) 15.].

205 B Broms "States" in M Bedjaoui (ed) International Law: Achievements and Prospects (1991) 41 45.

206 B Broms "States" in M Bedjaoui (ed) International Law: Achievements and Prospects (1991) 41 45.

question has the capacity to enter into treaties.[207] For the purposes of the research project it is sufficient to understand the term 'state' as qualified by the Montevideo Convention.

For purposes of completeness, the meaning of the state (and accordingly the meaning of the principle of state sovereignty over natural resources) can also be understood in reference to the concepts of territorial sovereignty and territorial integrity. Without a defined territory, there can be no state (in terms of the Montevideo Convention described above) and accordingly the term state sovereignty is to be understood *inter alia* in relation to territory.[208] Territorial sovereignty concerns the rights of the state over a defined territory, and different methods exist in international law by which to acquire or prove state territory.[209] Furthermore, the principle of respect for territorial integrity remains a so-called linchpin of the international legal system and an understanding of the state.[210] However, it should be noted that a contemporary understanding of the concepts of territorial sovereignty and territorial integrity also require due consideration of current concerns and corresponding legal developments. It is accepted that globalisation along with "transnational concerns"[211] such as issues of environmental degradation and human rights increase the interdependence between states. In this regard, the exercise of state sovereignty over natural resources (in particular where environmental degradation due to resource exploitation impact global climate change goals) may lead to international political pressure in the context of the international legal climate change regime, and in this way the international legal climate change

207 B Broms "States" in M Bedjaoui (ed) International Law: Achievements and Prospects (1991) 41 45.

208 MN Shaw *International Law* 6ed (2008) 487.

209 It would fall outside the ambit of the research project to discuss the international law which deal with the acquisition or proof of state territory. See in this regard: MN Shaw *International Law* 6ed (2008) 492-521 regarding the acquisition of state territory and the acquisition of additional state territory as largely dependent on effective control (in this regard Shaw includes a discussion on *inter alia* the operation of boundary treaties in the case of uncertain boundaries; the geographical formation of new land; the peaceful transfer of territory between sovereign states; the impact of occupation, the exercise of effective control, prescription, and the concept of the critical date; the doctrine of *uti possidetis* (which puts formard that "a new state has the boundaries of the predecessor entity"); and the relevance of subsequent conduct of states.).

210 MN Shaw *International Law* 6ed (2008) 488.

211 MN Shaw *International Law* 6ed (2008) 488.

regime may be considered a challenge to the territorial integrity of the state. In this context, section 2.5.2.3.3 discusses international environmental law (and by extension climate change law) as a limitation to state sovereignty (where state sovereignty is considered in its capacity as a constituent of PSNR).

2 5 2 2 Permanency as an attribute of state sovereignty

In addition to determining the meaning of the term 'state', terms such as 'permanent', 'full' or 'inalienable' are often used when referring to the state's sovereignty over natural resources. Academic opinions are most frequently expressed in regard to the meaning of the terms 'inalienable' and 'permanent', when used in conjunction with PSNR. Some of the most prominent opinions are mentioned below.

At face value the three adjectives, 'permanent', 'full' and 'inalienable', describe an attribute or component[212] of the principle or, more specifically, an attribute or component of the right to sovereignty over natural resources. These terms, especially the term 'permanent', are also used in the United Nations documents recording the inception of PSNR.[213] However, these terms are utilised inconsistently in United Nations publications and other relevant publications. Besides the inconsistency in the usage of these terms, the question can also be raised as to whether or not the state's (right to) sovereignty over natural resources is in fact 'permanent', 'full' or 'inalienable'. In any event, the exact meaning of these terms in the context of PSNR is to be determined below.

Chowdhury contends that the right to sovereignty over natural resources is an inalienable right, which cannot be "alienated or divested or ceded".[214] In other words, he argues that the right to permanent sovereignty over natural resources does not need to be secured by a treaty or a con-

---

212 W Scholtz "Custodial sovereignty: reconciling sovereignty and global environmental challenges amongst the vestiges of colonialism" (2008) Netherlands International Law Review 323 329.

213 See in particular UNGA resolution 1803 (XVII) of 14 December 1962.

214 SR Chowdhury "Permanent Sovereignty Over Natural Resources: Substratum of the Seoul Declaration" in P De Waart et al (eds) *International Law and Development* (1988) 59 62.

tract.[215] Chowdhury does however admit that a state may limit the right it has in this regard by taking on obligations in terms of its exercise of the right.[216]

In such a case, these obligations are limited in both scope and time, and cannot be permanent.[217] Abi-Saab submits concisely that "sovereignty is the rule and can be exercised at any time" and "limitations are the exception and cannot be permanent [as they] are limited in scope and time."[218] These obligations or limitations typically relate to foreign investment arrangements between states and foreign investors wishing to exploit the natural resources of that state.[219] The principle can also be called on when an agreement made by a state becomes "onerous and disadvantageous" to the point where it can be considered "a derogation of the sovereignty of that state" and not in the interest of the state's people.[220]

---

215 K Hossain "Introduction" in K Hossain & SR Chowdhury (eds) *Permanent Sovereignty over Natural Resources in International Law, Principle and Practice* (1984) ix xiv.

216 SR Chowdhury "Permanent Sovereignty Over Natural Resources: Substratum of the Seoul Declaration" in P De Waart et al (eds) *International Law and Development* (1988) 59 62.

217 UNITAR "Progressive Development of the Principles and Norms of International Law relating to the New International Economic Order" (1984) (UN Doc A/39/504/Add.1) as quoted by SR Chowdhury "Permanent Sovereignty Over Natural Resources: Substratum of the Seoul Declaration" in P De Waart et al (eds) *International Law and Development* (1988) 59 63.

218 G Abi-Saab "Progressive Development of the Principles and Norms of International Law Relating to the New International Economic Order" (1984) (UN Doc A/39/504/Add.1) as quoted by N Schrijver *Sovereignty Over Natural Resources, Balancing Rights and Duties* (1997) 263.

219 SR Chowdhury "Permanent Sovereignty Over Natural Resources: Substratum of the Seoul Declaration" in P De Waart et al (eds) *International Law and Development* (1988) 59 80.

220 SR Chowdhury "Permanent Sovereignty Over Natural Resources: Substratum of the Seoul Declaration" in P De Waart et al (eds) *International Law and Development* (1988) 59 64.

Abi-Saab contends that what is inalienable in terms of PSNR is a state's dominium[221] over, specifically, its "natural resources and wealth".[222] He continues by stating that what is permanent in terms of PSNR is a state's powers of imperium[223] over "economic assets and activities within its territorial jurisdiction".[224] He explains the term 'permanent' as denoting that the powers granted to the state are "continuous in character".[225]

Similarly, De Aréchaga is of the opinion that permanent sovereignty over natural resources depicts that the state can "never lose its legal capacity" – specifically not the legal capacity that allows the state to "change the destination or the method of exploitation of those resources, whatever arrangements have been made for their exploitation".[226]

Conversely, Hossain submits that the term 'inalienable' refers to the state's ability to transfer its sovereign rights in terms of PSNR to another state or person, and that this ability is what is in fact inalienable.[227]

The viewpoint of this research project is that a state's rights in terms of PSNR are inalienable, and the term 'inalienable' is understood to denote

---

221 Dominium is explained as "the patrimonial powers inherent in the institution of property in private law (which are in any case subject to the imperium of the State)". [G Abi-Saab "Permanent Sovereignty over Natural Resources and Economic Activities" in M Bedjaoui (ed) *International Law: Achievements and Prospects* (1991) 597 602.].

222 G Abi-Saab "Permanent Sovereignty over Natural Resources and Economic Activities" in M Bedjaoui (ed) *International Law: Achievements and Prospects* (1991) 597 607.

223 Imperium is explained as "the jurisdiction to prescribe and enforce over all persons, things and occurrences within the territorial ambit of the State". [G Abi-Saab "Permanent Sovereignty over Natural Resources and Economic Activities" in M Bedjaoui (ed) *International Law: Achievements and Prospects* (1991) 597 602.].

224 G Abi-Saab "Permanent Sovereignty over Natural Resources and Economic Activities" in M Bedjaoui (ed) *International Law: Achievements and Prospects* (1991) 597 607.

225 G Abi-Saab "Permanent Sovereignty over Natural Resources and Economic Activities" in M Bedjaoui (ed) *International Law: Achievements and Prospects* (1991) 597 607.

226 JE de Aréchaga "International Law in the Past Third of a Century" (1979) *Recueil des Cours* 159 as quoted by N Schrijver *Sovereignty Over Natural Resources, Balancing Rights and Duties* (1997) 263.

227 K Hossain "Introduction" in K Hossain & SR Chowdhury (eds) *Permanent Sovereignty over Natural Resources in International Law, Principle and Practice* (1984) ix xiv.

exactly the same characteristics as the term 'permanent' or 'full' when used in conjunction with the phrase 'the principle of sovereignty over natural resources'. In other words, the rights that are awarded to a state in terms of PSNR can never be taken from that state. However, it is possible that a state may enter into arrangements that limit these inalienable rights – as long as these arrangements are entered into freely and in the best interest and for the well-being of its people. This means that a state can make commitments in terms of climate change law and, by doing so, alienates or alters its right to freely use, dispose and exploit its natural resources by placing an encumbrance on its right.

It is clear that there is not a universally consistent understanding of what the usage of these descriptive terms, namely 'permanent', 'full' and 'inalienable', imply.[228] This is why this research project employs the term 'the principle of sovereignty over natural resources', or in short 'PSNR', rather than the three abovementioned, inconsistently used adjectives which carry different academic connotations.

2 5 2 3 State sovereignty

2 5 2 3 1 Introduction

There are a myriad of contexts within which to study sovereignty. Sovereignty has been described as one of the most frequently used, defended and contested terms of international law.[229] Accordingly, the full disposition of the concept would fall outside the scope of the research project. The discussion of sovereignty in this research project is limited to the context of state sovereignty as a foundational pillar of PSNR.

In the succeeding paragraphs, sovereignty is explored only in regard to serving the following three objectives. The first objective is to illustrate the evolutionary character of sovereignty as a foundational pillar of PSNR

228 In addition, Brownlie points out in discussing *jus cogens* that jurists have always tried to categorise rights by ascribing terms such as 'inalienable' to a right, but that these categorisations have been largely unsuccessful beyond the realm of influencing the interpretation of treaties. [I Brownlie *Principles of Public International Law* 5ed (1998) 514-515.].

229 FX Perrez Cooperative Sovereignty, From independence to interdependence in the structure of international environmental law (2000) 1.

by showing the varying understandings of sovereignty in regard to the existence of sovereignty, the denial of sovereignty, the criticisms against sovereignty, and the erosion or limitation of sovereignty. By illustrating the varied approaches that sovereignty allows, sovereignty as a foundational pillar of PSNR becomes able to accommodate a variety of interpretations. Accordingly, PSNR may be able to accommodate a variety of interpretations. The second objective is to illustrate the initial state-centric focus of PSNR, by illustrating the central role that states play in the understanding of sovereignty. The third objective is to gain an understanding of the term sovereignty as employed by PSNR.

2 5 2 3 2 Brief overview of the development of sovereignty

There is some debate as to whether or not the concept of sovereignty can be traced back to ancient empires such as that of Rome,[230] or only to the Peace of Westphalia.[231] In any event, by the end of the 15th century, Europe comprised about five hundred independent political entities, which claimed to derive their authority and power from the "law of God".[232] A shift occurred in the power of the monarchs, which gave them the power to rule over religion. Westphalian sovereignty was introduced in 1648 with the Peace of Westphalia ending the Thirty Years' War in Europe.[233] Westphalian sovereignty can be described as a horizontal system of states where all states are equal in terms of legal legitimacy and authority.[234] The power of the monarchy can be described as supreme in its territory during this time.[235] Therefore, the concept of sovereignty at this time is under-

---

230 For an exposition on sovereignty during the times of the ancient empires and the Hellenistic monarchies, see: FH Hinsley Sovereignty 2ed (1986) 27-44.
For an exposition on sovereignty during the decline and fall of the Roman Empire, see: FH Hinsley Sovereignty 2ed (1986) 45-60.

231 FX Perrez Cooperative Sovereignty, From independence to interdependence in the structure of international environmental law (2000) 18.

232 JA Camilleri & J Falk The End of Sovereignty? The Politics of a Shrinking and Fragmenting World" (1992) 12-13.

233 F Lenzertini "Sovereignty Revisited: International Law and Parallel Sovereignty of Indigenous Peoples" (2006) 42 Texas International Law Journal 155 157.

234 F Lenzertini "Sovereignty Revisited: International Law and Parallel Sovereignty of Indigenous Peoples" (2006) 42 Texas International Law Journal 155 157.

235 F Lenzertini "Sovereignty Revisited: International Law and Parallel Sovereignty of Indigenous Peoples" (2006) 42 Texas International Law Journal 155 157-158.

stood as expressing the idea of a "final and absolute authority in the political community".[236]

The concept of sovereignty came to the fore again during the 16th century owing to what Hinsley refers to as social disorder and political need.[237] Feudal absolutism theorists explored the concept of sovereignty, including its basis, definition and content, during the 16th and 17th centuries, including Bodin and Hobbes in particular.[238] A so-called "supreme authority"[239] existed which rested, firstly, in the right of the state to command and to be obeyed; secondly, in the fact that no other authority in the territory of the sovereign power would be superior to the sovereign power;[240] and, thirdly, in the existence of a defined territory.[241] Bodin considered sovereignty to express the state's political power, which was to be exercised by the king.[242] Hobbes considered sovereignty to mean that monarchs exercised unlimited powers with the ability to use the state's force as they saw fit.[243]

Hobbes succeeded in the first clear formulation of the concept of sovereignty in the period in which sovereignty took centre stage in European political thought.[244] Although his ideas were opposed, his work was able to bring about changes in thought at the time relating to the concept of sovereignty.[245] Unlike Bodin, Hobbes supported the idea that there are no limitations on sovereignty and subsequently removed all rights of the people from the concept of sovereignty.[246] In other words, Hobbes abandoned Bodin's idea of the so-called social contract that existed between

236 FH Hinsley Sovereignty 2ed (1986) 1.
237 FH Hinsley Sovereignty 2ed (1986) 125.
238 G Elian The Principle of Sovereignty over Natural Resources (1979) 1.
239 JA Camilleri & J Falk The End of Sovereignty? The Politics of a Shrinking and Fragmenting World (1992) 16.
240 FH Hinsley Sovereignty 2ed (1986) 122; F Lenzertini "Sovereignty Revisited: International Law and Parallel Sovereignty of Indigenous Peoples" (2006) 42 Texas International Law Journal 155 157-158.
241 F Lenzertini "Sovereignty Revisited: International Law and Parallel Sovereignty of Indigenous Peoples" (2006) 42 Texas International Law Journal 155 157.
242 G Elian The Principle of Sovereignty over Natural Resources (1979) 1.
243 G Elian The Principle of Sovereignty over Natural Resources (1979) 1.
244 FH Hinsley Sovereignty 2ed (1986) 141-144.
245 FH Hinsley Sovereignty 2ed (1986) 145.
246 A Camilleri & J Falk The End of Sovereignty? The Politics of a Shrinking and Fragmenting World (1992) 19.

the ruler and the ruled.[247] Hobbes replaced this idea with the idea that all individuals submit and surrender themselves to the state.[248]

Bodin supported the idea that, originally, sovereignty resided with the people, while anti-monarchists supported the idea that sovereignty resided permanently with the people.[249] Bodin's *Les six livres de la république* forms a vital component to the understanding and historical development of the theory of sovereignty.[250] Bodin contended, among others, that sovereignty is indivisible (a controversial contention and a contention that supported "a theory of ruler sovereignty"[251]).

Accordingly, the idea of sovereignty that existed in early modern Europe was that sovereignty was "a power concentrated in the hands of an authority bundled into a single entity, which governed a collectivity unified by the sharing of a single set of interests and confined within territorial borders.[252]

Grotius tried to bridge the gap between Bodin and the anti-monarchists. Grotius regarded sovereignty as a property, which could be held by a full property right (*jure pleno proprietas*) or a right of usufruct (*jure usufructario*).[253] In the 17th century, Grotius brought to light the idea that people were entitled to control the actions of the sovereign power, giving rise to the so-called sovereignty of the people.[254] More specifically, Grotius's *De Jure Belli ac Pacis*[255] acknowledged the king's sovereign power, but submitted that people had certain inalienable natural legal rights. It followed then that while a king held all sovereign power, he was guided in his exercise of sovereignty by natural law.[256] In *De Jure Belli ac Pacis* there was a

247 A Camilleri & J Falk The End of Sovereignty? The Politics of a Shrinking and Fragmenting World (1992) 19.
248 A Camilleri & J Falk The End of Sovereignty? The Politics of a Shrinking and Fragmenting World (1992) 19.
249 B Broms "States" in M Bedjaoui (ed) International Law: Achievements and Prospects (1991) 44 43.
250 Jean Bodin Les six livres de la république (1576).
251 JH Franklin (ed & translator) Jean Bodin On Sovereignty Four chapters from The Six Books of the Commonwealth (1992) xiii.
252 F Lenzertini "Sovereignty Revisited: International Law and Parallel Sovereignty of Indigenous Peoples" (2006) 42 Texas International Law Journal 155 157.
253 B Broms "States" in M Bedjaoui (ed) International Law: Achievements and Prospects (1991) 41 43.
254 G Elian The Principle of Sovereignty over Natural Resources (1979) 2.
255 First published in 1625.
256 G Elian The Principle of Sovereignty over Natural Resources (1979) 2.

clear effort to "reconcile the acceptance of the independence of the state with the old belief in the sway of the natural law".[257] There was no immediate attempt to follow up on the work of Grotius, as this would have been seen as recognition and approval of the "harmful doctrine" of so-called sovereignty of the people.[258]

During the era of Enlightenment, French philosophers, such as Rousseau, which excited revolutions throughout the late 18th and early 19th centuries, further developed the concept of sovereignty.[259] The French philosophers continued to develop the idea of sovereignty of the people and submitted that the term 'people' should include all those within the state's territory, with no influence by social class.[260] In his work *Le contrat social*,[261] Rousseau submitted the idea that the only legitimate sovereign was the people and that the state was the result of a social contract concluded between the people and those who exercised power within the state.[262] Rousseau agreed with Hobbes on the issue that state sovereignty is unlimited and indivisible, but differed from Hobbes by submitting that the state was only a commission of the people, formed by the abovementioned social contract.[263] Locke similarly contended that the people consent to the state merely exercising power on behalf of the people.[264]

---

257 FH Hinsley Sovereignty 2ed (1986) 139.

258 FH Hinsley Sovereignty 2ed (1986) 140.

259 G Elian The Principle of Sovereignty over Natural Resources (1979) 2.

260 G Elian The Principle of Sovereignty over Natural Resources (1979) 2.

261 1756.

262 G Elian The Principle of Sovereignty over Natural Resources (1979) 2.

263 A Camilleri & J Falk The End of Sovereignty? The Politics of a Shrinking and Fragmenting World (1992) 21; FH Hinsley Sovereignty 2ed (1986) 153.

264 A Camilleri & J Falk The End of Sovereignty? The Politics of a Shrinking and Fragmenting World (1992) 20; FH Hinsley Sovereignty 2ed (1986) 153.
Hinsley explains the difference in opinion between Rousseau and Locke on this matter as follows: "[I]n Locke's case remnants of the belief in a contract between the community and the government had restricted the right of revolution against the government to occasions when government had betrayed its trust or failed its proper task. For Rousseau the power to dismiss government, which was merely the non-sovereign executer of the legislative commands of the sovereign community, was permanently exercised by the community, which automatically suspended the government's commission whenever it assembled, because the unlimited sovereignty of the people could be bound by no law or constitution and could not be transferred even as to its exercise." [FH Hinsley Sovereignty 2ed (1986) 153-154.].

In the 18th and 19th centuries the concept of sovereignty was again placed under the spotlight. Montesquieu's work during this time focused on delegating power to the people and separating state powers.[265] Kant's work focused on political organisation, where all citizens were involved in making the law through a representative in the legislature, but the law was administered by the executive and interpreted by the judiciary equally on all citizens of the state.[266] Austin focused on explaining the law as resting entirely in positive law.[267] According to Austin, the state was sovereign with ultimate legal power and unlimited authority.[268] Hegel's work centred around the existence of the state as a result of the will of the citizens and denied the divine power of the monarch.[269] In the late 19th and early 20th century, Lasson, Zorn and Kaufmann expanded on Hegel's work and focused particularly on an understanding of sovereignty in terms of force, so that sovereignty was understood to embody force and in particular "the right of the strongest".[270] Following this line of thought where sovereignty was related to force, Treitschke, Lasson, Kaufmann and Jellinek supported the notion that sovereignty included the right to make war,[271] and Jellinek, in particular, went so far as to submit that the sovereignty of the state was absolute, to the degree that the state could ignore principles of international law.

Marx's theory of sovereignty can be differentiated from all other philosophical theories mentioned above, since his work is grounded in philosophy as well as economics, practice and political reality.[272] He considered it unrealistic that the state would promote and guard the interest of the people at all cost, especially in the context of a capitalist state.[273]

---

265 G Elian The Principle of Sovereignty over Natural Resources (1979) 3.
266 A Camilleri & J Falk The End of Sovereignty? The Politics of a Shrinking and Fragmenting World (1992) 22.
267 A Camilleri & J Falk The End of Sovereignty? The Politics of a Shrinking and Fragmenting World (1992) 22.
268 A Camilleri & J Falk The End of Sovereignty? The Politics of a Shrinking and Fragmenting World (1992) 22.
269 G Elian The Principle of Sovereignty over Natural Resources (1979) 3.
270 G Elian The Principle of Sovereignty over Natural Resources (1979) 3.
271 FH Hinsley Sovereignty 2ed (1986) 209.
272 A Camilleri & J Falk The End of Sovereignty? The Politics of a Shrinking and Fragmenting World (1992) 23-24.
273 A Camilleri & J Falk The End of Sovereignty? The Politics of a Shrinking and Fragmenting World (1992) 23.

The League of Nations, an aftermath of the First World War, aimed to prevent the repeat of such an event, which explains its objective "to maintain universal peace" and "develop cooperation among nations and to guarantee them peace and security".[274] Overall, the League of Nations is considered as somewhat of a political failure, being unable to prevent further conflicts.[275] Some have submitted that the League was destined to fail as it tried to suppress or limit sovereignty.[276] Leading into our present understanding of sovereignty, the United Nations, which was founded in 1945 after the Second World War to replace the ineffective League of Nations, extended the concept of sovereignty to all states and in particular also to non-European states.[277] Although, viewed by some as being "as ineffective as the League in its capacity as an enforcement machine",[278] the United Nations is still in force and arguably boasts many more successes than the League of Nations.[279]

---

274 United Nations Office at Geneva "The League of Nations (1919-1946)" (date of publication unavailable) (page numbers unavailable) United Nations Office at Geneva <http://www.unog.ch/80256EDD006AC19C/%28httpPages%29/17C8E6BCE10E3F4F80256EF30037D733?OpenDocument> (accessed 05-03-2015).

275 For example, the invasion of Austria by Hitler in 1938. However, limited successes were achieved by the League. [United Nations Office at Geneva "The League of Nations (1919-1946)" (date of publication unavailable) (page numbers unavailable) United Nations Office at Geneva <http://www.unog.ch/80256EDD006AC19C/%28httpPages%29/17C8E6BCE10E3F4F80256EF30037D733?OpenDocument> (accessed 05-03-2015).].

276 A Camilleri & J Falk The End of Sovereignty? The Politics of a Shrinking and Fragmenting World (1992) 28.

277 A Camilleri & J Falk The End of Sovereignty? The Politics of a Shrinking and Fragmenting World (1992) 28.

278 FH Hinsley Sovereignty 2ed (1986) 232.

279 Article 2 of the United Nations Charter, especially Article 2(1), 2(4) and 2(7) contain provisions relevant to state sovereignty.
Article 2(1)-(7) of the UN Charter reads as follows: "The Organization and its Members, in pursuit of the Purposes stated in Article 1, shall act in accordance with the following Principles:" (1) "The Organization is based on the principle of the sovereign equality of all its Members"; (2) "All Members, in order to ensure to all of them the rights and benefits resulting from membership, shall fulfil in good faith the obligations assumed by them in accordance with the present Charter"; (3) "All Members shall settle their international disputes by peaceful means in such a manner that international peace and security, and justice, are not endangered"; (4) "All Members shall refrain in their international relations from the threat or use of force against the territorial integrity or political independence of any state, or in any other manner inconsistent with the Purposes of the United

In the past, power was understood in terms of sovereignty, and sovereignty was understood as strengthening the state's claim to power.[280] Sovereignty meant that a "final and absolute authority" rested in the state.[281] A more contemporary understanding of sovereignty focuses on two primary elements of the concept, namely the independence of the state and the supremacy of the state, more specifically understood as independence of the state outside its borders and the supremacy of the state within its borders.[282] In other words, sovereignty is a supreme power which does not depend on another territory's power and enjoys independence within and outside its territory.[283] Contemporary sovereignty therefore comprises an external and an internal element – the external element referring to the independence that the state enjoys in relation to other states and the internal element referring to the domestic power of the state.[284]

The basic understanding of sovereignty today seems to reject the element of unlimited freedom and authority, but holds on to the element of independence.[285] Sovereignty is understood today as an attribute of the state.[286] Năstăsescu summarises the concept of sovereignty by referring to three features of sovereignty namely state power, indivisibility and inalienability.[287] State power refers to the state's power in performing state

---

Nations"; (5) "All Members shall give the United Nations every assistance in any action it takes in accordance with the present Charter, and shall refrain from giving assistance to any state against which the United Nations is taking preventive or enforcement action"; (6) "The Organization shall ensure that states which are not Members of the United Nations act in accordance with these Principles so far as may be necessary for the maintenance of international peace and security"; (7) "Nothing contained in the present Charter shall authorize the United Nations to intervene in matters which are essentially within the domestic jurisdiction of any state or shall require the Members to submit such matters to settlement under the present Charter; but this principle shall not prejudice the application of enforcement measures under Chapter VII.".

280 FH Hinsley Sovereignty 2ed (1986) 25.

281 FH Hinsley Sovereignty 2ed (1986) 1.

282 G Elian The Principle of Sovereignty over Natural Resources (1979) 5.

283 G Elian The Principle of Sovereignty over Natural Resources (1979) 6.

284 G Elian The Principle of Sovereignty over Natural Resources (1979) 6.

285 FX Perrez Cooperative Sovereignty, From independence to interdependence in the structure of international environmental law (2000) 16.

286 G Elian The Principle of Sovereignty over Natural Resources (1979) 1.

287 S Năstăsescu Suveranitatea si dinamica relatiilor internationale (1976) 22-23 as quoted by G Elian The Principle of Sovereignty over Natural Resources (1979) 5.

functions exclusively resting in the state.[288] Indivisibility refers to the "full freedom of choice in using the whole set of prerogatives of state power".[289] Inalienability refers to the fact that it would be impossible to transfer this state power to another entity.[290]

Hinsley contends that there can be no concept of sovereignty if there is no state.[291] The present understanding of sovereignty is therefore an understanding of so-called "relative sovereignty", which is limited by the freedom of others and the constraints of international law.[292] However, such an understanding of the concept of sovereignty may be considered as idealistic since governments still invoke sovereignty to withstand international pressures or compliance with international law.[293] It is not unheard of that states often use sovereignty and specifically sovereignty over natural resources as a "shield"[294] to control their own economic development at the cost of, for instance, the environment. This is clearly illustrated by certain states' failure to comply with climate commitments made in terms of the international legal climate change regime.

To define state sovereignty in detail would depend on the specific context in which state sovereignty is exercised. Moca provides a definition of sovereignty, which is useful in the context of this research project. Moca defines sovereignty as follows:

> "[Sovereignty is] the unique, full and indivisible supremacy of State power within the limits of the territorial frontiers and the independence of this power, in relation to any other power, which is expressed in the State's exclusive

---

288 S Năstăsescu Suveranitatea si dinamica relatiilor internationale (1976) 22-23 as quoted by G Elian The Principle of Sovereignty over Natural Resources (1979) 5.

289 S Năstăsescu Suveranitatea si dinamica relatiilor internationale (1976) 22-23 as quoted by G Elian The Principle of Sovereignty over Natural Resources (1979) 5.

290 S Năstăsescu Suveranitatea si dinamica relatiilor internationale (1976) 22-23 as quoted by G Elian The Principle of Sovereignty over Natural Resources (1979) 5.

291 FH Hinsley Sovereignty 2ed (1986) 22.

292 FX Perrez Cooperative Sovereignty, From independence to interdependence in the structure of international environmental law (2000) 16.

293 FX Perrez Cooperative Sovereignty, From independence to interdependence in the structure of international environmental law (2000) 17.

294 LA Miranda "The Role of International Law in Intrastate Natural Resource Allocation: Sovereignty, Human Rights, and Peoples-Based Development" (2012) 45 Vanderbilt Journal of Transnational Law 785 801; W Scholtz "Custodial sovereignty: reconciling sovereignty and global environmental challenges amongst the vestiges of colonialism" (2008) Netherlands International Law Review 323 329.

> and inalienable right to lay down and carry out its home and foreign policy independently, to discharge its functions, to implement the practical measures for organizing its social life at home and its foreign relations on the basis of respect for the sovereignty of other States, for the principles and norms of international law accepted of its own free will."[295]

In terms of this definition, a state would be able to limit its sovereignty by making commitments in terms of the international legal climate change regime. Furthermore, a state would have to respect the sovereignty of other states, and recognise that its contribution to global climate change could be seen as infringing on the sovereignty of other states. In this way the contemporary concern of climate change and the corresponding development of the international legal climate change regime has impacted the understanding of territorial sovereignty.

From the abovementioned theories of sovereignty, it is apparent that the understanding of sovereignty at any particular time and place is influenced by the social and economic circumstances of that particular time and place.[296] Climate change is an example of a social and economic circumstance impacting on our understanding of sovereignty today. In a broader context, the global recognition that the environment is one interconnected system which disregards the physical boundaries of states and state sovereignty is also considered as a limitation of sovereignty informing our current understanding of sovereignty, as will be explained in the following paragraphs.

2 5 2 3 3 International environmental law as a limitation of sovereignty

i Introduction

This section mainly serves to evaluate whether an argument can be made that sovereignty is limited or eroded specifically by international environmental law, and by climate change law, in particular. By grasping the possibility that environmental issues affecting the global interconnected envi-

---

295 G Moca Suveranitatea de stat si dreptul international contemporan (1970) 35 & G Moca Dreptul International Public (1977) 128 as quoted by G Elian The Principle of Sovereignty over Natural Resources (1979) 5.

296 JA Camilleri & J Falk The End of Sovereignty? The Politics of a Shrinking and Fragmenting World (1992) 12.

ronment[297] (such as climate change) can affect our understanding of sovereignty, it becomes possible to argue that the gravity of climate change has an impact on one of the foundational pillars of PSNR, namely sovereignty, and accordingly that the gravity of climate change has an impact on PSNR. There is a solid connection between sovereignty and PSNR. Sovereignty, including sovereignty over natural resources, represents the legal manifestation of the international reality that each state is the master of its own house.[298] The establishment of PSNR is more than a "terminological innovation" and the inception of PSNR meant that sovereignty was also officially linked to the utilisation of natural resources.[299] Elian considers PSNR to be a "constant and absolute component" of state sovereignty.[300] Sovereignty over natural resources is also founded within state supremacy and independence and is therefore an issue of domestic and foreign policy.[301] While sovereignty is understood to describe "the sum of all rights and obligations [that] the international system has allocated to a state",[302] PSNR is understood to describe "the sum of the rights and obligations [that] a state has with regard to its natural resources".[303]

Before considering the erosion of sovereignty argument specifically, it is useful to mention that opinions on the existence and continued existence of state sovereignty can be grouped into three categories. The first catego-

---

297 A Camilleri & J Falk The End of Sovereignty? The Politics of a Shrinking and Fragmenting World (1992) 173; C Redgwell "Life, the Universe and Everything: A Critique of Anthropocentric Rights" in A Boyle & M Anderson (eds) Human Rights Approaches to Environmental Protection (1996) 71 71, 73; S Bragdon "National Sovereignty and Global Environmental Responsibility: Can the Tension Be Reconciled for the Conservation of Biological Diversity?" (1992) 33(2) Harvard International Law Journal 381 381-392.

298 GM Abi-Saab "The Newly Independent Sates and the Rules of International Law: An Outline" (1962) 8 Harvard Law Journal 95 103 as quoted in A Anghie Imperialism, sovereignty and the making of international law (2004) 196.

299 G Abi-Saab "Permanent Sovereignty over Natural Resources and Economic Activities" in M Bedjaoui (ed) *International Law: Achievements and Prospects* (1991) 597 597.

300 G Elian The Principle of Sovereignty over natural resources (1979) 215.

301 G Elian The Principle of Sovereignty over Natural Resources (1979) 11.

302 FX Perrez Cooperative Sovereignty, From independence to interdependence in the structure of international environmental law (2000) 6.

303 FX Perrez Cooperative Sovereignty, From independence to interdependence in the structure of international environmental law (2000) 7.

ry of opinions is that sovereignty continues to exist, in an unaltered form, as a cornerstone of international law. The second category of opinions denies the existence or continued existence of sovereignty, criticises the existence of sovereignty, or predicts the extinction of sovereignty. The third category of opinions, or the so-called middle-ground between the first and second categories of opinions, is that an erosion or limitation of sovereignty is taking place. These three arguments are briefly set out in the succeeding paragraphs.

Points that should be noted before discussing state sovereignty include the fact that it is difficult to discuss state sovereignty as representing sovereignty for all states.[304] In addition, many so-called global environmental issues will not necessarily affect all states in the same way.[305] For instance, climate change affects developing and developed countries very differently because of the varying resources (for example capital and technology) that states and populations have at their disposal to adapt to climate change. Furthermore, climate change will have different impacts on states depending on certain factors, such as elevation above sea level or water-scarcity, which are related to the physical territory and location of states and make states predisposed to suffering the brunt of the effects of climate change.

ii Continued existence of sovereignty

It is possible to make an argument for the existence and continued existence of state sovereignty in an unaltered form. Those who hold this opinion do not admit to any limitation or erosion of sovereignty. According to such arguments, sovereignty is a fundamental or constitutive principle of international law;[306]sovereignty is a cornerstone of international law;[307]

304 K Conca "Rethinking the Ecology-Sovereignty Debate" (1994) 23 Millennium Journal of International Studies 701 703.

305 DA French "A reappraisal of sovereignty in the light of global environmental concerns" (2001) 21(3) Legal Studies 376 380.

306 FX Perrez "The Relationship between "Permanent Sovereignty" and the Obligation Not to Cause Transboundary Environmental Damage" (1996) 26 Environmental Law 1189 1188.

307 RC Gardner "Respecting Sovereignty" (1996) VIII Fordham Environmental Law Journal 133 133.

sovereignty shapes the content of international law;[308] the entire international legal order is an expression of external sovereignty;[309] and sovereignty is the means by which to become a member of international society.[310]In an older submission by Elian, he declares that sovereignty is a "permanent quality" of states.[311]

iii Denial or criticism of sovereignty

It is possible to challenge the existence of sovereignty. For example, Henkin prominently challenges the concept of sovereignty and submits that "sovereignty is a mistake, indeed a mistake built upon mistakes".[312] Furthermore, sovereignty has been described as "morally absolute" and "factually inaccurate".[313] Even the core constituent of sovereignty, namely the state, has been challenged and referred to as "a legal fiction" or "an artificial entity that cannot act by itself".[314]

Without denying sovereignty, the existence of sovereignty is often criticised. For instance, sovereignty is often seen as an impediment to solving international environmental problems. State decisions, relying on state sovereignty as a justification, can have far-reaching or global impacts on the environment that may not always have positive long-term results.[315]

308 A Bleckmann "Souveraineté étatique et protection internationale des minorities" in O Kimminich Einführung in das Völkerrecht 6ed (1997) 245 321 as quoted by FX Perrez Cooperative Sovereignty, From independence to interdependence in the structure of international environmental law (2000) 13.

309 A Bleckmann "Souveraineté étatique et protection internationale des minorities" in O Kimminich Einführung in das Völkerrecht 6ed (1997) 245 321 as quoted by FX Perrez Cooperative Sovereignty, From independence to interdependence in the structure of international environmental law (2000) 13.

310 A James Sovereign Statehood: The Basis of International Society (1986) 266 as quoted by FX Perrez Cooperative Sovereignty, From independence to interdependence in the structure of international environmental law (2000) 13.

311 G Elian The Principle of Sovereignty over Natural Resources (1979) 9.

312 L Henkin "Human Rights and State 'Sovereignty'" (1995/1996) 25 Georgia Journal of International and Comparative Law 31 31.

313 R Falk Law, Morality and War in the Contemporary World (1963) 29.

314 R McCorquodale "Beyond State Sovereigtny: The International Legal System and Non-State Participants" (2006) 8 International Law: Revista Colombiana de Derecho Internacional 103 109.

315 A Camilleri & J Falk The End of Sovereignty? The Politics of a Shrinking and Fragmenting World (1992) 185.

Finally, without denying sovereignty or criticising sovereignty, the future extinction of sovereignty is often argued. Even older literature has linked environmental interdependence with the complete disintegration of state sovereignty.[316] This disintegration, it is argued, occurs either from above in the form of international governance or from below where states transfer power to local communities or community-based, non-governmental organisations.[317] Where it is argued that the state's legitimacy is granted to the state by a form of international governance or supra-national order, it is usually suggested that sovereignty and instruments structured around sovereignty such as the UN Charter should be or might eventually be replaced.[318] The line between the extinction of sovereignty and the limitation of sovereignty is often blurred, especially when the argument is made that states may surrender certain aspects of their sovereignty to a supra-national organisation,[319] and this would be a limitation on their sovereignty and not an extinction of their sovereignty, since their sovereignty allows them to make this self-imposed limitation. Therefore it can be argued that a state entering freely without coercion into an agreement of a certain kind, as per the example below, would limit its own sovereignty and that this is nothing more than the ultimate exercise of its sovereignty. A practical example of the aforementioned is the state action of member states to join the United Nations, become a party to the UNFCCC, or the Kyoto Protocol. This action would be a self-imposed limitation on sovereignty.

iv Erosion of sovereignty

A middle-ground approach to sovereignty is to argue that sovereignty is being eroded – which involves neither the continued unaltered support for sovereignty, nor the denial or criticism of sovereignty. In view of the objective of the research project, international environmental law is used in

316 KT Liftin "Sovereignty in Ecopolitcs" (1997) 41 Mershon International Studies Review 167 170.

317 KT Liftin "Sovereignty in Ecopolitcs" (1997) 41 Mershon International Studies Review 167 170.

318 P Clark & L Sohn World peace through World Law (1960); C Eagleton International Government (1957); A Jessup A Modern Law of Nations (1948) as quoted by G Elian The Principle of Sovereignty over Natural Resources (1979) 8.

319 G Elian The Principle of Sovereignty over Natural Resources (1979) 8.

particular as a means to illustrate the argument for the erosion of sovereignty.[320]

The argument for the erosion of sovereignty is especially relevant and applicable in reference to "interlocking crises" that have been observed for some time.[321] These crises are explained by the World Commission on Environment and Development (Brundtland Commission) as follows:

> "Until recently, the planet was a large world in which human activities and their effects were neatly compartmentalized within nations, within sectors (energy, agriculture, trade), and within broad areas of concern (environmental, economic, social). These compartments have begun to dissolve. This applies in particular to the various global 'crises' that have seized public concern, particularly over the past decade. These are not separate crises: an environmental crisis, a developmental crisis, an energy crisis. They are all one."[322]

Accordingly, the tension between state sovereignty and international environmental law (instead of erosion of sovereignty by international environmental law) is not a new topic of debate, as the abovementioned submission was made in 1987. Following the statement by the Brundtland Commission in 1987, Principle 21 of the Stockholm Declaration also applies the principle of state sovereignty to the environmental realm.[323] Principle 21 of the Stockholm Declaration reads as follows:

> "States have, in accordance with the Charter of the United Nations and the principles of international law, the sovereign right to exploit their own resources pursuant to their own environmental policies, and the responsibility to ensure that activities within their jurisdiction or control do not cause damage to the environment of other states or of areas beyond the limits of national jurisdiction."

Principle 21 encapsulates two fundamental, albeit conflicting, objectives of international environmental law: firstly, the sovereign right of states to

---

320 In regard to the limitation of sovereignty by sustainable development, see: MO Hinz "Agenda 21 and Climate Protection: The Development of Global and Local Governance for Environment and Development - Obeservations from Research in Namibia" in OC Ruppel et al (eds) Climate change: International law and global governance. Vol. 1 Policy, diplomacy and governance in a changing environment. (2013) 621 621-659.

321 World Commission on Environment and Development Our Common Future (1987) 4.

322 World Commission on Environment and Development Our Common Future (1987) 4.

323 A Kiss & D Shelton *International Environmental Law* 3ed (2004) 27.

exploit and utilise their natural resources according to their own national policies; and, secondly, the obligation upon states not to cause environmental damage to other states or areas outside their national jurisdiction.[324] In other words, principle 21 encapsulates two principles of international law: the first being the principle of sovereignty over natural resources, and the second being the doctrine of state responsibility.[325]

It is stated that the meaning of sovereignty has changed.[326] Shelton states that is clear that environmental protection has had an impact on the "traditional role" of state sovereignty today.[327] It is this impact that is being referred to as a limitation on or an erosion of sovereignty. Aiding the argument for erosion of state sovereignty (and blurring the lines between an argument for the erosion of sovereignty and a criticism of sovereignty) is the fact that the sovereign state is often to blame for its role in international environmental degradation. The sovereign state has been described as "an agent or accomplice in ecological devastation across the globe",[328] and an "obstacle"[329] or "hindrance"[330] to the effective response in international environmental law and the realisation of environmental values.

State sovereignty rests on the idea of separate areas or states governed by separate governments, each according to their own set of laws. Each state exercises sovereignty over a defined territory and accordingly over a defined environment according to its own set of laws. The international environment and ecosystem of the Earth is an interrelated system which does not recognise these man-made boundaries, and cannot prosper without being managed in a globally oriented way that looks past state bound-

324 P Sands *Principles of International Environmental Law* 2ed (2003) 234.

325 J Schneider *World public order of the environment: Towards an international ecological law and organization* (1979) 21.

326 C Nyst "Solidarity in a Disaggregated World, Universal Jurisdiction and the Evolution of Sovereignty" (2012) 8 Journal of International Law and International Relations 36 48.

327 D Shelton "Common Concern of Humanity" (2009) 1 Iustum Aequum Salutare 33 40.

328 KT Liftin "Sovereignty in Ecopolitcs" (1997) 41 Mershon International Studies Review 167 168.

329 RC Gardner "Respecting Sovereignty" (1996) VIII Fordham Environmental Law Journal 133 133; KT Liftin "Sovereignty in Ecopolitcs" (1997) 41 Mershon International Studies Review 167 168.

330 RC Gardner "Respecting Sovereignty" (1996) VIII Fordham Environmental Law Journal 133 133; KT Liftin "Sovereignty in Ecopolitcs" (1997) 41 Mershon International Studies Review 167 168.

ary lines. As concisely stated by Camilleri and Falk, the "sovereign domain is confronted by the biosphere".[331]

The argument for the erosion of sovereignty is organised by arranging impacting factors into internal and external factors which are believed to diminish a state's power and authority.[332] Examples of internal factors would include the form of government, such as a federal system, or means of providing checks and balances.[333] Other examples of internal factors can include the realisation of the people's rights to their natural environment,[334] which may take the form of the right of people to hold a state accountable for climate commitments made in terms of the international legal climate change regime, as will be discussed in Chapter 3 and Chapter 4 of the research project. Examples of external factors include the expansion of international law and international organisations,[335] the development of world trade, and regional economic integration and cooperation.[336]

According to Elliot, there can be no doubt that environmental degradation challenges state sovereignty.[337] Similarly, Eckstein submits the following in the context of climate change:

> "Moreover, the international community must accept the reality that climate change is a global phenomenon that necessitates a collective and coordinated response, and that principles of sovereignty over territory and natural resources must yield to that reality.[338]

---

331 A Camilleri & J Falk The End of Sovereignty? The Politics of a Shrinking and Fragmenting World (1992) 192.

332 A Camilleri & J Falk The End of Sovereignty? The Politics of a Shrinking and Fragmenting World (1992) 31.

333 A Camilleri & J Falk The End of Sovereignty? The Politics of a Shrinking and Fragmenting World (1992) 31.

334 N Schrijver Sovereignty Over Natural Resources, Balancing Rights and Duties (1997) 378.

335 A Camilleri & J Falk The End of Sovereignty? The Politics of a Shrinking and Fragmenting World (1992) 31.

336 N Schrijver Sovereignty Over Natural Resources, Balancing Rights and Duties (1997) 377-379.

337 L Elliot "Sovereignty and the Global Politics of the Environment: Beyond Westphalia?" in T Jacobsen, C Sampford & R Thakur (eds) Re-envisioning sovereignty, The end of Westphalia (2008) 193 195.

338 G Eckstein "Water Scarcity, Conflict, and Security in a Climate Change World: Challenges and Opportunities for International Law and Policy" (2009) 27(3) Wisconsin International Law Journal 409 461.

These challenges to or limitations of sovereignty exist on a global scale, for example as seen above in the context of climate change, or on a smaller scale, for example when air or water pollution is carried from one state to another.[339] Whatever the scale of the environmental impact, it is clear that national state decisions can have international environmental impacts, and Bragdon contends that this is why "new meaning must be given to the concept of national sovereignty and state responsibility".[340]

Some remain doubtful that sovereignty is being eroded and argue that, owing to the multifaceted nature of sovereignty, it is impossible to claim with certainty that it is being eroded.[341] Somewhere within this argument, state sovereignty is considered as part of the solution to address international environmental issues effectively.[342] This entails that state sovereignty be strengthened by placing states at the centre of international responses to environmental issues and allowing them to act collectively in solving international problems.[343] States have the authority, resources and territorial control to implement and enforce treaties that would contribute to solving international environmental problems.[344] States become parties to these treaties by their own choosing (and by exercising their sovereignty). In this context it may be useful to note the distinction drawn by Levy, Keohane and Haas between 'operational sovereignty', which is "their legal freedom of action under international law", and 'formal sovereignty', which is "a state's legal supremacy and independence," not threatened "by international environmental interdependence nor by the agreements a state makes to regulate it".[345] By this understanding, formal sovereignty does

---

339 A Camilleri & J Falk The End of Sovereignty? The Politics of a Shrinking and Fragmenting World (1992) 178.

340 SH Bragdon "National Sovereignty and Global Environmental Responsibility: Can the Tension Be Reconciled for the Conservation of Biological Diversity?" (1992) 33(2) Harvard International Law Journal 381 384.

341 KT Liftin "Sovereignty in Ecopolitcs" (1997) 41 Mershon International Studies Review 167 171.

342 K Conca "Rethinking the Ecology-Sovereignty Debate" (1994) 23 Millennium Journal of International Studies 701 703.

343 K Conca "Rethinking the Ecology-Sovereignty Debate" (1994) 23 Millennium Journal of International Studies 701 702.

344 KT Liftin "Sovereignty in Ecopolitcs" (1997) 41 Mershon International Studies Review 167 168.

345 MA Levy et al "Improving the Effectiveness of International Environmental Institutions" in PM Haas et al (eds) Institutions for the Earth, Sources of Effective International Environmental Protection (1993) 397 417.

not prevent climate change, and states must allow for a collective state effort to curb climate change by eroding their operational sovereignty.[346] From this viewpoint, states are bound to international agreements as an "effect" of their sovereignty and not as a "derogation" from their sovereignty.[347] This manner of working in international relations can also be explained as follows:

> "[S]tates engage in sovereignty bargains in which they voluntarily accept some limitations in exchange for certain benefits. The cumulative effect of these trade-offs, however, may be to alter the norms and practices of sovereignty by reconfiguring expectations regarding state autonomy, control, and legitimacy."[348]

In terms of this understanding, state autonomy is not limited, and instead the problem-solving capacity of states is expanded.[349]

According to Conca, it is possible that the two contrasting viewpoints (sovereignty is eroded by international environmental law versus internal environmental law is an effect of sovereignty) regarding the erosion of sovereignty can be compatible, and proves nothing more than that sovereignty is a multifaceted concept.[350] Liftin, too, refers to sovereignty as having a "multifaceted character".[351] Regardless, sovereignty must be treated not as a concept which is "static and immovable", but rather as a "flexible tool" by which states act in response to environmental concerns.[352]

It is not the objective of the research project to prove that sovereignty is eroding. It is important to reiterate the reason for including a discussion on sovereignty in this research project. The rationale is that state sovereignty

---

346 MA Levy et al "Improving the Effectiveness of International Environmental Institutions" in PM Haas et al (eds) *Institutions for the Earth, Sources of Effective International Environmental Protection* (1993) 397 417.

347 M Koskenniemi "What Use for Sovereignty Today?" (2011) 1 Asian Journal of International Law 61 62.

348 KT Liftin "Sovereignty in Ecopolitcs" (1997) 41 Mershon International Studies Review 167 170.

349 K Conca "Rethinking the Ecology-Sovereignty Debate" (1994) 23 Millennium Journal of International Studies 701 703.

350 K Conca "Rethinking the Ecology-Sovereignty Debate" (1994) 23 Millennium Journal of International Studies 701 703, 711.

351 KT Liftin "Sovereignty in Ecopolitcs" (1997) 41 Mershon International Studies Review 167 169.

352 DA French "A reappraisal of sovereignty in the light of global environmental concerns" (2001) 21(3) Legal Studies 376 376-377.

forms a foundational pillar of PSNR, and it needs to be shown that sovereignty is a concept capable of evolving or being able to be subjected to differing interpretations. This, in turn, will allow PSNR to be considered as a concept capable of evolving or being able to be subjected to differing interpretations. All the arguments above that can be made in response to the concept of sovereignty prove that sovereignty is, at its core, a concept which is able to lend itself to a multitude of interpretations and has evolved over time. Accordingly, since sovereignty forms the foundational pillar of PSNR, it is responsible for PSNR being a concept that is able to evolve and be reinterpreted and reconfigured.

v The principle of sovereignty over natural resources and global climate justice

Climate justice is understood as expanding climate change (which is chiefly considered as an environmental and ecological issue) towards a social, economic and political issue as well. It can be submitted that climate justice aims to foster the connection between climate change and the effect of climate change on people and access to justice (which would include for example their right to water). In this regard the discussion in section 3.5, which considers *inter alia* the environmental, social and economic effects of climate change on the well-being of people, is relevant.

As stated in section 2.5.2.3.3 (iii), it is often accepted that states use their sovereignty over natural resources as a shield[353] of protection against allegations of environmental degradation, excessive natural resource exploitation or accompanied human rights violations within their territory, which would impinge global climate justice. Allegations (albeit formal or informal allegations) of environmental degradation, excessive natural resource exploitation or accompanied human rights violations within the territory of a particular state can be received from a number of sources, including *inter alia* indigenous communitites of the contravening state. The

353 LA Miranda "The Role of International Law in Intrastate Natural Resource Allocation: Sovereignty, Human Rights, and Peoples-Based Development" (2012) 45 Vanderbilt Journal of Transnational Law 785 801; W Scholtz "Custodial sovereignty: reconciling sovereignty and global environmental challenges amongst the vestiges of colonialism" (2008) Netherlands International Law Review 323 329. Also see discussion at 2.5.2.3.2.

Saramaka case[354] constitutes an example where the state made contentions which can be understood as employing state sovereignty over natural resources as a shield of protection. By way of example, Suriname contended in this case that "all land ownership including all natural resources, vests in the State, and that, as such, the State may grant logging and mining concessions within alleged Saramaka territory, while respecting as much as possible Saramaka customs and traditions."[355] Furthermore, the state contended that "should the Court recognize a right of the members of the Saramaka people to the natural resources found within traditionally owned lands, this right must be limited to those resources traditionally used for their subsistence, cultural and religious activities [...] [and accordingly] the alleged land rights of the Saramakas would not include any interests on forests or minerals beyond what the Tribe traditionally possesses and uses for subsistence (agriculture, hunting, fishing etc.) and the religious and cultural needs of its people".[356]

The Social and Economic Rights Action Center (SERAC) case,[357] which is also discussed in section 4.3.3.2, is mostly considered for its contribution to promoting the human right to a satisfactory environment as contained in section 24 of the Banjul Charter.[358] In the SERAC case it was alleged that the dealings of the Nigerian government with oil consortiums led to oil exploitation and environmental degradation in Ogoniland.[359] The

354 Inter-American Court of Human Rights: Case of the Saramaka People v Suriname (Preliminary Objections, Merits, Reparations, and Costs) 28 November 2007 Inter-American Court of Human Rights Series C No 172.

355 Inter-American Court of Human Rights: Case of the Saramaka People v Suriname (Preliminary Objections, Merits, Reparations, and Costs) 28 November 2007 Inter-American Court of Human Rights Series C No 172 par 124.

356 Inter-American Court of Human Rights: Case of the Saramaka People v Suriname (Preliminary Objections, Merits, Reparations, and Costs) 28 November 2007 Inter-American Court of Human Rights Series C No 172 par 125.

357 ACHPR: Social and Economic Rights Action Center (SERAC) and the Center for Economic and Social Rights v Nigeria 27 October 2001 ACHPR 155/96.

358 OC Ruppel "Climate Change Law Policy Positions in the African Union and Related Developments in Selected African Countries" in OC Ruppel et al (eds) Climate Change: International Law and Global Governance, Volume II: Policy, Diplomacy and Governance in a Changing Environment (2013) 411 418.

359 ACHPR: Social and Economic Rights Action Center (SERAC) and the Center for Economic and Social Rights v Nigeria 27 October 2001 *ACHPR 155/96* par 55; N Schrijver "Self-determination of peoples and sovereignty over natural wealth and resources" in United Nations Human Rights Office of the High Commissioner (ed) *Realizing the Right to Development* (2013) 95 99.

African Commission on Human and Peoples' Rights (ACHPR) determined that section 24 of the African Charter "imposes clear obligations upon a government" and that it "requires the state to take reasonable and other measures to prevent pollution and ecological degradation, to promote conservation, and to secure an ecologically sustainable development and use of natural resources".[360] In this sense a balance was struck by the ACHPR between the right to a satisfactory environment (which requires a state to "secure an ecologically sustainable development and use of natural resources"[361]) and oil exploitation which leads to environmental degradation (which forms part of the exercise of sovereignty over natural resources). Indirectly, the opinion of the ACHPR therefore promotes climate justice by "imposing clear obligations upon a government"[362] regarding human rights in the exploitation of natural resources (and accordingly the exercise of the state's sovereignty over natural resources).

As also discussed in section 4.3.3.2, the Endorois case,[363] which involved the displacement of the indigenous Endorois community from their ancestral lands in Kenya, also resulted in human rights violations of the

---

360 ACHPR: Social and Economic Rights Action Center (SERAC) and the Center for Economic and Social Rights v Nigeria 27 October 2001 ACHPR 155/96 par 52. Also see: OC Ruppel "Climate Change Law Policy Positions in the African Union and Related Developments in Selected African Countries" in OC Ruppel et al (eds) Climate Change: International Law and Global Governance, Volume II: Policy, Diplomacy and Governance in a Changing Environment (2013) 411 418.

361 ACHPR: Social and Economic Rights Action Center (SERAC) and the Center for Economic and Social Rights v Nigeria 27 October 2001 ACHPR 155/96 par 52. Also see: OC Ruppel "Climate Change Law Policy Positions in the African Union and Related Developments in Selected African Countries" in OC Ruppel et al (eds) Climate Change: International Law and Global Governance, Volume II: Policy, Diplomacy and Governance in a Changing Environment (2013) 411 418.

362 ACHPR: Social and Economic Rights Action Center (SERAC) and the Center for Economic and Social Rights v Nigeria 27 October 2001 ACHPR 155/96 par 52. Also see: OC Ruppel "Climate Change Law Policy Positions in the African Union and Related Developments in Selected African Countries" in OC Ruppel et al (eds) Climate Change: International Law and Global Governance, Volume II: Policy, Diplomacy and Governance in a Changing Environment (2013) 411 418.

363 ACHPR: Centre for Minority Rights Development (Kenya) and Minority Rights Group International (on behalf of Endorois Welfare Council) v Kenya 25 November 2009 ACHPR 276/03.

Banjul Charter. Of particular concern to the current context, the case involved mining concessions which were granted in regard to the ancestral land without giving the indigenous community a share in these resources.[364] The ACHPR had to decide *inter alia* whether article 21 of the Charter, regarding the right to dispose freely of natural resources, was violated. The Endorois community's displacement from their ancestral lands meant that they were not allowed access to vital resources such as traditional medicine, salt licks for cattle, and fertile soil.[365] The ACHPR found that the inaccessibility to natural resources constituted a violation of article 21 of the Banjul Charter.[366] The ACHPR found that article 21 protects the right of the community to potential wealth from their land, and a violation of article 21 occurred when the community did not benefit from their land's potential wealth.[367] In this way the exercise of sovereignty over natural resources by a state (in this case involving the granting of mining concessions) is also restricted by the ACHPR's consideration of human rights.

It can therefore be submitted that the exercise of sovereignty over natural resources is restricted by concerns for climate justice and *vice versa*, meaning that global climate justice is impinged by the exercise of sovereignty over natural resources.

2 5 3 People as the subjects of sovereignty over natural resources

2 5 3 1 Meaning of people

People are considered to be the last of the terminology employed by PSNR to be examined in Chapter 2 and people also form a subject of

---

364 ACHPR: Centre for Minority Rights Development (Kenya) and Minority Rights Group International (on behalf of Endorois Welfare Council) v Kenya 25 November 2009 ACHPR 276/03 par 120.

365 ACHPR: Centre for Minority Rights Development (Kenya) and Minority Rights Group International (on behalf of Endorois Welfare Council) v Kenya 25 November 2009 ACHPR 276/03 par 120.

366 ACHPR: Centre for Minority Rights Development (Kenya) and Minority Rights Group International (on behalf of Endorois Welfare Council) v Kenya 25 November 2009 ACHPR 276/03 par 120.

367 ACHPR: Centre for Minority Rights Development (Kenya) and Minority Rights Group International (on behalf of Endorois Welfare Council) v Kenya 25 November 2009 ACHPR 276/03 par 124.

PSNR. As mentioned in Section 1.5, the author makes the deliberate decision to refer to 'people' (as opposed to 'peoples') throughout the research project, except where a quote specifically refers to the term 'peoples'. As mentioned in Section 1.5, the term 'people' is understood by the author as encompassing the whole of the state population, but also being able to refer to a specific group of people forming a part of the state population such as indigenous communities.[368] The term 'people' is therefore understood by the author as a term that includes 'peoples' (regardless of the different interpretations of the term 'peoples' in international law). Accordingly, in order to ascertain the meaning of the term 'people' in this section, it is beneficial to consider the meaning of the term 'peoples' (although the extensive debate on the meaning of the term 'peoples' falls outside the scope of the research project).

A legal study conducted by the Secretariat of the International Commission of Jurists set out to examine the term 'a people' as used in the United Nations Charter and other relevant international law instruments in the context of the right to self-determination. The Secretariat concluded that "[w]hat is plain is that there is no single, authentic answer to the question 'what is a people'? All the official texts ignore it, presumably owing to the difficulty of definition."[369] Today, a general, internationally accepted definition for the term 'peoples' still does not exist.[370] As a result, peoples remains, for the most part, a term in international law which is unclear and disputed.[371]

---

368 This is a possibility according to: RM Kiwanuka "The meaning of 'People' in the African Charter on Human and Peoples' Rights" (1988) 82 American Journal of International Law 80.

369 International Commission of Jurists "The Events in East Pakistan, 1971" (1972) (71) International Commission of Jurists <http://icj.wpengine.netdna-cdn.com/wp-content/uploads/1972/06/Bangladesh-events-East-Pakistan-1971-thematic-report-1972-eng.pdf> (accessed 24-02-2015).

370 T van Boven "The Relations Between People's Rights and Human Rights in the African Charter" 1986 (7) Human Rights Law Journal 183 191 as quoted by E Duruigbo "Permanent Sovereignty and Peoples' Ownership of Natural Resources in International Law" (2006) 38 The George Washington International Law Review 33 51.

371 P Malanczuk *Akehurst's Modern Introduction to International Law* 7ed (1997) 107.
For an introduction to the debate, see: B Algan "Rethinking 'Third Generation' Human Rights" (2004) 1(1) *Ankara Law Review* 121 131; TT Onifade "Peoples-Based Permanent Sovereignty over Natural Resources: Toward Functional Dis-

In addition to states, people form the subjects of PSNR. The International Commission of Jurists found that the term 'peoples' include human communities whose members share common features or characteristics, which can include the following: historical, racial or ethnic, cultural or linguistic, religious or ideological, geographical or territorial, and economic or quantitative features.[372] However, the Commission also found that such a list of features is not very helpful, and an idealistic (rather than physical) characteristic should rather be considered, namely that "a people begins to exist only when it becomes conscious of its own identity and asserts its will to exist".[373]

Instead of focusing on characteristics or features, Kinakawu submits that the term 'peoples' can refer to the following four groups: persons under colonial domination; persons forming a part of the state population such as indigenous peoples; persons making up the whole of the state population;[374] or, lastly, 'peoples' can be a synonym for the state.[375] When considering only the historical development of PSNR, it makes sense to explain the term 'peoples' in this specific context as historically encompassing "people under colonial domination who could not fit into the definition of states or nations".[376] However, a more contemporary consideration of the definition of people as the subject of PSNR, would interpret the

---

tributive Justice?" (2015) 16 *Human Rights Review* 343 344; D Makinson "On Attributing Rights to all Peoples: Some Logical Questions" (1989) 8(1) *Law and Philosophy* 53 62-63.

372 International Commission of Jurists "The Events in East Pakistan, 1971" (1972) (70) International Commission of Jurists <http://icj.wpengine.netdna-cdn.com/wp-content/uploads/1972/06/Bangladesh-events-East-Pakistan-1971-thematic-report-1972-eng.pdf> (accessed 24-02-2015).

373 International Commission of Jurists "The Events in East Pakistan, 1971" (1972) (70) International Commission of Jurists <http://icj.wpengine.netdna-cdn.com/wp-content/uploads/1972/06/Bangladesh-events-East-Pakistan-1971-thematic-report-1972-eng.pdf> (accessed 24-02-2015).

374 R Araujo "Sovereignty, Human Rights, and Self-Determination: The Meaning of International Law" (2000) 24(5) Fordham International Law Journal 1476 1477.

375 RM Kiwanuka "The meaning of 'People' in the African Charter on Human and Peoples' Rights" (1988) 82 American Journal of International Law 80 as quoted by E Duruigbo "Permanent Sovereignty and Peoples' Ownership of Natural Resources in International Law" (2006) 38 The George Washington International Law Review 33 52.

376 E Duruigbo "Permanent Sovereignty and Peoples' Ownership of Natural Resources in International Law" 2006 (38) *The George Washington International Law Review* 33 52.

term 'peoples' as encompassing the whole of the state population. Accordingly, the term 'people', in its capacity as a subject of PSNR and as employed in the research project, is also understood by the author to include at least the whole of the state population. This understanding can be subjected to the limitation that people encompassing the whole of the state population are able to "participate in the conduct of their common affairs within the State".[377] In addition and for the sake of completeness, people as the subjects of PSNR not only encompass the whole of the state population, but also include future generations.[378]

2 5 4 Final note on the subjects of the principle of sovereignty over natural resources

In summation of the topic of the subjects of PSNR, the UNGA resolutions of the 1950s consider people, states and underdeveloped countries to be the subjects of PSNR. The decolonisation period bring about a decrease in the focus on people as subjects of PSNR and increase the focus on developing states as subjects of PSNR. During the 1970s and 1980s the focus on states as the subjects of PSNR increase, and people are mentioned in this regard only when a state is under foreign control. From the 1990s onwards, only certain Palestinian people are mentioned as subjects of PSNR.

---

377 M Koskenniemi "National self-determination today: problems of legal theory and practice" (1994) 43 International and Comparative Law Quarterly 241 249 as quoted by A Farmer "Towards a meaningful rebirth of economic self-determination: Human rights realization in resource-rich countries" (2006) 39 International Law and Politics 417 441.

378 Such a view corresponds to the views put forth in the following literature: A Farmer "Towards a meaningful rebirth of economic self-determination: Human rights realization in resource-rich countries" (2006) 39 *International Law and Politics* 417 438, 452; E Duruigbo "Permanent Sovereignty and Peoples' Ownership of Natural Resources in International Law" (2006) 38 *The George Washington International Law Review* 33 52-54; A Cassese "The self-determination of People" in L Henkin (ed) *The International Bill of Human Rights: The Covenant on Civil and Political Rights* (1981) 92 94; A Kiss "The Peoples' Right to Self-Determination" (1986) *Human Rights Law Journal* 165 171.
Camilleri and Falk submit that the truly sovereign state will have a supreme sovereignty not limited in time – reaching into the past and into the future. [A Camilleri & J Falk *The End of Sovereignty? The Politics of a Shrinking and Fragmenting World* (1992) 31.].

According to human rights instruments, in particular the International Covenants on Human Rights and the Banjul Charter, people naturally form the subjects of PSNR. The Vienna Conventions on Succession of States, consider both people and states as the subjects of PSNR. According to the Convention on the Law of the Sea, the UNFCCC, the Biodiversity Convention, the African Convention on the Conservation of Nature and Natural Resources), and the Energy Charter Treaty, states are the subjects of PSNR.

It is illustrated in UNGA resolutions that, even if people are theoretically just as important as a subject of PSNR as states, they are most recently only considered as subjects of PSNR by resolutions when people in question are under colonial or alien control or living in an occupied territory. Thus, according to the UNGA resolutions and most relevant international instruments, states are key subjects of international law[379] and key subjects of PSNR.[380] The aforementioned also supports the contention that PSNR has enjoyed an application focused on states, rather than people. However, an important qualification to make when discussing the role of states, is one based on the degree of power a state is able to exercise in international relations in the context of PSNR.[381] States, as the subjects of PSNR, are expected to exercise their rights in terms of PSNR for the benefit of their people. UNGA resolution 1803 (XVII)[382] of 1962 specifically states that "the right of people and nations to permanent sovereignty over their natural wealth and resources must be exercised in the interest of their national development and of the well-being of the people of the State concerned". It is not uncommon to describe the state in such an instance as a trustee or custodian of the country's natural resources, where the state administers and cares for its natural resources solely to benefit its people. If states were to be construed as the only subjects of PSNR without any reference to the interests of its people, it could imply that governments alone

379 MN Shaw *International Law* 6ed (2008) 197.

380 See N Schrijver *Sovereignty Over Natural Resources, Balancing Rights and Duties* (1997) 7-8; LS Clark "International Law and Natural Resources" (1976-1977) 4 *Syracuse Journal of International Law and Commerce* 377 380; E Duruigbo "Permanent Sovereignty and Peoples' Ownership of Natural Resources in International Law" (2006) 38 *The George Washington International Law Review* 33 45, 49.

381 E Blanco & J Razzaque *Globalisation and Natural Resources Law: Challenges, Key Issues and Perspectives* (2011) 173-174.

382 UNGA resolution 1803 (XVII) of 14 December 1962.

are able to exercise the rights in terms of PSNR and that all benefits arising from the exercise of these rights would be designated to a minor proportion of the state's population (often government officials) instead of the entire state population.[383]

This research project therefore considers both states and people as subjects of PSNR. However, the research project highlights people as the most important subject in terms of PSNR, in order to enable the principle to constitute a valuable addition to the fight against climate change.

2 6 Reinterpreting and reconfiguring the principle of sovereignty over natural resources as a means to combat climate change

As determined above, both states and people are considered the subjects of PSNR. It can be argued that the historic focus of PSNR has been on states as the subjects of the principle and the principle has been employed mainly to promote their interests. Accordingly, the principle has not been employed to promote the rights of the people of the state in terms of the principle. It is suggested by the author that the gravity of climate change warrants a reinterpretation and reconfiguration of PSNR that would enable the principle to be employed in the fight against climate change. This can be achieved by promoting people's rights in terms of the principle, as will be explained in Chapter 4. The customary state-centric application of the principle does not aid in the fight against climate change, because state rights in terms of PSNR constitute a barrier to effective climate change responses and state duties in terms of PSNR are difficult to enforce. Therefore, this research project aims to reinterpret PSNR so that people are promoted as subjects of PSNR, and people are reminded that, as right-bearers in terms of the principle, they can institute action based on their rights in terms of the principle. Accordingly, throughout the research project, this act will be referred to by the phrase 'reconfiguration or reinterpretation of PSNR as a principle focused on people'. This phrase should be clarified: firstly, by 'reconfiguring' the principle, constitutive elements of the principle are organised in a different form, meaning in this case that people's rights as a constitutive element of PSNR are organised in such a way as to

383 E Duruigbo "Permanent Sovereignty and Peoples' Ownership of Natural Resources in International Law" (2006) 38 *The George Washington International Law Review* 33 44.

promote people's rights, rather than states' rights; and, secondly, by 'reinterpreting' the principle, the principle is interpreted or construed in a different light, meaning in this case that the principle is interpreted in light of climate change.

As will be shown in the next chapter, there is a lack of state compliance with climate commitments (in terms of the Kyoto Protocol and the Paris Agreement) in terms of the international legal climate change regime. As Chapter 3 will show, this lack of state compliance is considered a root cause of the ineffectiveness of the international legal climate change regime. As will be set out below, a sophisticated international legal climate change regime already exists, meaning that the science behind the regime is certain and the legal formulation of what needs to happen in order to mitigate the effects of climate change is clear. The key to unlocking effectiveness within the international legal climate change regime is state compliance with the climate commitments made in terms of the regime. A possible solution to the lack of state compliance witnessed in the regime could be founded in PSNR. Although PSNR has been state-centric in its application thus far, the construed subjects of the principle have also been people, even if they were not always practically recognised as the most important subjects of PSNR. Reinterpreting and reconfiguring PSNR so that people are promoted as subjects of PSNR, means that people's rights in terms of PSNR would be promoted. People's rights in terms of PSNR, as discussed in Chapter 4, stem from self-determination and include specifically the right to dispose freely of natural resources.[384] This right can constitute a basis upon which people can hold states accountable for their climate commitments, as will be discussed in Chapter 4. This aforementioned action would constitute a negative incentive for state compliance to climate commitments and is discussed further in Chapter 3.

---

384 HM Haugen "Peoples' right to self-determination and self-governance over natural resources: Possible and desirable" (2014) 8(1) *Nordic Journal of Applied Ethics* 3 3-21; A Farmer "Towards a meaningful rebirth of economic self-determination: human rights realization in resources-rich countries" (2006) 39 *International Law and Politics* 417 417-473; D Cambou & S Smis "Permanent sovereignty over natural resources from a human rights perspective: natural resources exploitation and indigenous peoples' rights in the Arctic" (2013) 22(1) *Michigan State International Law Review* 347 347-376; N Schrijver "Self-determination of peoples and sovereignty over natural wealth and resources" in United Nations Human Rights Office of the High Commissioner (ed) *Realizing the Right to Development* (2013) 95 95-102.

2 7 Conclusion

The main objective of Chapter 2 is to determine what PSNR entails. This chapter examines PSNR by considering its origin and legal nature and the terminology it employs. It has been found that the origin of PSNR lies in the economic development of underdeveloped states and the self-determination of people. Also, the discussion of the origin and development of PSNR has illustrated the evolutionary character of the principle, which means that PSNR is capable of having key characteristics that shift, such as focusing on states and people as subjects of PSNR. This means that it is possible for climate change to have an impact on the principle and to affect how we reinterpret or reconfigure the principle (such as, in this case, interpreting and configuring the principle as a principle focused on people instead of a principle focused on states). As to the legal nature of PSNR, it is submitted that the principle finds its legal roots in UNGA resolutions and is cemented today as a fundamental principle of international law, a right of international law, and a part of treaty and customary international law.

Chapter 2 establishes PSNR as a concept which is able to evolve along with the phenomenon of climate change, which has significant environmental, political and legal implications. This is an important step in establishing that it is possible for the gravity of climate change to have an impact on the evolution of PSNR. The impact of the gravity of the effects of climate change on PSNR is explored further in Chapter 3.

In considering the terminology employed by PSNR, the objects and subjects of PSNR are also determined. The main object of PSNR is found to be natural resources, and the definition and precise scope of this term (specifically in the context of the research project) are determined. It is also concluded in this context that the atmosphere forms part of natural resources, the main object of PSNR, which is valuable in illustrating the nexus between PSNR and climate change, as is continued in Chapter 3.

The main subjects of the principle are found to be states and people. In examining the meaning of the term 'state', the concept of sovereignty is brought to light and examined. It is found that sovereignty, as a foundational pillar of PSNR, further contributes to giving PSNR its evolutionary character, since sovereignty itself is a concept able to evolve. Furthermore, in examining the meaning of the term 'state', it becomes apparent that the principle was originally state-oriented. Chapter 2 establishes people as subjects of PSNR and accordingly right-bearers in terms of PSNR. The chapter concludes by introducing the idea of reinterpreting and reconfiguring PSNR as a principle focused on people that can increase state compliance with the international legal climate change regime. The manifestation and implementation of people's rights in terms of PSNR is further explored in Chapter 4.

CHAPTER 3: The international legal climate change regime and the question of whether the gravity of climate change warrants a reinterpretation and reconfiguration of the principle of sovereignty over natural resources

3 1 Introduction

This chapter commences by introducing the nexus between climate change and PSNR, which forms the basis of the possibility for investigating the impact of climate change on the principle. The nexus between climate change and PSNR will continually be explored throughout the research project. Bear in mind that it was submitted in Chapter 1 that the reference to the term 'climate change' throughout the research project can include the environmental, political and the legal component of climate change. The research project is entitled "the impact of climate change law on the principle of state sovereignty over natural resources", as it highlights the legal component of climate change. This chapter defines the term 'international legal climate change regime', a term employed throughout the research project. This chapter also examines the international legal climate change regime, and embeds PSNR throughout the regime. In examining the international legal climate change regime, the development of the regime will be outlined, illuminating the most important developmental milestones such as the emergence of climate change as a political and policy issue separate from a scientific issue; the metamorphosis of climate change into an intergovernmental issue; the adoption of the UNFCCC as the cornerstone of the regime; the adoption of the Kyoto Protocol; and the adoption of the Paris Agreement.

In the process of examining the international legal climate change regime, important principles upon which the regime is founded are also discussed. These principles include equity; common but differentiated responsibilities; the precautionary principle; the polluter-pays principle; and the principle of sustainable development. Other aspects of the regime related to the functionality of the UNFCCC and the Kyoto Protocol, such as targets and timetables and market-based mechanisms, are also explained.

The Paris Agreement will be examined throughout the chapter and also in a separate section, as the latest significant addition and constituent of the regime.[385]

This chapter will illustrate why it is vital to explore innovative ways in which to combat climate change. The chapter will accordingly set out the scientific basis of climate change and outline the environmental, social and economic effects of climate change, and the resulting effects on the well-being of people, all in an effort to illustrate the urgency, scale and gravity of climate change. This chapter shows that the gravity of the effects of climate change are so significant that it warrants the reinterpretation and reconfiguration of any principle of international law (in this case PSNR) that can contribute to a more effective international legal climate change regime. This chapter of the research project will also briefly address issues of mitigation and adaptation as relevant actions in response to climate change and objectives of the international legal climate change regime, in order to have a complete understanding of the many different components comprising the complex international legal climate change regime. This first section of Chapter 3 will therefore aim to determine whether the gravity of climate change has an impact on PSNR (or warrants an impact on PSNR).

This chapter will illustrate where a cause of ineffectiveness of the international legal climate change regime lies, by setting out the legal framework in which the regime functions. In this regard, state compliance with climate commitments will be identified as a key component in ensuring a more effective international legal climate change regime. It will be shown that the Kyoto Protocol, as a current constituent of the regime, is a key component in ensuring the effectiveness of the regime and builds on the foundation established by the UNFCCC. The chapter considers whether the Kyoto Protocol ensures the effectiveness of the regime by considering the Kyoto Protocol's functionality and its record in ensuring state compliance with climate commitments. It will be determined that in order to increase state compliance with the regime, a viable solution is the implementation of the rights of people in terms of PSNR, which if implemented can act as a basis upon which to hold states accountable for climate commitments made in terms of the international legal climate change regime.

385 Note that the international legal regime is considered only including developments up until the end of June 2016.

At this point of the chapter it should also be clear that the gravity of the effects of climate on the well-being of people, together with the ineffectiveness of the regime due to a lack of state compliance, warrant a reinterpretation and reconfiguration of PSNR.

3 2 Nexus between climate change and the principle of sovereignty over natural resources

This research project reconfigures and reinterprets PSNR so that people's rights in terms of the principle can be utilised in the fight against climate change, by contributing to increasing state compliance with climate commitments. The reason why PSNR, as the specific international environmental law principle, is selected for examination in this context is not arbitrary. There is a considerable link between PSNR and climate change, which makes the study of the impact of climate change on PSNR a natural and meaningful choice.

The nexus between PSNR and climate change is significant. As submitted in Chapter 1, PSNR, like climate change, spans ecological, political and legal realms. As determined in Chapter 2 of the research project, the object of PSNR is natural resources, which include the atmosphere (according to Cano's definition of natural resources, as examined in Chapter 2). The object of the international legal climate change regime is also the atmosphere. Chapter 3 will illustrate how the state activities pertaining to natural resources cause an increase in greenhouse gases, which are responsible for altering the composition of the atmosphere and causing climate change. This means there is an overlap in the object of PSNR and the object of the international legal climate change regime. As will be explored in Chapter 4 by reinterpreting and reconfiguring PSNR to focus on people as the subjects of PSNR, people have rights over their natural resources – meaning that people have rights over the atmosphere and that it may be possible that people can hold states accountable for their actions in managing the atmosphere in terms of the international legal climate change regime. It is reasonable therefore to link the right of people in terms of PSNR with the international legal climate change regime.

Furthermore, state sovereignty lies at the core of PSNR and the climate change regime as an attribute and also as an impediment to the effectiveness of both the principle (from the viewpoint of promoting state duties and not state rights) and the regime. Firstly, PSNR was built on a foundation of *inter alia* state sovereignty and the struggle of newly independent states to gain control over their natural resources. However, this also means that state sovereignty has presented an impediment to PSNR functioning equally successfully in regard to executing state rights as well as state duties in regard to the management of the state's natural resources. Secondly, and similarly to PSNR, state sovereignty also lies at the heart of the international legal climate change regime, as the functionality of the regime is built on state action and compliance. However, this also affects the functioning of the regime since the lack of state compliance (strengthened by reliance on state sovereignty) is considered to be a considerable cause of ineffectiveness of the regime. In both the case of PSNR and the international legal climate change regime, state sovereignty can therefore act as a shield against a state's mismanagement of its natural resources and disregard for climate commitments respectively.

This leads to the next point of connection between PSNR and climate change. The subjects of PSNR have traditionally been considered to be most importantly states, although people are also considered to be subjects of the principle. States are also subjects of the international legal climate change regime, since it is states that commit to the legal instruments of the regime and undertake to fulfil obligations in terms of the regime. In addition, the United Nations (a state-based system) is the principal driving force behind the international legal climate change regime. Therefore PSNR and the international legal climate change regime also share states as their subjects.

This section does not exhaustively list the connection between PSNR and climate change. The nexus between climate change and PSNR will also be further explored at various points throughout the research project. For instance, Chapter 3 will show how the activities of states in regard to their natural resources (and this includes the activities that states allow their citizens to perform), which are firmly planted in PSNR, are also the main sources of greenhouse gas emissions. Therefore, while considering the most damaging greenhouse gases and the corresponding anthropocentric activities leading to their emission, the nexus between PSNR and climate change will be illustrated. Chapter 3 will also consider whether human well-being can be regarded as an objective of the international legal climate change regime and PSNR, in order to explore the nexus between the regime and the principle. In this context, it will also be discussed how a strengthening of the implementation of a state duty in terms of PSNR can strengthen the implementation of the international legal climate change regime and vice versa. By considering the content of the UNFCCC and certain principles which the international legal climate change regime promotes, such as sustainable development, the link between PSNR and the principles of the regime will also be explored. The example of the United Nations Programme on Reducing Emissions from Deforestation and Forest Degradation in Developing Countries (REDD) as a mitigation strategy of the international legal climate change regime will further illustrate the connection between PSNR and climate change, since states must manage their forests (as natural resources in terms of PSNR) in accordance with both PSNR and the international legal climate change regime. The embedding of PSNR within the UNFCCC, Kyoto Protocol and Paris Agreement will also be explained in this chapter.

3 3 Defining the international legal climate change regime

3 3 1 Definition of climate change

Climate change is an interdisciplinary and complex problem. It relates to a vast range of subject areas, such as science, law, economics, health, energy and human security.[386] Within the United Nations alone, there are currently more than 20 agencies concerned with the issue of climate change.[387] It is accordingly no surprise that climate change is described as "a second coming for international law".[388] This section defines the term 'climate change' in the context of the international legal climate change regime.

Climate change can be described in simple terms as the "response of the planet's climate system to altered concentrations of carbon dioxide and other 'greenhouse gases' in the atmosphere".[389] It is also necessary to consider more authoritative definitions of the term as found in the assessment

386 Council on Foreign Relations "The Climate Change Regime" (19-06-2013) (page numbers not available) *Council on Foreign Relations* <http://www.cfr.org/climate-change/global-climate-change-regime/p21831> (accessed 17-04-2015).

387 Council on Foreign Relations "The Climate Change Regime" (19-06-2013) (page numbers not available) *Council on Foreign Relations* <http://www.cfr.org/climate-change/global-climate-change-regime/p21831> (accessed 17-04-2015).

388 D Badrinarayana "Global Warming: A Second Coming for International Law?" (2010) 85 *Washington Law Review* 253 253-292.

389 C Wold et al Climate Change and the Law (2009) 1.

reports of the Intergovernmental Panel on Climate Change (IPCC) and also in the text of the UNFCCC. The fifth assessment report of the IPCC defines climate change as follows:

> "[A] change in the state of the climate that can be identified (e.g., by using statistical tests) by changes in the mean and/or the variability of its properties and that persists for an extended period, typically decades or longer. Climate change may be due to natural internal processes or external forcings such as modulations of the solar cycles, volcanic eruptions and persistent anthropogenic changes in the composition of the atmosphere or in land use."[390]

The UNFCCC defines climate change as follows:

> "[A] change of climate which is attributed directly or indirectly to human activity that alters the composition of the global atmosphere and which is in addition to natural climate variability observed over comparable time periods."[391]

It should be noted that the definition of the IPCC and the UNFCCC includes climate change due to anthropogenic activities and due to natural causes. This research project is not concerned with climate change due to natural causes as the international legal climate change regime is concerned with addressing anthropogenic activities which are altering the atmosphere. However, for the sake of completeness, it is submitted that a definition for climate change must be understood to contain both natural and anthropocentric activities. The definition of climate change adhered to in this research project corresponds with the meaning of climate change as described by the UNFCCC.

The utilisation and degradation of natural resources by states under the mantle of PSNR are key contributors to the increasing greenhouse gases in the atmosphere responsible for climate change.[392] This further underlines the nexus between PSNR and climate change as discussed in Section 3.2. It is therefore a natural inference that it should be possible to fight climate change by using PSNR so that it becomes more difficult for states to utilise natural resources in a way that would contribute excessively to climate change and without compliance with the climate commitments states

---

390 IPCC [KJ Mach et al (eds)] Annex II: Glossary. In Climate Change 2014: Synthesis Report (2014) 120.

391 Article 1(2), UNFCCC.

392 RR Judkins & W Fulkerson "The dilemma of fossil fuel use and global climate change" (1993) 7 Energy & Fuels 14 16; C Wold et al Climate Change and the Law (2009) 5.

make in terms of the international legal climate change regime. The possibility for people to hold states accountable for their commitments made under the international legal climate change regime, in terms of people's rights under PSNR, is explored in Chapter 4.

3 3 2 Definition of regime

In the context of international relations, Krasner defines the term 'regime' as "implicit or explicit principles, norms, rules, and decision-making procedures around which actors' expectations converge".[393] However, in the context of the international legal climate change regime, Yamin and Depledge define the term 'regime' as referring simply to the rules, regulations and institutions relevant to a specific subject area.[394] In order to understand the terms employed by these definitions, it is useful to consider the difference between the terms 'rules' and 'principles', which is that rules are considered to be more specific, while principles are considered to be general standards of behaviour.[395] Furthermore, 'norms' are defined as "a broad category covering any provision with prescriptive content"[396] and 'decision-making procedures' are defined as "prevailing practices for making and implementing collective choice"[397]. An important characteristic of a 'regime' is that it denotes something less permanent in nature, and can be altered by a "shift in power or interest".[398] Ultimately, 'regimes' can also be referred to as man-made arrangements or grouped activities that serve certain objectives.[399] In the case of the international legal cli-

393 SD Krasner "Structural causes and regime consequences: regimes as intervening variables" in SD Krasner (ed) *International Regimes* (1983) 1 2.

394 F Yamin & J Depledge *The international climate change regime: A guide to rules, institutions and procedures* (2004) 6.

395 F Yamin & J Depledge *The international climate change regime: A guide to rules, institutions and procedures* (2004) 7.

396 F Yamin & J Depledge *The international climate change regime: A guide to rules, institutions and procedures* (2004) 7.

397 SD Krasner "Structural causes and regime consequences: regimes as intervening variables" in SD Krasner (ed) *International Regimes* (1983) 1 2.

398 SD Krasner "Structural causes and regime consequences: regimes as intervening variables" in SD Krasner (ed) *International Regimes* (1983) 1 2.

399 DJ Puchala & RF Hopkins "International regimes: lessons from inductive analysis" in SD Krasner (ed) *International Regimes* (1983) 61 61; SD Krasner "Struc-

mate change regime, one such objective of the regime would be to mitigate the effects of climate change.

It is sufficient for the purposes of this research project to understand the term 'regime' as defined above by Yamin and Depledge. Furthermore, it is possible to understand and define the term 'climate change regime' by considering the meaning of the term 'climate change' and 'regime', as was done in preceding paragraphs. The term 'climate change regime' can be further specified by being confined to the international legal realm. Accordingly, references in the research project to the climate change regime will include the preceding term 'international legal'. Accordingly, the term 'international legal climate change regime' is defined within the research project as the international law (including rules, principles and norms) that govern all aspects related to climate change as well as national, regional and international institutions that regulate and administer according to the international law on the subject.

Even though PSNR is a principle of international law, it is also observed that some may refer to a "PSNR regime'".[400] It should be noted that this research project understands any reference to a PSNR regime as referring to the rights and duties flowing from the principle and the various institutions, such as the United Nations, which may have an impact on its operation. However, in order to avoid confusion, this research project will refer to PSNR as a principle and not a regime.

3 4 Scientific basis of climate change: linking activity in terms of sovereignty over natural resources with greenhouse gas emissions

3 4 1 Link between greenhouse gas emissions and climate change

The process by which greenhouse gases warm the planet is referred to as 'the greenhouse effect' or 'global warming' and has been documented from as early as 1896.[401] It is the unbalanced increase in the concentration

---

tural causes and regime consequences: regimes as intervening variables" in SD Krasner (ed) *International Regimes* (1983) 1 28.

400 For example, see: I Mgbeoji "Beyond Rhetoric: State Sovereignty, Common Concern, and the Inapplicability of the Common Heritage Concept to Plant Genetic Resources" (2003) 16(4) *Leiden Journal of International Law* 821 835.

401 C Wold et al Climate Change and the Law (2009) 2.

of greenhouse gases in the atmosphere, causing a global increase in average temperatures and associated changes to weather patterns on a global scale, that is referred to as global warming or climate change.[402] It has been determined to be more than 90% likely that the observed increase in global temperature since the mid-20th century is due to the increase in anthropocentric emissions of greenhouse gases.[403] Accordingly, the IPCC has laid to rest the questions regarding the causal link between global warming and anthropocentric emissions of greenhouse gases.[404] Accordingly, the science on which the international legal climate change regime is based forms a solid foundation of the regime. The most important questions within the regime at present, in regard to climate change, do not concern the existence of climate change but concern the rate of warming and the impact of the warming that needs to be addressed.[405]

According to the IPCC the increase in emissions of carbon dioxide and other greenhouse gases have increased to "unprecedented levels" from fossil fuel emissions and net land-use change emissions.[406] Carbon dioxide emissions in particular have increased by 40% since pre-industrial times.[407] Besides an increase in temperature, anthropocentric emissions of greenhouse gases cause other changes in climate such as ocean warming, temperature extremes and changes in wind patterns.[408] According to the IPCC, there is a greater than 90% likelihood that the continuation of greenhouse gas emissions at the current rate or higher will cause further warming and will affect the global climate system to a greater extent than has already been observed.[409] The IPCC's research makes it clear that the

---

402 C Wold et al Climate Change and the Law (2009) 2.

403 IPCC [S Solomon et al (eds)] Climate Change 2007: The Physical Science Basis (2007) 10.

404 C Wold et al Climate Change and the Law (2009) 12.

405 C Wold et al Climate Change and the Law (2009) 15.
The existence of climate change is not questioned in this research project and this research project does not set out to prove that climate change is real. Accordingly, this research project is written on the assumption that climate change exists.

406 IPCC [TF Stocker et al (eds)] *Summary for Policymakers in Climate Change 2013: The Physical Science Basis* (2013) 11.

407 IPCC [TF Stocker et al (eds)] *Summary for Policymakers in Climate Change 2013: The Physical Science Basis* (2013) 11.

408 IPCC [S Solomon et al (eds)] Climate Change 2007: The Physical Science Basis (2007) 10.

409 IPCC [S Solomon et al (eds)] Climate Change 2007: The Physical Science Basis (2007) 10.

scale and gravity of climate change are a formidable challenge to address within the international legal climate change regime, and it is in this light that the roping in of PSNR as a positive contribution to the regime is warranted.

A consideration of the most prevalent greenhouse gases, and the corresponding anthropocentric activities leading to their emission, further illustrates the gravity, scale and complexity of that which the international legal climate regime seeks to address. A consideration of the most prevalent greenhouse gases, and the corresponding anthropocentric activities listed below, illustrates that the activities of states in terms of their natural resources are firmly planted in PSNR (including the activities that states allow their citizens to perform in this regard) and are also the main source of greenhouse gas emissions. In this regard, the nexus between PSNR and climate change is strengthened.

The most prevalent anthropocentrically produced greenhouse gas to take note of is carbon dioxide, for reasons that will be set out below. The other relevant greenhouse gases anthropocentrically produced are methane, nitrous oxide, hydrofluorocarbons, perfluorocarbons and sulphur hexafluoride.[410] These greenhouse gases are specifically mentioned by the Kyoto Protocol, unlike gases of which the impacts are less well known such as sulphur dioxide, nitrogen oxides, carbon monoxide, hydrogen sulphide, ozone and volatile organic compounds.[411] The greenhouse gases mentioned in the Kyoto Protocol and the corresponding anthropocentric activities contributing to their emissions are considered below.

Global greenhouse gas emissions can be broken down according to emission of greenhouse gases by economic sector. The following sectors and percentage of greenhouse gas emissions is a good indication of the contribution of anthropocentric activity to greenhouse gas emissions: electricity and heat (involving the burning of coal natural gas and oil) account for 25% of emissions; agriculture (pertaining to crops and livestock), forestry (including deforestation) and other land-use account for 24% of

410 C Wold et al Climate Change and the Law (2009) 5.
Chlorofluorocarbons, although constituting an anthropocentric greenhouse gas, are regulated under the Montreal Protocol. [C Wold et al Climate Change and the Law (2009) 5.].

411 C Wold et al Climate Change and the Law (2009) 6.

emissions;[412] industry (involving mainly the burning of fossil fuels) accounts for 21% of emissions; transportation (involving burning of fossil fuels for various means of transportation) accounts for 14% of emissions; other energy (related to activities such as fuel refining, extraction, or processing) accounts for 10% of emissions; and buildings (involving on-site energy generation and burning of fossil fuels to create electricity) account for 6% of emissions.[413] The abovementioned emissions, according to the economic sector to which they pertain, illustrate the complexity of addressing the issue of climate change. Many different sectors of a state's economy would have to undergo transformation in order to conform to the emission levels advanced by the international legal climate change regime. The succeeding paragraphs of this chapter serve to illustrate the complexity, scale and gravity of the problem of climate change in greater detail, by considering the main greenhouse gases and the accompanying anthropocentric activity that leads to an increase in the emission of a specific greenhouse gas.

3 4 2 Anthropocentric activity in terms of sovereignty over natural resources and the resulting increase in a specific greenhouse gas emission

3 4 2 1 Carbon dioxide and anthropocentric activity

Carbon dioxide composes more than half of greenhouse gas emissions anthropocentrically produced.[414] More specifically, carbon dioxide from fossil fuel and industrial usage forms 65% of global greenhouse gas emis-

---

412 This percentage does not include the offset of greenhouse gases removed from the atmosphere. [United States Environmental Protection Agency "Global Greenhouse Gas Emissions Data" (10-06-2016) (page numbers unavailable) United States Environmental Protection Agency <https://www3.epa.gov/climatechange/ghgemissions/global.html> (accessed 04-07-2016).].

413 United States Environmental Protection Agency "Global Greenhouse Gas Emissions Data" (10-06-2016) (page numbers unavailable) United States Environmental Protection Agency <https://www3.epa.gov/climatechange/ghgemissions/global.html> (accessed 04-07-2016).

414 IPCC [S Solomon et al (eds)] Climate Change 2007: The Physical Science Basis (2007) 2; C Wold et al Climate Change and the Law (2009) 5.

sions.[415] The atmospheric concentration of carbon dioxide has increased since the pre-industrial period mainly owing to fossil fuel use.[416] The burning of fossil fuels accounts for more than 80% of the carbon dioxide concentration in the atmosphere.[417] The sources of the burning of fossil fuels range from large-scale operations involving power plants or cement manufacturing to smaller individual activities such as the use of motor vehicles.[418] The atmospheric concentration of carbon dioxide has also increased since the pre-industrial period owing to factors such as land-use change to about 11% of global greenhouse gas emissions.[419]

The role of black carbon (also known as soot) in the climate change process is relevant in this context. To date, black carbon is believed to be second only to carbon dioxide in its role in climate change. Black carbon is technically a solid and not a gas.[420] Nevertheless, black carbon may account for nearly a third of the warming in the Arctic.[421] Simply put, black carbon is formed in flames.[422] The most noteworthy sources of black carbon production are the incomplete combustion of coal, diesel, wood and biomass fuel.[423] Activities which produce the largest sources of black car-

---

415 United States Environmental Protection Agency "Global Greenhouse Gas Emissions Data" (10-06-2016) (page numbers unavailable) United States Environmental Protection Agency <https://www3.epa.gov/climatechange/ghgemissions/global.html> (accessed 04-07-2016).

416 IPCC [S Solomon et al (eds)] Climate Change 2007: The Physical Science Basis 2007 2.

417 C Wold et al Climate Change and the Law (2009) 5.

418 IPCC [S Solomon et al (eds)] Climate Change 2007: The Physical Science Basis (2007) 2; C Wold et al Climate Change and the Law (2009) 5.

419 P Taylor An ecological approach to international law, Responding to challenges of climate change (1998) 12; IPCC [S Solomon et al (eds)] Climate Change 2007: The Physical Science Basis (2007) 2; C Wold et al Climate Change and the Law (2009) 5; United States Environmental Protection Agency "Global Greenhouse Gas Emissions Data" (10-06-2016) (page numbers unavailable) United States Environmental Protection Agency <https://www3.epa.gov/climatechange/ghgemissions/global.html> (accessed 04-07-2016).

420 C Wold et al Climate Change and the Law (2009) 6.

421 C Wold et al Climate Change and the Law (2009) 6.

422 TC Bond et al "Bounding the role of black carbon in the climate system: A scientific assessment" (2013) 18 Journal of Geophysical Research 5380 5384; C Wold et al Climate Change and the Law (2009) 6.

423 TC Bond et al "Bounding the role of black carbon in the climate system: A scientific assessment" (2013) 18 Journal of Geophysical Research 5380 5385; C Wold et al Climate Change and the Law (2009) 6.

bon include the burning of forests and savannas.[424] A noteworthy difference between black carbon and greenhouse gas emissions lies in the rough differentiation that the effects of black carbon are localised while the effects of greenhouse gas emissions are not.[425]

3 4 2 2 Methane and anthropocentric activity

The atmospheric concentration of methane has increased since the pre-industrial period.[426] Methane forms about 16% of global greenhouse gas emissions.[427] More specifically, methane is produced by anthropocentric activities relating to coal mining, waste decomposition, plant decay, agriculture involving livestock, the increase in the area or productivity of rice paddies, and draining or flooding of wetlands.[428] It has also been noted that the increase in temperature may cause methane to seep from ocean beds and frozen lake beds.[429]

3 4 2 3 Nitrous oxide and anthropocentric activity

Nitrous oxide forms about 6% of global greenhouse gas emissions.[430] The increase in the atmospheric concentration of nitrous oxide is due to certain anthropocentric activities mainly involving agricultural activities, such as

---

424 TC Bond et al "Bounding the role of black carbon in the climate system: A scientific assessment" (2013) 18 Journal of Geophysical Research 5380 5385.

425 C Wold et al Climate Change and the Law (2009) 6.

426 IPCC [S Solomon et al (eds)] Climate Change 2007: The Physical Science Basis (2007) 3.

427 United States Environmental Protection Agency "Global Greenhouse Gas Emissions Data" (10-06-2016) (page numbers unavailable) United States Environmental Protection Agency <https://www3.epa.gov/climatechange/ghgemissions/global.html> (accessed 04-07-2016).

428 P Taylor An ecological approach to international law, Responding to challenges of climate change (1998) 13; C Wold et al Climate Change and the Law (2009) 5-6.

429 C Wold et al Climate Change and the Law (2009) 6.

430 United States Environmental Protection Agency "Global Greenhouse Gas Emissions Data" (10-06-2016) (page numbers unavailable) United States Environmental Protection Agency <https://www3.epa.gov/climatechange/ghgemissions/global.html> (accessed 04-07-2016).

the production of livestock, and the burning of biomass and fossil fuels, which are related to the use of motor vehicles and industrial activity.[431]

3 4 2 4 Fluorinated gases and anthropocentric activity

Fluorinated gases form about 2% of global greenhouse gas emissions.[432] As a greenhouse gas, chlorofluorocarbons have an effect which is 10 000 times greater than the effect of carbon dioxide.[433] Chlorofluorocarbons and hydrochlorofluorocarbons, which were used in products such as air conditioners and refrigerators, are being phased out by the Montreal Protocol on Substances that Deplete the Ozone Layer 1987 (Montreal Protocol) and the Vienna Convention for the Protection of the Ozone Layer 1985 (Vienna Convention for the Protection of the Ozone Layer). However, the emission of hydrofluorocarbons, perfluorocarbons and sulphur hexafluoride (also known as fluorinated gases) are emitted through different industrial processes and are potent alternatives to chlorofluorocarbons that currently contribute to climate change.[434]

431 P Taylor An ecological approach to international law, Responding to challenges of climate change (1998) 13; IPCC [S Solomon et al (eds)] Climate Change 2007: The Physical Science Basis (2007) 3; C Wold et al Climate Change and the Law (2009) 6.

432 United States Environmental Protection Agency "Global Greenhouse Gas Emissions Data" (10-06-2016) (page numbers unavailable) United States Environmental Protection Agency <https://www3.epa.gov/climatechange/ghgemissions/global.html> (accessed 04-07-2016).

433 P Taylor An ecological approach to international law, Responding to challenges of climate change (1998) 13.

434 United States Environmental Protection Agency "Overview of Greenhouse Gases" (26-05-2016) (page numbers unavailable) United States Environmental Protection Agency <https://www3.epa.gov/climatechange/ghgemissions/gases.html> (accessed 30-06-2016); C Wold et al Climate Change and the Law (2009) 6.

3 5 Illustrating the gravity of climate change in terms of its environmental, social and economic impacts on the well-being of people

3 5 1 Introduction

Climate change is the cause of a myriad of "problems without passports"[435] – it is an issue that also touches on the erosion of state sovereignty, as discussed in Chapter 2. Whether these problems without passports are classified as environmental, social or economic problems, the common denominator seems to be that the well-being of people is affected by climate change, and it is necessary to forge a more effective response to climate change.

This section of the research project sets out the meaning of the term 'well-being' as used in the research project, by considering physiological and psychological components of human well-being. In addition, this section will also consider whether human well-being can be construed as an objective of both the international legal climate change regime and PSNR, in order to explore the connection between the regime and the principle further.

Lastly, this section considers the environmental, social and economic effects of climate change and their impact on the well-being of people.[436] By understanding the effect of climate change on the well-being of people, the gravity of the problem of climate change and the urgent need to explore innovative ways, such as the reinterpretation and reconfiguration of PSNR, in which to combat climate change is emphasised.

---

435 Worldwatch Institute "State of the World 2005 Trends and Facts – Security Redefined" (30-04-2015) (page numbers unavailable) *Worldwatch Institute* <http://www.worldwatch.org/node/65> (accessed 30-04-2015).

436 Importantly, this section is not a philosophical study of the term 'well-being' as this would fall outside the scope of the research project. For a philosophical exposition of 'well-being', see: WJ Talbott Human Rights and Human Well-Being (2010).

3 5 2 Understanding the term 'well-being'

The 'well-being of people' is a general and broad term, applicable throughout a wide array of subject fields, and consequently the term is difficult to define. Most frequently, the concept of well-being is discussed in the context of determining quality of life in disciplines such as economics, environmental science, psychology, medicine and sociology.[437] In a legal context, well-being is most frequently discussed within the realm of human rights and environmental law. Even though it seems that well-being ought to be an obvious objective of certain laws, legal expositions of the term are rare and there remains no single accepted legal definition of the term 'well-being'.

An example of the positioning of the term 'well-being' within the legal realm is found in the UNGA resolution 45/94 of 1990.[438] The resolution, entitled "Need to ensure a healthy environment for the well-being of individuals", considers a healthy environment as a basic component for the fulfilment of human rights.[439] Although the term 'well-being' is not defined in the resolution, it is employed alongside dignity, equality and health as conditions for the fulfilment of human rights.

For the purposes of understanding the term 'well-being', as used in the context of the research project, it is useful to understand the well-being of people as comprising two components, namely physiological well-being and psychological well-being.[440] Physiological well-being would concern considerations of income, food and health. Psychological well-being would concern a less-defined set of considerations, which are set out below.

Income and food are prominent components of physiological well-being, especially in poorer communities or less-developed countries.[441] For example, people in a rural community collect their food from the environ-

437 R Costanza et al "Quality of life: An approach integrating opportunities, human needs, and subjective well-being" (2007) 61 *Ecological Economics* 268.

438 UNGA resolution 45/94 of 14 December 1990.

439 UNGA resolution 45/94 of 14 December 1990.

440 KS Meijer *Human Well-Being Values of Environmental Flows, Enhancing Social Equity in Integrated Water Resources Management* PhD thesis, Technische Universiteit Delft (2007) 33.

441 KS Meijer *Human Well-Being Values of Environmental Flows, Enhancing Social Equity in Integrated Water Resources Management* PhD thesis, Technische Universiteit Delft (2007) 33.

ment, take what they need to survive from their harvest, sell the rest of their harvest, and buy what they need from other people's harvests with income obtained from selling their own harvest.[442] This example shows the interdependence that can exist between income and food, and it is clear to see how a change in the environment due to climate change may affect people's harvests and their income, thus affecting their physiological well-being.

Health is also a component of the physiological well-being of people.The World Health Organization defines human health as "a state of complete physical, mental and social well-being" and goes further by including that health is "not merely the absence of disease or infirmity".[443] Therefore, health can pertain to physiological well-being, and to social and mental (or psychological) well-being. This means that there is a degree of overlap between physiological and psychological well-being. Social and mental well-being will be discussed under psychological components of well-being. The component of psychological well-being (including social and mental well-being) can refer to issues of belonging, identity, community, tradition, culture, religion, and recreation.[444] The psychological well-being of people is especially relevant in cases involving the relationship between indigenous people and their right to natural resources. This relationship is addressed further in Chapter 4 of the research project.

When considering physiological and psychological well-being, it is clear how the environment can impact on this component of well-being, and accordingly how environmental changes due to climate change can affect this component of well-being. Environmental impacts of climate change on the well-being of people are discussed in greater detail below. However, it can be mentioned briefly at this point that climate change can have an impact on physiological and psychological human health in a variety of ways, such as by causing heat stress, floods, drought, intense

442 KS Meijer *Human Well-Being Values of Environmental Flows, Enhancing Social Equity in Integrated Water Resources Management* PhD thesis, Technische Universiteit Delft (2007) 33.

443 Preamble, Constitution of the World Health Organization.

444 KS Meijer *Human Well-Being Values of Environmental Flows, Enhancing Social Equity in Integrated Water Resources Management* PhD thesis, Technische Universiteit Delft (2007) 34.

storms, spread of disease vectors, food insecurity and malnutrition, displacement, and mental ill-health.[445]

Therefore, in short, this research project considers the term 'well-being of people' as referring to physiological components (which include income, food and physical health) and psychological components (which include social and mental health). For the purposes of this research project, this understanding of the term 'well-being' is sufficient, since it is only necessary to understand well-being in order to show that the effects of climate change gravely impact on well-being of people and accordingly warrant a reconfiguration of PSNR in order to increase the effectiveness of the international legal climate change regime.

3 5 3 Well-being of people as an objective of the international legal climate change regime and as an objective of the principle of sovereignty over natural resources

The position or relevance of human well-being within the international legal change regime is most vital when considering the environmental, social and economic impacts of climate change, as will be done in the following paragraphs of this section. The international legal climate change regime – by addressing the impacts of climate change through mitigation of impacts and adaption to impacts – therefore automatically addresses and promotes human well-being. However, it is not suspected that the well-being of people is a central and main concern of the international legal climate change regime, as the regime is a state-centric regime. As has already been submitted at the beginning of Chapter 3, the state forms the core of the international legal climate change regime, as the functionality of the regime is built on state participation and action, and organised by the United Nations which is ultimately also a state-based system. In support of the state-centric arrangement of the international legal climate change regime, discussions of the regime which refer to the term 'well-being' usually refer to the term in the context of the economic well-being of the state (as opposed to people); this focus on the economic well-being of the state explains why a state which relies heavily on fossil fuels for eco-

445 N Watts et al "Health and climate change: policy responses to protect public health" (23-06-2015) (1) *The Lancet* <http://www.thelancet.com/pdfs/journals/lancet/PIIS0140-6736(15)60854-6.pdf> (accessed 30-06-2015).

nomic growth chooses not to participate in the regime.[446] Therefore, while human well-being is very relevant to the international legal climate change regime and is greatly affected by the operation of the regime, it can be submitted that human well-being does not form an objective of the regime.

The well-being of people is a more suitable objective within PSNR than it is within the international legal climate change regime. From the historical exposition of PSNR in Chapter 2, it is evident that the well-being of people is construed in two ways. In the first instance, the well-being of people can be construed as a limitation on the exercise of the state's rights in terms of PSNR. Two UNGA resolutions stipulate that states should exercise their sovereignty over natural resources for the well-being of people.[447] The landmark General Assembly resolution 1803 (XVII) of 14 December 1962 declares that the "right of peoples and nations to permanent sovereignty over their natural wealth and resources must be exercised in the interest of their national development and of the well-being of the peoples of the State concerned."[448] General Assembly resolution 2692(XXV) reaffirms this statement of resolution 1803 (XVII).[449] In the second instance, the well-being of people can also be construed as a duty of states in exercising PSNR. More accurately, Schrijver considers it a well-established duty of states to exercise sovereignty over natural resources for national development and the well-being of the people.[450] This state duty, and the balance which is struck between national development and well-being, is discussed in greater detail in Chapter 4. Accordingly, the well-being of people, although not a main objective of PSNR, is settled as a

---

446 J von Stein "The International Law and Politics of Climate Change, Ratification of the United Nations Framework Convention and the Kyoto Protocol" (2008) 52(2) 243 253.

447 D Cambou & S Smis "Permanent sovereignty over natural resources from a human rights perspective: Natural resources exploitation and indigenous peoples' rights in the arctic" (2013) 22(1) Michigan State International Law Review 347 359.
These UNGA resolutions are resolution 1803 (XVII) entitled 'Permanent Sovereignty over Natural Resources' of 14 December 1962 and UNGA resolution 2692 (XXV) entitled 'Permanent Sovereignty over Natural Resources of Developing Countries and Expansion of Domestic Sources of Accumulation for Economic Development' of 11 December 1970.

448 UNGA resolution 1803 (XVII) of 14 December 1962.

449 UNGA resolution 2692 (XXV) of 11 December 1970.

450 N Schrijver Sovereignty Over Natural Resources, Balancing Rights and Duties (1997) 309.

definite facet of PSNR. As was submitted above, the same does not hold true for the international legal climate change regime.

If the state duty in terms of PSNR to promote the well-being of people is exercised and the international legal climate change regime objective to mitigate the impacts of climate change is realised (and by effect this promotes the well-being of people), it can be forwarded that the well-being of people is promoted by both PSNR and the international legal climate change regime. In addition, a strengthening of the implementation of this state duty in terms of PSNR will strengthen the implementation of the international legal climate change regime and vice versa. This not only shows a nexus between PSNR and the international legal climate change regime, but also shows how PSNR and the regime, when implemented, can strengthen each other.

3 5 4 Environmental effects of climate change on the well-being of people

According to the IPCC, it is clear that there are a variety of environmental effects or perceived changes in the global climate system caused by climate change.[451] These changes illustrate the gravity and seriousness of climate change and relate mainly to the atmosphere, the ocean, the cryosphere,[452] the sea level, and to carbon and biogeochemical cycles.[453]

Notable changes in the atmosphere due to climate change have caused some of the following changes in regard to temperature and precipitation: there has been a global decrease in cold nights and days, as well as a global increase in warm nights and days; there has been an increase in the frequency of heat waves in parts of Europe, Asia and Australia; and regions

451 IPCC [TF Stocker et al (eds)] *Summary for Policymakers in Climate Change 2013: The Physical Science Basis* (2013) 4.

452 According to the National Snow & Ice Data Center, the cryosphere refers to areas where water is found in its solid form. [National Snow & Ice Data Center "What is the Cryosphere?" (date of publication unavailable) (page numbers unavailable) *National Snow & Ice Data Center* <https://nsidc.org/cryosphere/what-is-the-cryosphere.html> (accessed 08-04-2015).].

453 An example of a biochemical cycle is the carbon cycle. For an explanation of biochemical cycles see: The Editors of Encyclopedia Britannica "Biochemical cycle" (04-01-2015) (page numbers unavailable) *Encyclopedia Britannica* <https://global.britannica.com/science/biogeochemical-cycle> (accessed 19-07-2016).

such as North America and Europe experience a heavier, more frequent and more intense precipitation.[454]

The environment is also impacted by climate change in regard to changes in snow, ice and frozen ground (including permafrost) as follows: glacial lakes have enlarged and their number has increased; permafrost regions show an increase in ground instability; rock avalanches occur more frequently in mountainous regions; and ecosystems of the Arctic and Antarctic are impacted in various ways.[455] In the last 20 years, ice sheets found in Greenland and the Antarctic have lost mass and globally glaciers are shrinking.[456] In addition, spring snow cover continues to decrease in the Arctic sea and Northern Hemisphere.[457]

The IPCC has also observed that climate change has impacted on global hydrological systems in various ways, which include changes in snow-fed and glacier-fed rivers owing to increased run-off and changes in the times of run-off during the year.[458] In addition, many lakes and rivers are warming, which has various effects on, for instance, water quality.[459] The ocean has absorbed about 30% of the anthropogenically emitted carbon dioxide, which is causing the ocean to become more acidic.[460] In addition, the ocean is warming, and the evaporation of and precipitation over the oceans have changed.[461] Furthermore, the sea level has risen more since the mid-19th century than during the previous two millennia combined.[462]

---

454 IPCC [TF Stocker et al (eds)] *Summary for Policymakers in Climate Change 2013: The Physical Science Basis* (2013) 5.

455 IPCC [ML Parry et al (eds)] Summary for Policymakers. In: Climate Change 2007: Impacts, Adaptation and Vulnerability (2007) 8.

456 IPCC [TF Stocker et al (eds)] *Summary for Policymakers in Climate Change 2013: The Physical Science Basis* (2013) 9.

457 IPCC [TF Stocker et al (eds)] *Summary for Policymakers in Climate Change 2013: The Physical Science Basis* (2013) 9.

458 IPCC [ML Parry et al (eds)] Summary for Policymakers. In: Climate Change 2007: Impacts, Adaptation and Vulnerability (2007) 8.

459 IPCC [ML Parry et al (eds)] Summary for Policymakers. In: Climate Change 2007: Impacts, Adaptation and Vulnerability (2007) 8.

460 IPCC [TF Stocker et al (eds)] *Summary for Policymakers in Climate Change 2013: The Physical Science Basis* (2013) 11.

461 IPCC [TF Stocker et al (eds)] *Summary for Policymakers in Climate Change 2013: The Physical Science Basis* (2013) 6.

462 IPCC [TF Stocker et al (eds)] *Summary for Policymakers in Climate Change 2013: The Physical Science Basis* (2013) 11.

The impact that climate change has on hydrological systems can lead to water scarcity and water damage by means of floods.[463]

Climate change also impacts on terrestrial biological systems. Usual spring events such as leaf-unfolding, bird or other species migration, and egg-laying are now occurring at different times.[464] The IPCC also presents evidence that links the changes in seawater and freshwater biological systems to the increased water temperatures, as well as associated changes in ice cover, salinity, oxygen levels and circulation of water to climate change.[465] These changes include changes in algal, plankton and fish abundance in the high-latitude oceans, high-latitude and high-altitude lakes show an increase in algal and zooplankton abundance, and fish in rivers show range changes and earlier migration patterns.[466]

To summarise, climate change can be said to have environmental impacts in regard to melting ice, rising sea levels, changing ocean ecology, escalating weather events, declining forest areas, increasing desertification, ecosystems and wildlife.[467] It is possible to argue that the well-being of people is being impacted upon and will be impacted upon by the environmental effects of climate change. The well-being of people is deemed to be affected by the environmental impacts of climate change if the environmental impacts affect the physiological component of well-being by affecting food security, health security or the psychological component of well-being. The environmental impacts of climate change listed above affect physiological components of human health, especially in respect of food and health security. In addition, it may be argued that these effects of climate change have an impact on psychological components of well-being, either directly or as a result of the impact on physiological components of well-being.

---

463 IPCC [DJ Arent et al (eds)] Key Economic Sectors and Service in Climate Change 2014: Impacts, Adaptation, and Vulnerability. Part A: Global and Sectoral Aspects (2014) 659 662.

464 IPCC [ML Parry et al (eds)] Summary for Policymakers. In: Climate Change 2007: Impacts, Adaptation and Vulnerability (2007) 8.

465 IPCC [ML Parry et al (eds)] Summary for Policymakers. In: Climate Change 2007: Impacts, Adaptation and Vulnerability (2007) 8.

466 IPCC [ML Parry et al (eds)] Summary for Policymakers. In: Climate Change 2007: Impacts, Adaptation and Vulnerability (2007) 8.

467 WE Easterling "Climate Trends and Prospects" in RN Sampson & D Hair (eds) *Natural Resources for the 21st Century* (1990) 32 47; C Wold et al *Climate Change and the Law* (2009) 18-26.

Well-being can also be discussed within the sphere of "ecosystem services".[468] Ecosystem functions become ecosystem services when certain environmental occurrences are utilised by people to generate valuable processes, materials or commodities.[469] For example, nutrient recycling in soil (an ecosystem function) is utilised so as to produce crops (an ecosystem service).[470] Another example of an ecosystem service isforest system services, which include the provision of raw material, space for human recreation, and natural habitats for certain animals.[471] Climate change will significantly impact on certain ecosystem services by influencing certain ecosystem functions, which in turn will impact the physiological component of human well-being.

3 5 5 Social effects of climate change on the well-being of people

From the abovementioned environmental impacts of climate change, social and economic impacts of climate change develop. Food security may be impacted by climate change by way of the environmental effects of climate change on agriculture that include increasing water scarcity and in-

---

468 C Corvalan et al "Ecosystems and Human Well-Being" (2005) (2) *WHO* <http://www.who.int/globalchange/ecosystems/ecosys.pdf> (accessed 19-07-2016); EJ Nelson et al "Climate change's impact on key ecosystem services and the human well-being they support in the US" (2013) 11(9) *Frontiers in Ecology and the Environment* 483 483.

469 EJ Nelson et al "Climate change's impact on key ecosystem services and the human well-being they support in the US" (2013) 11(9) *Frontiers in Ecology and the Environment* 483 483.

470 EJ Nelson et al "Climate change's impact on key ecosystem services and the human well-being they support in the US" (2013) 11(9) *Frontiers in Ecology and the Environment* 483 483.

471 EJ Nelson et al "Climate change's impact on key ecosystem services and the human well-being they support in the US" (2013) 11(9) *Frontiers in Ecology and the Environment* 483 486-487.

fluencing soil quality and land availability.[472] In addition, this effect on the agricultural sector can increase cases of malnutrition.[473]

Specific health risks of climate change include: high temperature-related illness and death; air pollution-related health issues; extreme weather-related events which could cause drowning and exposure; increasing water- and food-borne diseases; increasing vector-borne and rodent-borne diseases; and mental health effects which can occur after extreme weather events due to personal or economic loss.[474] The impact that climate change will have on human health will also cause an increase in the demand for qualified healthcare personnel, medical supplies and medical infrastructure, and an increase in activities related to disease prevention and confinement of disease.[475] In turn, these effects lead to economic impacts, as will be discussed in the succeeding section.

The social impacts of climate change can further become an issue of national security when they directly result in, or contribute to causing, conflict or war. Naturally, war and conflict present risks to human health, as they can lead to physical injury and death, or they can present risks to psychological health. The effect which climate change can have on natural resources (whether in the form of, for example, increasing water scarcity[476] or a decrease in mineral resource abundance) can exacerbate geopolitical friction; cause confrontation with indigenous people; and precipitate civil wars.[477] Therefore, in the same way that natural resources have been a cause of conflict over millennia, the effects of climate change can cause conflict. The social effects of climate change therefore present new chal-

472 T Wheeler & J von Braun "Climate Change Impacts on Global Food Security" (2013) 341 *Science* 508 508; Worldwatch Institute "State of the World 2005 Trends and Facts – Security Redefined" (30-04-2015) (page numbers unavailable) *Worldwatch Institute* <http://www.worldwatch.org/node/65> (accessed 30-04-2015).

473 C Wold et al *Climate Change and the Law* (2009) 33; T Wheeler & J von Braun "Climate Change Impacts on Global Food Security" (2013) 341 *Science* 508 508.

474 M Zsamboky et al *Impacts of climate change on disadvantaged UK coastal communities* (2011) 31.

475 IPCC [DJ Arent et al (eds)] Key Economic Sectors and Service in Climate Change 2014: Impacts, Adaptation, and Vulnerability. Part A: Global and Sectoral Aspects (2014) 659 663.

476 In this context, see: JR Starr "Water wars" (1991) 82 *Foreign Policy* 17-36.

477 Worldwatch Institute "State of the World 2005 Trends and Facts – Security Redefined" (30-04-2015) (page numbers unavailable) *Worldwatch Institute* <http://www.worldwatch.org/node/65> (accessed 30-04-2015).

lenges to national security of states, as it becomes clear that some problems caused by climate change cannot be solved by conventional methods of national security (such as dispatching military troops).[478] Climate change has been labelled "the mother of all security problems",[479] and the threats posed by climate change have brought new aspects required to ensure national security to the fore. Firstly, contemporary national security is not necessarily dependent on weaponry, and non-military aspects of security are revealed as important.[480] Secondly, security cannot be sufficiently provided for on a purely national basis and some kind of international cooperation is required to ensure national security.[481] Finally, the problems caused by climate change are solved more easily if these problems are understood as "shared risks and vulnerabilities" of all states.[482] Similarly, the international legal climate change regime depends on increased collective participation by all states, whereby participation also implies compliance.

The social impacts of climate change, such as conflict, war, food scarcity and health risks, may also lead to cases of mass migration. In this context, there is an emerging problem of "climate refugees".[483] A good example of the urgency with which to address the problem of climate change is found in the fact that as many as 20 to 200 million climate refugees may be created because of the impact of climate change on the agricultural sec-

478 Worldwatch Institute "State of the World 2005 Trends and Facts – Security Redefined" (30-04-2015) (page numbers unavailable) *Worldwatch Institute* <http://www.worldwatch.org/node/65> (accessed 30-04-2015).

479 O Brown, A Hammill & R Mcleman "Climate change as the 'new' security threat: implications for Africa" (2007) 83 *International Affairs* 1141 1142.

480 Worldwatch Institute "State of the World 2005 Trends and Facts – Security Redefined" (30-04-2015) (page numbers unavailable) *Worldwatch Institute* <http://www.worldwatch.org/node/65> (accessed 30-04-2015).

481 Worldwatch Institute "State of the World 2005 Trends and Facts – Security Redefined" (30-04-2015) (page numbers unavailable) *Worldwatch Institute* <http://www.worldwatch.org/node/65> (accessed 30-04-2015).

482 Worldwatch Institute "State of the World 2005 Trends and Facts – Security Redefined" (30-04-2015) (page numbers unavailable) *Worldwatch Institute* <http://www.worldwatch.org/node/65> (accessed 30-04-2015).

483 OC Ruppel & S van Wyk "Climate-change-induced Movement of Persons in Africa: Human Rights Responses to Aspects of Human Security" in OC Ruppel et al (eds) *Climate Change: International Law and Global Governance* (2013) 799-826.

tor, increased desertification, increased flooding and the rise in sea level.[484]

The well-being of people will be affected by the social impacts of climate change if these influence the physiological or psychological components of well-being. It is clear that social impacts of climate change affecting human health will affect the well-being of people, as human health forms part of the physiological component of human well-being. It is also clear that the effect of climate change on food security will affect the well-being of people, as food security forms part of the physiological component of human well-being.

3 5 6 Economic effects of climate change on the well being of people

As mentioned above, the social impacts of climate change (such as health impacts) can lead to economic impacts. However, there are some economic impacts that can be separated from social impacts, as listed below. Similarly, certain environmental impacts of climate change lead to various economic impacts.[485] For instance, more frequent and intense weather events due to climate change will have a significant impact on the insurance industry on a national and international basis.[486] The insurance industry will also be challenged in providing affordable insurance cover in those particular countries which are less able to afford it,[487] but prone to rely on it.

Climate change can also have an impact on the housing infrastructure, and economic considerations in this regard include changes to property values and property insurance premiums. As was explained when dis-

---

484 C Wold et al *Climate Change and the Law* (2009) 33.

485 For a more detailed consideration of the economics of climate change, see the following landmark review: N Stern "Stern Review: The Economics of Climate Change" (2006) (page numbers unavailable) *Centro de Ciência do Sistema Terrestre* <http://mudancasclimaticas.cptec.inpe.br/~rmclima/pdfs/destaques/sternreview_report_complete.pdf> (accessed 30-04-2015)].

486 IPCC [DJ Arent et al (eds)] Key Economic Sectors and Service in Climate Change 2014: Impacts, Adaptation, and Vulnerability. Part A: Global and Sectoral Aspects (2014) 659 662.

487 IPCC [DJ Arent et al (eds)] Key Economic Sectors and Service in Climate Change 2014: Impacts, Adaptation, and Vulnerability. Part A: Global and Sectoral Aspects (2014) 659 662.

cussing the environmental effects of climate change, the impact of climate change on hydrological systems can lead to floods, causing physical damage to infrastructure. Damage to transport infrastructure and the cost of repairs, for example, can have a significant impact on the economic sector of the country. Transport infrastructure can be further affected by other kinds of environmental impacts caused by climate change, such as temperature extremes, as most roads and bridges are vulnerable to such changes.[488] In addition, as explained above, environmental impacts of climate change can lead to water scarcity, and securing the demand and supply infrastructure for water will have various economic impacts on a country.[489]

Furthermore, economic sectors such as tourism, fishing and agriculture will be affected by climate change.[490] It is easy to understand why the tourism sector may be affected by extreme weather events: in the case of ski resorts, beach resorts and nature resorts, the feasibility or enjoyment of the resorts is permitted by specific weather conditions.[491] The fishing industry will be affected in terms of providing a livelihood and employment to a certain part of the population.[492] Temperature and precipitation changes, as well as introduction of different weeds, pests and diseases, and issues of soil degradation will affect the agricultural sector with an ensuing effect on crop yields and overall productivity.[493]

The impact of climate change on the energy sector of the economy will be varied and significant. Firstly, climate change may increase the energy needed to heat and cool living spaces, depending on its specific effects in

---

488 IPCC [DJ Arent et al (eds)] Key Economic Sectors and Service in Climate Change 2014: Impacts, Adaptation, and Vulnerability. Part A: Global and Sectoral Aspects (2014) 659 662.

489 IPCC [DJ Arent et al (eds)] Key Economic Sectors and Service in Climate Change 2014: Impacts, Adaptation, and Vulnerability. Part A: Global and Sectoral Aspects (2014) 659 662.

490 M Zsamboky et al *Impacts of climate change on disadvantaged UK coastal communities* (2011) 6 & 32; WE Easterling "Climate Trends and Prospects" in RN Sampson & D Hair (eds) *Natural Resources for the 21st Century* (1990) 32 47.

491 IPCC [DJ Arent et al (eds)] Key Economic Sectors and Service in Climate Change 2014: Impacts, Adaptation, and Vulnerability. Part A: Global and Sectoral Aspects (2014) 659 663.

492 M Zsamboky et al *Impacts of climate change on disadvantaged UK coastal communities* (2011) 32.

493 C Wold et al *Climate Change and the Law* (2009) 31-32; M Zsamboky et al *Impacts of climate change on disadvantaged UK coastal communities* (2011) 32.

specific locations.[494] Furthermore, energy sources and technologies relating to energy generation will be affected. For example, climate change will affect availability and temperature of water needed for cooling in regard to thermal and nuclear power plants.[495] Furthermore, climate change can affect the structure and functionality of pipelines and electricity grids in various ways,[496] for instance floods may cause significant damage to this infrastructure.

The well-being of people will be affected by the economic impacts of climate change,[497] because economic hardships of states ultimately trickle down to the citizens of the state, affecting the income-component of the well-being of people. A clear example of how the well-being of people is affected by the economic impact of climate change on the state is found in the imposition of a carbon tax.[498]

The various economic impacts of climate change illustrate the urgency of the problem of climate change especially well. Nicholas Stern produced a landmark review of the economics of climate change, in which he submits that early action in reducing greenhouse gas emissions to avoid the worst impacts of climate change is paramount as immediate responses would economically amount to about 1% of global GDP each year, while later responses could amount to as much as 20% of global GDP each

494 IPCC [DJ Arent et al (eds)] Key Economic Sectors and Service in Climate Change 2014: Impacts, Adaptation, and Vulnerability. Part A: Global and Sectoral Aspects (2014) 659 662.

495 IPCC [DJ Arent et al (eds)] Key Economic Sectors and Service in Climate Change 2014: Impacts, Adaptation, and Vulnerability. Part A: Global and Sectoral Aspects (2014) 659 662.

496 IPCC [DJ Arent et al (eds)] Key Economic Sectors and Service in Climate Change 2014: Impacts, Adaptation, and Vulnerability. Part A: Global and Sectoral Aspects (2014) 659 662.

497 IPCC [DJ Arent et al (eds)] Key Economic Sectors and Service in Climate Change 2014: Impacts, Adaptation, and Vulnerability. Part A: Global and Sectoral Aspects (2014) 659 663.

498 For more information on a carbon tax, see: GE Metcalf & DA Weisbach "The Design of a Carbon Tax" (2009) Public Law and Legal Theory Working Papers 1-62.

year.[499] The effort to increase the effectiveness of the international legal climate change regime therefore remains a matter of urgency.

At this point of the research project, the gravity of climate change and its environmental, social and economic impacts on the well-being of people should be evident. It is therefore vital to explore innovative ways in which to combat climate change. As submitted in Chapter 1, the primary research question of the research project is to determine if climate change has an impact on the principle of sovereignty over natural resources. It has already been demonstrated how climate change and PSNR are intrinsically connected, and because of the nexus between climate change and PSNR it might be possible to submit that the gravity of climate change has an impact on PSNR. However, even if the connection between PSNR and climate change does not exist, the gravity and scale of the environmental (and social and economic) impacts of climate change in themselves warrant the consideration of any principle of international environmental law that is able to contribute positively to the international legal climate change regime.

If the gravity of climate change warrants a reinterpretation and reconfiguration of PSNR, how is PSNR to be reinterpreted and reconfigured? In other words, what objective should the reinterpretation and reconfiguration of the principle serve? This question will be explored in the remaining part of Chapter 3. Firstly, by considering the international legal climate change regime as a whole, which is done below, it will become clear that state compliance seems to be the most significant hurdle to an effective regime. Secondly, it will be shown below that the gravity of climate change necessitates that PSNR is reinterpreted and reconfigured to promote people's rights in terms of the principle, which can in turn contribute to increase state compliance with climate commitments in terms of the international legal climate change regime. Chapter 4 will go on to explore the specific rights of people in terms of PSNR and the implementation and enforcement of their rights, which can improve state compliance with climate commitments and increase the effectiveness of the international legal climate change regime.

499 N Stern "Stern Review: The Economics of Climate Change" (2006) (page numbers unavailable) *Centro de Ciência do Sistema Terrestre* <http://mudancasclimaticas.cptec.inpe.br/~rmclima/pdfs/destaques/sternreview_report_complete.pdf> (accessed 30-04-2015) vi.

3 6 Development of the international legal climate change regime: pre-United Nations Framework Convention on Climate Change

3 6 1 Introduction

Arrhenius introduced the theory of global warming as early as 1896.[500] However, scientific concern regarding climate change only started gaining momentum around 1958, when technology made it possible to better monitor the increase of carbon dioxide concentration in the atmosphere and provide evidence that this increase was linked to the increased anthropocentric use of fossil fuel.[501]

The First World Climate Conference (FWCC) was held in Geneva in 1979, and was convened by the World Meteorological Organization (WMO), the United Nations Environment Programme (UNEP), the Food and Agriculture Organization of the United Nations (FAO), the United Nations Educational, Scientific and Cultural Organization (UNESCO) and the World Health Organization (WHO). The Conference was convened in order to discuss the state of knowledge on climate change as well as the effect of climate change on "human society".[502] The Second World Climate Conference (SWCC) was held in Geneva in 1990 and the Third World Climate Conference (TWCC) followed in Geneva in 2009.

As far as the development in formulating the legal framework by which to address climate change is concerned, the existing and accruing scientific knowledge was only transformed and incorporated into policy in the 1980s.[503] It can be argued that the mid 1980s marks the legal origin of the

500 E Crawford "Arrhenius' 1896 Model of the Greenhouse Effect in Context" (1997) 26(1) Ambio 6 6; J Uppenbrink "Arrhenius and Global Warming" (1996) 272 Science 1122 1122; D Bodansky "The History of the Global Climate Change Regime" in U Luterbacher & DF Sprinz (eds) International Relations and Global Climate Change (2001) 23 24.

501 This information is displayed by the famous Keeling curve, named after Charles David Keeling. [CD Keeling "The concentration and isotopic abundances of carbon dioxide in the atmosphere" (1960) 12(2) Tellus 200.].

502 WMO "First World Climate Conference" (date of publication unavailable) (page numbers unavailable) WMO <http://www.wmo.int/pages/themes/climate/international_wcc.php> (accessed 20-05-2013).

503 SC Mosner "Communicating climate change: history, challenges, process and future directions" (2010) 1 WIREs Climate Change 31 31-32; D Bodansky "The History of the Global Climate Change Regime" in U Luterbacher & DF Sprinz (eds) International Relations and Global Climate Change 2001 23 23-24.

international legal climate change regime.[504] In the 1980s there was a significant increase in the growing international environmental concern for issues such as deforestation, biodiversity loss, ocean pollution and hazardous waste trade,[505] and the Brundtland Commission report was published.[506] In addition, with the discovery the damage to the ozone and the negotiation of the Vienna Convention for the Protection of the Ozone Layer and the Montreal Protocol, the effect of emissions of carbon dioxide and other gases on the atmosphere came to the international forefront once again.[507]

3 6 2 The emergence of climate change as an intergovernmental issue and the international legal climate change regime as a state-centric regime

In the twenty years which followed, great leaps were made to combat climate change on the international stage. Climate change first became an intergovernmental issue in 1988.[508] In that year, the UNGA passed resolution 43/53, entitled "Protection of Global Climate for Present and Future Generations of Mankind", which refers to climate change as a "common

504 SC Mosner "Communicating climate change: history, challenges, process and future directions" (2010) 1 WIREs Climate Change 31 31-32; D Bodansky "The History of the Global Climate Change Regime" in U Luterbacher & DF Sprinz (eds) International Relations and Global Climate Change (2001) 23 23.

505 D Bodansky "The History of the Global Climate Change Regime" in U Luterbacher & DF Sprinz (eds) International Relations and Global Climate Change (2001) 23 23-24.

506 World Commission on Environment and Development Our Common Future (1987).

507 D Freestone "The UN Framework Convention on Climate Change, the Kyoto Protocol and the Kyoto Mechanisms" in D Freestone & C Streck (eds) Legal Aspects of Implementing the Kyoto Protocol Mechanisms, Making Kyoto Work (2005) 3 3-24; SC Mosner "Communicating climate change: history, challenges, process and future directions" (2010) 1 WIREs Climate Change 31 31-32; D Bodansky "The History of the Global Climate Change Regime" in U Luterbacher & DF Sprinz (eds) International Relations and Global Climate Change (2001) 23 23.

508 D Bodansky "The History of the Global Climate Change Regime" in U Luterbacher & DF Sprinz (eds) International Relations and Global Climate Change (2001) 23 28.

concern for mankind".[509] In 1988, the UNEP and the WMO brought into existence the IPCC – an action supported by the UNGA through resolution 43/53.[510] The IPCC is the leading international body for the assessment of climate change, created to provide clear scientific knowledge on climate change and the environmental and socioeconomic impacts it may have.[511]

Along with the aforementioned UNGA resolution 43/53 and the creation of the IPCC, other events are also important to the legal development of the international climate change regime. These events include the Hague Summit held in 1989, the Noordwijk Ministerial Conference on Atmospheric Pollution and Climatic Change (Noordwijk Conference) held in 1989, the SWCC and Bergen Ministerial Conference on Sustainable Development (Bergen Conference) held in 1990.[512] The Noordwijk Conference resulted in the declaration entitled "Noordwijk Declaration Atmospheric Pollution and Climatic Change 1989" (Noordwijk Declaration), which can be considered a noteworthy step in the establishment of emission targets. This declaration reads as follows:

> "The Conference recognizes the need to stabilize, while ensuring stable development of the world economy, CO2 emissions and emissions of other greenhouse gases not controlled by *the Montreal Protocol*. Industrialized nations agree that such stabilization should be achieved by them as soon as possible, at levels to be considered by *the IPCC* and *the Second World Climate Conference* of November 1990. In the view of many industrialized nations such stabilization of CO2 emissions should be achieved as a first step at the latest by the year 2000."[513]

The Noordwijk Declaration also already hints at the differentiation that would occur between countries in the international legal climate change

---

509 UNGA resolution 43/53 of 6 December 1988.

510 IPCC "Organization" (date of publication unavailable) (page numbers unavailable) IPCC <http://www.ipcc.ch/organization/organization.shtml#.URKQ3KW5cII> (accessed 06-02-2013).

511 IPCC "Organization" (date of publication unavailable) (page numbers unavailable) IPCC <http://www.ipcc.ch/organization/organization.shtml#.URKQ3KW5cII> (accessed 06-02-2013).

512 D Bodansky "The History of the Global Climate Change Regime" in U Luterbacher & DF Sprinz (eds) International Relations and Global Climate Change (2001) 23 28.

513 UNFCCC "The Noordwijk Ministerial Declaration on climate change" (01-05-1993) (page numbers unavailable) UNFCCC <http://unfccc.int/resource/ccsites/senegal/fact/fs218.htm> (accessed 02-07-2016).

regime. Until about 1990, the issue of climate change was mostly an issue researched and debated by the governments of Western industrialised countries. At the Noordwijk Conference there was a definite difference of opinion between two groups concerning the action to be taken in response to climate change: the first group being a large part of Europe, joined by Canada, Australia and New Zealand, and the second group being the United States joined by Japan and the former Soviet Union.[514] The "Noordwijk divide" grew greater at the Bergen Conference and at the SWCC.[515] The first group wanted to establish quantitative limitations to national greenhouse gas emissions where, initially, carbon dioxide would be stabilised at the current level,[516] a line of action based on the precautionary principle.[517] The second group adopted a more nationalistic approach to the problem of climate change and supported the idea that further scientific research would have to be conducted concerning the national economic implications of stabilisation of greenhouse gases and the uncertainties of climate change before it would be agreed on an international level to stabilise greenhouse gases.[518] The reasons for the divide between the two groups were the promotion of economic self-interest and benefit of each country, differing domestic politics in the branches of government in-

514 W Lang "Is the Ozone Depletion Regime a Model for an Emerging Regime on Global Warming?" (1991) 9(2) UCLA Journal of Environmental Law & Policy 161196; K Fisher-Vanden "Debate: A Case Study of the United States" (1997) (6) ENRP Discussion Paper 1 5; D Bodansky "The History of the Global Climate Change Regime" in U Luterbacher & DF Sprinz (eds) International Relations and Global Climate Change (2001) 23 29.

515 K Fisher-Vanden "Debate: A Case Study of the United States" (1997) (6) ENRP Discussion Paper 1 5.

516 D Bodansky "The History of the Global Climate Change Regime" in U Luterbacher & DF Sprinz (eds) International Relations and Global Climate Change (2001) 23 29.

517 K Fisher-Vanden "Debate: A Case Study of the United States" (1997) (6) ENRP Discussion Paper 1 5.
The reluctance of the USA, in particular, to adopt the precautionary principle is well documented. For example, see: J Cameron & W Wade-Cery "Addressing Uncertainty: Law, Policy and the Development of the Precautionary Principle" (1992) 43 CSERG Working Paper GEC 1 27.

518 J Cameron & W Wade-Cery "Addressing Uncertainty: Law, Policy and the Development of the Precautionary Principle" (1992) 43 CSERG Working Paper GEC 1 27; D Bodansky "The History of the Global Climate Change Regime" in U Luterbacher & DF Sprinz (eds) International Relations and Global Climate Change (2001) 23 29.

volved in forming the climate change policy of each country, and national tactics targeted at retaining power in future negotiations.[519]

In 1990, UNGA resolution 45/212 was passed, sharing the title of the resolution of 1988, namely "Protection of Global Climate for Present and Future Generations of Mankind".[520] This resolution is an important part of the development of climate change from a concern of developed countries to a concern of developed and developing countries. This movement includes a shift for the issue of climate change from a severely technical environmental issue spearheaded by the IPCC towards an environmental and developmental issue spearheaded by the UNGA. This movement is evident in the UNGA resolution 45/212, which reads as follows:

> "[The General Assembly] decides to establish a single intergovernmental negotiating process under the auspices of the General Assembly, supported by the United Nations Environment Programme and the World Meteorological Organization, for the preparation by an Intergovernmental Negotiating Committee of an effective framework convention on climate change, containing appropriate commitments, and any related instruments as might be agreed upon, taking into account proposals that may be submitted by States participating in the negotiating process, the work of the Intergovernmental Panel on Climate Change and the results achieved at international meetings on the subject, including the Second World Climate Conference"[521]

In the climate change discussions that followed, developing countries were able to show more unity in their opinion than developed countries; however, developing countries mainly insisted only on financial assistance and technology transfer from developed countries, which would enable them to implement the UNFCCC.[522] Financial assistance was a very controversial issue in the climate change negotiations.[523] The controversy rested in either using an established fund, namely the Global Environment

519 D Bodansky "The History of the Global Climate Change Regime" in U Luterbacher & DF Sprinz (eds) International Relations and Global Climate Change (2001) 23 28-29.

520 UNGA resolution 45/212 of 6 December 1990.

521 UNGA resolution 45/212 of 6 December 1990.

522 D Bodansky "The History of the Global Climate Change Regime" in U Luterbacher & DF Sprinz (eds) International Relations and Global Climate Change (2001) 23 30.

523 AK Johnson "The Influence of Institutional Culture on the Formation of Pre-regime Climate Change Policies in Sweden, Japan and the United States" (1998) 7(2) Environmental Values 223 235.

Facility (GEF), or creating a new fund.[524] Developing countries supported the creation of a new fund, while developed countries wanted to use the GEF.[525]

3 6 3 Targets and timetables negotiations

There were a few clear positions concerning the establishment of targets and timetables in the period leading up to the formulation and adoption of the UNFCCC. At one end, the European Union (EU), which was supported by the Alliance of Small Island States (AOSIS), supported the creation of targets and timetables to combat climate change.[526] At the other end, the United States of America (USA) and the oil-producing states did not support the creation of targets and timetables.[527] This showed *inter alia* that even more developed states, which relied heavily on fossil fuels to ensure economic growth, were hesitant to limit the right to dispose freely of, explore and exploit their natural resources in terms of PSNR, regardless of the amount of greenhouse gas emissions it would contribute. The developing countries, including China, India and Brazil, submitted that they did not want climate change measures to have an impact on their right to use natural resources for economic development, which is their right in terms

---

524 D Bodansky "The History of the Global Climate Change Regime" in U Luterbacher & DF Sprinz (eds) International Relations and Global Climate Change (2001) 23 33.

525 R Clémençon
"What future for the Global Environment Facility?" (2006) 15 The Journal of Environmental Development 50 52.

526 S Andresen & S Agrawala "Leaders, pushers and laggards in the making of the climate change regime" (2012) 12 Global Environmental Change 41 46; D Bodansky "The History of the Global Climate Change Regime" in U Luterbacher & DF Sprinz (eds) International Relations and Global Climate Change (2001) 23 30, 33.

527 D Jamieson "Climate Change, Consequentialism, and the Road Ahead" (2013) 13(2) Chicago Journal of International Law 439 448; S Andresen & S Agrawala "Leaders, pushers and laggards in the making of the climate change regime" (2012) 12 Global Environmental Change 41 46; D Bodansky "The History of the Global Climate Change Regime" in U Luterbacher & DF Sprinz (eds) International Relations and Global Climate Change (2001) 23 30-31.

of PSNR.[528] Overall, developing countries supported the establishment of targets and timetables, as long as these were only binding on developed countries.[529] In order to come to fruition the UNFCCC could not provide clear answers to all contentious issues facing the international legal climate change regime and deliberately made use of "highly ambiguous language"[530].

Two years after the UNGA passed resolution 45/212 of 1990, entitled "Protection of Global Climate for Present and Future Generations of Mankind",[531] the UNFCCC was adopted. At this point, existing documents dealing with stratospheric ozone depletion (the Vienna Convention for the Protection of the Ozone Layer and the Montreal Protocol) and transboundary air pollution (Convention on Long-Range Transboundary Air Pollution 1979[532]) were unable to deal effectively with the growing international problem of climate change. An option at this point – an option which was not accepted – was to handle the issue of climate change in a more comprehensive regime dealing with all interlinked international atmospheric issues in one document, similarly to the approach followed in the Convention of the Law of the Sea.[533] Eventually, the international climate change regime was established as a problem-specific regime, modelled on the approach followed by the Vienna Convention for the Protection of the Ozone Layer where there is a clear "step-by-step approach" to solving a specific problem.[534] By this time customary international law principles already existed, such as principle 21 of the Stockholm Declara-

528 D Bodansky "The History of the Global Climate Change Regime" in U Luterbacher & DF Sprinz (eds) International Relations and Global Climate Change (2001) 23 31.

529 D Bodansky "The History of the Global Climate Change Regime" in U Luterbacher & DF Sprinz (eds) International Relations and Global Climate Change (2001) 23 33.

530 D Jamieson "Climate Change, Consequentialism, and the Road Ahead" (2013) 13(2) Chicago Journal of International Law 439 448.

531 UNGA resolution 45/212 of 6 December 1990.

532 Hereafter referred to as 'Convention on Long-Range Transboundary Air Pollution'.

533 D Bodansky "The History of the Global Climate Change Regime" in U Luterbacher & DF Sprinz (eds) International Relations and Global Climate Change (2001) 23 31.

534 D Bodansky "The History of the Global Climate Change Regime" in U Luterbacher & DF Sprinz (eds) International Relations and Global Climate Change (2001) 23 31.

tion, which would have to be taken into consideration during the establishment of an international legal regime for climate change.[535] Principle 21 of the Stockholm Declaration reads as follows:

> "States have, in accordance with the Charter of the United Nations and the principles of international law, the sovereign right to exploit their own resources pursuant to their own environmental policies, and the responsibility to ensure that activities within their jurisdiction or control do not cause damage to the environment of other States or of areas beyond the limits of national jurisdiction."

Interestingly, principle 21 encapsulates PSNR. Accordingly, certain duties of states in terms of PSNR are relevant to the establishment of an international legal climate change regime, such as the duty to cooperate for the international development and sustainable use of natural resources.[536]

3 7 The United Nations Framework Convention on Climate Change as the cornerstone of the international legal climate change regime and the embedding of the principle of sovereignty over natural resources

3 7 1 Introduction to the United Nations Framework Convention on Climate Change

The UNGA established the Intergovernmental Negotiating Committee for a Framework Convention on Climate Change (INC/FCCC), which conducted its first meeting to negotiate the UNFCCC in 1991.[537] The UNFCCC was adopted on 9 May 1992 and entered into force on 21 March 1994 with the ratification of 50 states. Currently, the UNFCCC has 197 parties.[538] The UNFCCC is a crucial component of the international legal cli-

535 K Hossain "Introduction" in K Hossain & SR Chowdhury (eds) Permanent Sovereignty Over Natural Resources in International Law, Principle and Practice (1984) ix ix.

536 The rights and duties are of states are set out in Chapter 4.

537 UN "Towards a climate agreement" (date of publication unavailable) (page numbers unavailable) UN <http://www.un.org/climatechange/towards-a-climate-agreement/ > (accessed 02-07-2016).

538 UNFCCC "Status of Ratification of the Convention" (date of publication unavailable) (page numbers unavailable) UNFCCC <http://unfccc.int/essential_background/convention/status_of_ratification/items/2631.php > (accessed 30-06-2016). This chapter contains all updates up until the end of June 2016.

mate change regime and forms the "constitution"[539] or cornerstone of the regime.

The preamble of the UNFCCC reads as follows:

> "States have, in accordance with the Charter of the United Nations and the principles of international law, the sovereign right to exploit their own resources pursuant to their own environmental and developmental policies, and the responsibility to ensure that activities within their jurisdiction or control do not cause damage to the environment of other States or of areas beyond the limits of national jurisdiction".[540]

The preamble of the UNFCCC thus embeds PSNR in the international legal climate change regime, by including the right of states to exploit their natural resources pursuant to their own domestic policies, which is a right of states in terms of PSNR. It can also be submitted that the state responsibility included in the preamble, to ensure that the exercise of this right does not cause damage to the environment of other states or areas beyond the limits of national jurisdiction, is in line with the state duties in terms of PSNR that demands, for instance, that the state exercises PSNR in a way that conserves the environment and promotes the sustainable use of natural resources.[541] This limitation on the state right included in the preamble of the UNFCCC can also be considered as recognition that state sovereignty over natural resources is not absolute,[542] an idea that is explored in Chapter 2. The embedding of PSNR in the UNFCCC, which is the cornerstone of the international legal climate change regime, is accordingly a solid indication of the nexus between PSNR and climate change.

As was included in Section 2.4.2.3.2, the preamble of the UNFCCC specifically acknowledges that a "change in the Earth's climate and its adverse effects are a common concern of humankind". Since PSNR is included in the preamble of the UNFCCC as submitted above, it can be further submitted that PSNR must be implemented as a common concern of humankind within the international legal climate change regime. It is re-

539 D Bodansky & L Rajamani "The Evolution and Governance Architecture of the Climate Change Regime" (2013) (2) SSRN <http://papers.ssrn.com/sol3/papers.cfm?abstract_id=2168859> (accessed 12-05-2015).

540 Preamble, UNFCCC.

541 The rights and duties of states in terms of PSNR are discussed in Chapter 4.

542 W Scholtz "Greening Permanent Sovereignty through the Common Concern in the Climate Change Regime: Awake Custodial Sovereignty!" in OC Ruppel et al (eds) Climate Change: International Law and Global Governance. Volume II: Policy, Diplomacy and Governance in a Changing Environment (2013) 201 208.

peated at this juncture that this research project does not consider the common concern of humankind to run counter to PSNR, since the state duties in terms of PSNR include *inter alia* exercising PSNR for the well-being of people and in terms of the principle of sustainable development.[543] However, the scope of the research project does not allow for a full exposition of the principle of common concern of humanity in this context.

The objective of the UNFCCC is *inter alia* to stabilise the concentration of greenhouse gases in the atmosphere "at a level that would prevent dangerous anthropogenic interference with the climate system",[544] and according to the IPCC this means no more than a 2 degrees Celsius increase in temperature above pre-industrial levels. Article 2 of the UNFCCC also stipulates the time frame in which this stabilisation of greenhouse gases should occur, and provides that the time frame must allow for natural adaptation of ecosystems to climate change, must not threaten food production and must enable sustainable economic development. Although unrealistic, a possible criticism against the UNFCCC is that it aims merely to stabilise greenhouse gas concentrations, and not reverse it.[545]

Other articles relevant when discussing the content of the UNFCCC include article 7 to 11 of the Convention, which set up the institutional framework of the Convention; the financial mechanism, in particular, is included in article 11.[546] In a financial context, article 21(3) stipulates that

---

543 N Schrijver Sovereignty Over Natural Resources, Balancing Rights and Duties (1997) 308, 324.
The state's rights and duties in terms of PSNR are discussed in Chapter 4.

544 Article 2 of the UNFCCC reads as follows: "The ultimate objective of this Convention and any related legal instruments that the Conference of the Parties may adopt is to achieve, in accordance with the relevant provisions of the Convention, stabilization of greenhouse gas concentrations in the atmosphere at a level that would prevent dangerous anthropogenic interference with the climate system. Such a level should be achieved within a time frame sufficient to allow ecosystems to adapt naturally to climate change, to ensure that food production is not threatened and to enable economic development to proceed in a sustainable manner.".

545 D Freestone "The UN Framework Convention on Climate Change, the Kyoto Protocol and the Kyoto Mechanisms" in D Freestone & C Streck (eds) Legal Aspects of Implementing the Kyoto Protocol Mechanisms, Making Kyoto Work (2005) 3-24 4-5; P Birnie & A Boyle International Law and the Environment 2ed (2002) 527.

546 The Green Climate Fund was established at COP 16 as an operating entity of the financial mechanism in terms of article 11 of the UNFCCC. [UNFCCC "Green Climate Fund" (date of publication unavailable) (page numbers unavailable) *UN-*

the GEF shall be designated as a financial mechanism. For the sake of completeness, it should be noted that article 14 deals with the settlement of disputes in the context of the Convention and article 15 deals with amendments to the Convention.

Article 3 of the UNFCCC sets out the guidelines by which parties to the Convention should pursue the Convention's objective as stipulated in article 2. These guidelines include well-known principles of international law, point to the specific characteristics of the international legal climate change regime, and are vital to inform the decisions which are made by

---

FCCC <http://unfccc.int/cooperation_and_support/financial_mechanism/green_climate_fund/items/5869.php> (accessed 03-07-2016).]

Article 11(1)-(5) of the UNFCCC reads as follows: (1) "A mechanism for the provision of financial resources on a grant or concessional basis, including for the transfer of technology, is hereby defined. It shall function under the guidance of and be accountable to the Conference of the Parties, which shall decide on its policies, programme priorities and eligibility criteria related to this Convention. Its operation shall be entrusted to one or more existing international entities"; (2) "The financial mechanism shall have an equitable and balanced representation of all Parties within a transparent system of governance"; (3) "The Conference of the Parties and the entity or entities entrusted with the operation of the financial mechanism shall agree upon arrangements to give effect to the above paragraphs, which shall include the following: (a) Modalities to ensure that the funded projects to address climate change are in conformity with the policies, programme priorities and eligibility criteria established by the Conference of the Parties; (b) Modalities by which a particular funding decision may be reconsidered in light of these policies, programme priorities and eligibility criteria; (c) Provision by the entity or entities of regular reports to the Conference of the Parties on its funding operations, which is consistent with the requirement for accountability set out in paragraph 1 above; and (d) Determination in a predictable and identifiable manner of the amount of funding necessary and available for the implementation of this Convention and the conditions under which that amount shall be periodically reviewed"; (4) "The Conference of the Parties shall make arrangements to implement the above-mentioned provisions at its first session, reviewing and taking into account the interim arrangements referred to in Article 21, paragraph 3, and shall decide whether these interim arrangements shall be maintained. Within four years thereafter, the Conference of the Parties shall review the financial mechanism and take appropriate measures"; (5) "The developed country Parties may also provide and developing country Parties avail themselves of, financial resources related to the implementation of the Convention through bilateral, regional and other multilateral channels.".

parties in the context of the international legal climate change regime.[547] The guidelines include *inter alia* equity, the principle of common but differentiated responsibility, the precautionary principle, and the principle of sustainable development; and are discussed in the succeeding paragraphs. The main objective of the discussion of these principles is to understand the foundation of the international legal climate change regime and to illustrate the complexities and scale of the legal effort required to address climate change, which justifies the usage of other principles of international environmental law (in this case, specifically PSNR) to assist the international legal climate change regime, if possible. By considering the principles of the international legal climate change regime in the succeeding section, the most prominent aspects and content of the UNFCCC, as the cornerstone of the international legal climate change regime, are also examined further. Finally, the discussion of these principles also contributes to solidifying the nexus between PSNR and the international legal climate change regime.

3 7 2 Principles of the international legal climate change regime

3 7 2 1 Principle of equity

The principle of equity is a central principle within the international legal climate change regime.[548] In general terms, equity can be considered as an issue of or ideal of fairness or justice.[549] There are many reasons why it is

547 B Metz "International equity in climate change policy" (2000) 1 *Integrated Assessment* 111 125; A Kiss & D Shelton *International Environmental Law* 3ed (2004) 203.

548 M Paterson "Principles of Justice in the Context of Global Climate Change" (2001) (121) Graduate Institute Geneva <http://graduateinstitute.ch/files/live/sites/iheid/files/shared/iheid/800/luterbacher/luterbacher%20chapter%206%20106.pdf> (accessed 09-09-2016); A Hurrell and S Sengupta "Emerging powers, North-South relations and global climate politics" (2012) 88(3) International Affairs 463 472; L Ringius , A Torvanger & A Underdal "Burden Sharing and Fairness principles in International Climate Policy" (2002) 2 International Environmental Agreements: Politics, Law and Economics 1-22.

549 B Metz "International equity in climate change policy" (2000) 1 *Integrated Assessment* 111 111; J Ashton & X Wang "Equity and climate, In principle and practice" in JE Aldy et al (eds) *Beyond Kyoto, Advancing the international effort against climate change* (2003) 61 62.

vitally important to incorporate equity into the international legal climate change regime, if it is to be an internationally supported regime among states. Many countries which are set to endure the worst impacts of climate change and which are the least equipped to respond to these impacts are less developed and have historically contributed the least to greenhouse gas emissions responsible for climate change.[550] Accordingly, there has been much needed debate between countries regarding what is deemed equitable in addressing climate change. The UNFCCC, the Kyoto Protocol and the Paris Agreement contain evidence of the attempt by the international community to give effect to the principle of equity, while addressing the problem of climate change. In this context it should be noted that the international legal climate change regime, which developed after the establishment of PSNR, is linked to a state duty in terms of PSNR, namely the duty to share transboundary natural resources (the atmosphere in particular) in a manner which is equitable.[551]

It is possible to understand equity as an umbrella term, as many principles can be designated as equity principles. It is, for example, possible to classify the polluter-pays principle and the principle of common but differentiated responsibility as equity principles. Whatever the exact understanding of equity, it is important that it be balanced with environmental efficiency, cost efficiency and political considerations,[552] as is illustrated within the international legal climate change regime.

As shown by Ashton and Wang, it is possible to deconstruct equity in the context of climate change, revealing different dimensions of the term.[553] Depending on whether a country is classified as a developed or developing country in terms of the international legal climate change regime, it may favour a specific dimension of equity. Therefore, in trying to understand the meaning of the term equity within the international legal climate change regime, it is useful to consider the different dimensions of the term, which include: responsibility, equal entitlement, the capacity to

550 S McInerney-Lankford "Climate Change and Human Rights: An Introduction to Legal Issues" (2009) 33 *Harvard Environmental Law Review* 431 431-437.

551 The state duties in terms of PSNR are discussed more fully in Chapter 4.

552 B Metz "International equity in climate change policy" (2000) 1 *Integrated Assessment* 111 125.

553 J Ashton & X Wang "Equity and climate, In principle and practice" in JE Aldy et al *Beyond Kyoto, Advancing the international effort against climate change* (2003) 61 64-66.

act (or the ability to pay), basic needs, and comparability of effort.[554] Firstly, the concept of responsibility ensures equity by asking who is responsible for the damage in the first place.[555]This dimension of equity, namely responsibility, is illustrated by the principle of equity known as the polluter-pays principle. Secondly, the concept of equal entitlement is seen as an important dimension of equity within the international legal climate change regime and refers to equal entitlement of all states and all people to the atmosphere, as the atmosphere is considered a common good in addition to being considered a natural resource in terms of PSNR.[556] Thirdly, the capacity to act (or the ability to pay) refers to a country's capacity to act or the country's capability to contribute, and the country that is able to contribute the most should do so.[557] This ability to contribute can pertain to many different areas, including for instance technology and finance. The fourth dimension of equity, as deconstructed by Ashton and Wang, is in respect of basic needs and refers to a country's ability to meet the basic needs of its population.[558] An equitable agreement would therefore not impede the actions of poorer countries to meet the basic needs of their population.[559] Comparability of effort is the fifth dimension of equity, and refers to the effort that is expected from a country, which must be fair and

---

554 J Ashton & X Wang "Equity and climate, In principle and practice" in JE Aldy et al *Beyond Kyoto, Advancing the international effort against climate change* (2003) 61 64-66.

555 J Ashton & X Wang "Equity and climate, In principle and practice" in JE Aldy et al *Beyond Kyoto, Advancing the international effort against climate change* (2003) 61 64; Neumayer E "In defence of historical accountability for greenhouse gas emissions" (2000) 33(2) *Ecological Economics* 185 185-192.

556 J Ashton & X Wang "Equity and climate, In principle and practice" in JE Aldy et al *Beyond Kyoto, Advancing the international effort against climate change* (2003) 61 65.

557 M Carzola & M Toman "International Equity and Climate Change Policy" (2000) 27 *Climate Issue Brief 1* 7; J Ashton & X Wang "Equity and climate, In principle and practice" in JE Aldy et al *Beyond Kyoto, Advancing the international effort against climate change* (2003) 61 65.

558 J Ashton & X Wang "Equity and climate, In principle and practice" in JE Aldy et al *Beyond Kyoto, Advancing the international effort against climate change* (2003) 61 66.

559 J Ashton & X Wang "Equity and climate, In principle and practice" in JE Aldy et al *Beyond Kyoto, Advancing the international effort against climate change* (2003) 61 66.

must be determined according to every country's particular circumstances.[560] Therefore, according to the approach followed by Ashton and Wang, it is possible to understand the principle of equity as part of the international legal climate change regime by understanding the different dimensions of equity, as outlined above.

It is also possible to understand the principle of equity by considering equity as a principle functioning on three different levels, namely an international level (this is referred to as international equity); a national level (this is referred to as national or social equity); and an intergenerational level (this is referred to as intergenerational equity).[561] International equity is understood as the equity between countries.[562] National or social equity is understood as the equity between different social groups within a country.[563] Intergenerational equity is understood as the equity or justice between different generations.[564] For the sake of clarity, intragenerational equity is differentiated from intergenerational equity and is understood as the equity or justice between those people living today.[565] All levels of equity must be balanced within the international climate change regime. Also, in light of ensuring sustainable development (another principle of

560 J Ashton & X Wang "Equity and climate, In principle and practice" in JE Aldy et al *Beyond Kyoto, Advancing the international effort against climate change* (2003) 61 66.

561 B Metz "International equity in climate change policy" (2000) 1 *Integrated Assessment* 111 111.

562 B Metz "International equity in climate change policy" (2000) 1 *Integrated Assessment* 111 111.

563 B Metz "International equity in climate change policy" (2000) 1 *Integrated Assessment* 111 111.

564 M Paterson "Principles of Justice in the Context of Global Climate Change" (2001) (121) *Graduate Institute Geneva* <http://graduateinstitute.ch/files/live/sites/iheid/files/shared/iheid/800/luterbacher/luterbacher%20chapter%206%20106.pdf> (accessed 09-09-2016); B Metz "International equity in climate change policy" (2000) 1 *Integrated Assessment* 111 111.

565 E B Weiss "In Fairness To Future Generations and Sustainable Development" (1992) 8(1) *American University International Law Review* 19 21.
The actions needed to achieve intergenerational equity and intragenerational equity are in large consistent with each other. [E B Weiss "In Fairness To Future Generations and Sustainable Development" (1992) 8(1) *American University International Law Review* 19 21-22.].

the international legal climate change regime discussed below), intergenerational equity needs to be ensured.[566]

The principle of equity is clearly implemented throughout the UNFCCC, as clearly illustrated by article 3(1) of the UNFCCC, which refers in particular to intergenerational equity between present and future generations and equity through the principle of common but differentiated responsibilities. Article 3 reads as follows:

> "The Parties should protect the climate system for the benefit of present and future generations of humankind, on the basis of equity and in accordance with their common but differentiated responsibilities and respective capabilities. Accordingly, the developed country Parties should take the lead in combating climate change and the adverse effects thereof."

The principle of equity is also included in the UNFCCC on a procedural level. The financial mechanism of the UNFCCC, as set out in article 11, includes the stipulation that the mechanism shall represent the parties in a way that is "equitable and balanced". It can also be submitted that the principle of equity is incorporated into the amendment procedure set out by article 15 of the UNFCCC. Article 15 states that "the amendment shall as a last resort be adopted by a three-fourths majority vote of the Parties present and voting at the meeting".

In accordance with the UNFCCC, equity is also a principle employed by the Kyoto Protocol, as is especially evident in article 2, article 3 and article 12 of the Protocol, and the detail contained in these articles are included below when discussing the principle of common but differentiated responsibility.

Equity can also be highlighted as an important principle in the Paris Agreement 2015. The preamble of the Paris Agreement sets out *inter alia* that it is guided by principles of equity and specifically also refers to intergenerational equity. Other specific references to equity in the Agreement include article 2(2), which stipulates that the Agreement will be "implemented to reflect equity",[567] and article 4, which stipulates that the various

566 E B Weiss "Intergenerational equity: a legal framework for global environmental change" in E B Weiss (ed) *Environmental change and international law: New challenges and dimensions* (1992) (page number of chapter unavailable).

567 Article 2(2) of the Paris Agreement reads as follows: "This Agreement will be implemented to reflect equity and the principle of common but differentiated responsibilities and respective capabilities, in the light of different national circumstances.".

objectives of the Agreement will be achieved “on the basis of equity”.[568] In addition, article 14 describes the action of the parties to the Convention as occurring “in light of equity”[569].

The Paris Agreement also addresses the “equity conditions”[570] proposed by Ashton and Wang for inclusion in any future climate agreement. The first equity condition that was set for a future climate agreement involves action by and the cooperation of the USA.[571] So far, the Paris Agreement has been able to satisfy this condition as it achieved the participation of the USA. The second equity condition that was set for a future climate agreement was that industrialised countries should continue to lead the way in the efforts to combat climate change, as stipulated by the UNFCCC.[572] Although the Paris Agreement incorporates nationally deter-

---

568 Article 4(1) of the Paris Agreement reads as follows: “In order to achieve the long-term temperature goal set out in Article 2, Parties aim to reach global peaking of greenhouse gas emissions as soon as possible, recognizing that peaking will take longer for developing country Parties, and to undertake rapid reductions thereafter in accordance with best available science, so as to achieve a balance between anthropogenic emissions by sources and removals by sinks of greenhouse gases in the second half of this century, on the basis of equity, and in the context of sustainable development and efforts to eradicate poverty. ”.

569 Article 14(1) of the Paris Agreement reads as follows: “The Conference of the Parties serving as the meeting of the Parties to this Agreement shall periodically take stock of the implementation of this Agreement to assess the collective progress towards achieving the purpose of this Agreement and its long-term goals (referred to as the "global stocktake"). It shall do so in a comprehensive and facilitative manner, considering mitigation, adaptation and the means of implementation and support, and in the light of equity and the best available science.”.

570 J Ashton & X Wang “Equity and climate, In principle and practice” in JE Aldy et al *Beyond Kyoto, Advancing the international effort against climate change* (2003) 61 76.

571 W Scholtz “Equity as the basis for a future international climate change agreement: between pragmatic panacea and idealistic impediment. The optimization of the CBDR principle via realism” (2009) 42(2) *The Comparative and International Law Journal of South Africa* 166 173; J Ashton & X Wang “Equity and climate, In principle and practice” in JE Aldy et al *Beyond Kyoto, Advancing the international effort against climate change* (2003) 61 76.

572 W Stark “Phoenix from the Ashes – An Analysis of the Paris Agreement to the United Nations Framework Convention on Climate Change” (2016) (41) *Wuppertal Institut* <http://wupperinst.org/fa/redaktion/downloads/publications/Paris_Results.pdf> (accessed 10-07-2016); J Ashton & X Wang “Equity and climate, In principle and practice” in JE Aldy et al *Beyond Kyoto, Advancing the international effort against climate change* (2003) 61 76.

mined contributions (NDCs) from all countries, it does still differentiate between the capabilities of developed and developing countries. This feature is elucidated below, where the Paris Agreement is discussed in greater detail. The third equity condition that was set for a future climate agreement involved imposing emission cuts also on developing countries, and not only on developed countries.[573] This condition was satisfied entirely by the Paris Agreement's implementation of NDCs for both developed and developing countries. The fourth and fifth equity conditions that were set for a future climate agreement involved assistance to and between countries in mitigating and adapting to climate change and reaching emission targets.[574] The Paris Agreement performs well in terms of this equity condition as it actually strengthens support from developed countries to developing countries in the form of financial assistance and technology transfer, as is evidenced in the full discussion on the Agreement in this chapter. Accordingly, it can be submitted that the Paris Agreement has successfully incorporated the most prominent equity conditions set by Ashton and Wang for the formulation of any climate agreement of the future.

573 W Scholtz "Equity as the basis for a future international climate change agreement: between pragmatic panacea and idealistic impediment. The optimization of the CBDR principle via realism" (2009) 42(2) *The Comparative and International Law Journal of South Africa* 166 173-174; J Ashton & X Wang "Equity and climate, In principle and practice" in JE Aldy et al *Beyond Kyoto, Advancing the international effort against climate change* (2003) 61 77.

574 J Ashton & X Wang "Equity and climate, In principle and practice" in JE Aldy et al *Beyond Kyoto, Advancing the international effort against climate change* (2003) 61 78.
Guillaumont follows an interesting approach here based on vulnerability as a determinant of developmental assistance in the context of climate change. [P Guillaumont "Measuring Structural Vulnerability to Allocate Development Assistance and Adaptation Resources" (05-03-2012) (1-22) *Fondation Pour Les Études Et Recherches Sur Le Développement International* <http://www.ferdi.fr/sites/www.ferdi.fr/files/publication/fichiers/wp68_guillaumont_web_0.pdf> (accessed 10-07-2016).].

3 7 2 2 Principle of common but differentiated responsibility

3 7 2 2 1 The principle

As mentioned above, the principle of common but differentiated responsibility can be described as a principle of equity. The principle of common but differentiated responsibility was already part of international environmental law before the adoption of the UNFCCC, as evidenced by Article 7 of the Rio Declaration, which reads as follows:

> "States shall cooperate in a spirit of global partnership to conserve, protect and restore the health and integrity of the Earth's ecosystem. In view of the different contributions to global environmental degradation, States have common but differentiated responsibilities. The developed countries acknowledge the responsibility that they bear in the international pursuit of sustainable development in view of the pressures their societies place on the global environment and of the technologies and financial resources they command."

The international legal climate change regime aims to compel states to react to climate change through mitigation or adaptation, pursuant to the principle of common but differentiated responsibility. Within the UNFCCC and Kyoto Protocol countries are differentiated according to whether they are categorised as an Annex I country,[575] an Annex II country,[576] a least developed country (LDC), a small island country, or a country with a

575 See UNFCCC website for a list of the Annex I countries. [UNFCCC "List of Annex I Parties to the Convention" (date unavailable) (page numbers unavailable) *UNFCCC* <http://unfccc.int/parties_and_observers/parties/annex_i/items/2774.php> (accessed 05-09-2016).]
Annex I countries include industrialised countries that were members of the OECD in 1992 and countries with EIT (including the Russian Federation, the Baltic States, and several Central and Eastern European States). [UNFCCC "Parties & Observers" (2014) (page numbers unavailable) *UNFCCC* <http://unfccc.int/parties_and_observers/items/2704.php> (accessed 05-09-2016).].

576 Annex II countries include the OECD members of Annex I, but not the EIT parties. [UNFCCC "Parties & Observers" (2014) (page numbers unavailable) *UNFCCC* <http://unfccc.int/parties_and_observers/items/2704.php> (accessed 05-09-2016).]
See UNFCCC website for a list of Annex II countries. [UNFCCC "ANNEX II Countries" (date unavailable) (page numbers unavailable) *UNFCCC* <http://unfccc.int/cop3/fccc/climate/annex2.htm> (accessed 06-09-2016).].

particularly high dependence on fossil fuels.[577] Therefore, while all countries share the responsibility of combating climate change within the regime, their responsibility in this regard is differentiated on the basis of category.

Article 3(1) of the UNFCCC establishes the position of the principle of common but differentiated responsibility within the international legal climate change regime and reads as follows:

> "The Parties should protect the climate system for the benefit of present and future generations of humankind, on the basis of equity and in accordance with their common but differentiated responsibilities and respective capabilities. Accordingly, the developed country Parties should take the lead in combating climate change and the adverse effects thereof."

Article 4 of the UNFCCC sets out the commitments of developed and developing countries within the framework of common but differentiated responsibility. Article 4(1) specifically sets out obligations by taking into account the principle of common but differentiated responsibility and "national and regional development priorities, objectives and circumstances". Article 4(2) sets out the commitments of the developed country parties and other parties included in Annex I. Article 4(3) pertains to financial resources provided by developed country parties and other developed parties included in Annex II. Article 4(4) deals with assistance from developed countries and other developed countries included in Annex II to developing countries that are particularly vulnerable, in the context of adaptation. Article 4(5) deals with the issue of financing and technology transfer between developed countries (including developed countries listed in Annex II) and developing countries. Article 4(6) of the UNFCCC mentions specifically that "flexibility" should be allowed in the specific circumstance where Annex I countries are transitioning to a market economy and takes into consideration the country's historic level of anthropogenic greenhouse gas emissions in this regard. Article 4(7) of the UNFCCC takes into consideration the specific priorities of developing countries, such as economic and social development as well as poverty eradication. Article 4(8) recognises the principle of common but differentiated responsibility by stating that necessary actions, such as funding, insurance and

---

577 D Bodansky et al "Issues for a 2015 climate agreement" (May 2015) (2) *Center for Climate and Energy Solutions* <http://www.c2es.org/docUploads/issues-for-a-2015-climate-agreement.pdf> (accessed 04-05-2015).

technology transfer, must take into consideration the specific needs of developing countries or countries that are particularly vulnerable, while listing characteristics which may contribute to the vulnerability of a country. In addition, article 4(9) reiterates that funding and technology transfer should take into consideration the specific needs and situations of the least developed countries. The final sub-section of article 4, namely article 4(10), stipulates that developing countries must receive consideration during the implementation of measures to respond to climate change, especially in light of the fact that their economies might be vulnerable to the implementation of such measures. Other articles of the UNFCCC also illustrate the employment of the principle of common but differentiated responsibility throughout the Convention. For instance, article 5 of the UNFCCC pertains to research and systematic observation, while article 6 concerns education, training and public awareness, but both article 5 and 6 also take cognisance of the position of developing countries in particular.

In accordance with the UNFCCC, the principle of common but differentiated responsibility is also a principle employed by the Kyoto Protocol. This is particularly evident in article 2, article 3, article 10 and article 12 of the Protocol. Article 2(3) states that Annex I countries should ensure that their implementation of policies and measures under article 2 have minimal adverse effect on developing countries, illustrating the differentiation between countries to achieve equity. Article 3(14) reiterates the aforementioned idea and states that Annex I countries must implement their commitments in terms of article 3(1) so that the social, environmental and economic impacts on developing countries are minimal. Article 10 of the Kyoto Protocol also takes into account the principle of common but differentiated responsibility in addition to "national and regional development priorities, objectives and circumstances". Furthermore, article 12 of the Protocol deals with the clean development mechanism and article 12(8) stipulates that a share of the proceeds of the mechanism should be used to cover administrative expenses of developing countries and to assist developing countries that are particularly vulnerable in adapting to climate change.

At this point, the articulations of the principle of common but differentiated responsibility in the international legal climate change regime show that the principle is made up of two elements. The first element consists of the common responsibility shared by all states, namely the duty to protect the environment (albeit on a national, regional, or international scale). The second element concerns the issue of differentiation in responsibility between states, and specifically involves the consideration of a country's contribution towards the problem of climate change and a country's capability to contribute to addressing the problem of climate change.[578] Practically, the principle allows developed and developing countries to be treated according to their developmental status when international environmental law relating to the international legal climate change regime is applied. In accordance with the provisions of the Kyoto Protocol, developed countries bear the brunt of the responsibility for greenhouse gas emissions responsible for causing climate change.[579] In so doing, one of the greatest barriers to the success of the international legal climate change regime arises, namely the hesitation of developing counties to limit their greenhouse gas emissions,[580] and the refusal of developed countries, which are massive emitters of greenhouse gases, to ratify the Kyoto Protocol because of *inter alia* the lack of limitation placed on developing countries. Accordingly, the Kyoto Protocol is criticised on the basis that it is not fair to allow developing countries to emit greenhouse gases without binding them, like developed countries, to emission caps.[581] This is also considered the main reason why the USA did not ratify the Protocol.[582] One of the obstacles in regulating emission caps of developing countries thus far rested in the fact that some developing countries constitute some of the world's largest greenhouse gas emitters, while other developing countries hardly contribute to the emission of greenhouse gases. In this sense the principle of common but differentiated responsibility can be considered as an "ob-

---

578 P Sands *Principles of International Environmental Law* 2ed (2003) 286.

579 M Getu "Accommodating the Interests of Developing Countries n the Climate Change Regime: Lessons from the Ozone Layer Regime" (2012) 6(1) Mizan Law Review 1 13.

580 D Bodansky & L Rajamani "The Evolution and Governance Architecture of the Climate Change Regime" (2013) (2) SSRN <http://papers.ssrn.com/sol3/papers.cfm?abstract_id=2168859> (accessed 12-05-2015).

581 P Baer "Equity, Greenhouse Gas Emissions, and Global Common Resources" in SH Schneider et al (eds) *Climate Change Policy, A Survey* (2002) 393 394; Getu M "Accommodating the Interests of Developing Countries n the Climate Change Regime: Lessons from the Ozone Layer Regime" (2012) 6(1) *Mizan Law Review* 1 1-44.

582 S Barrett & R Stavins "Increasing Participation and Compliance in International Climate Change Agreements" (2002) 9-13 *Fondazione Eni Enrico Mattei* <http://www.feem.it/web/activ/_wp.html> (accessed: 04-07-2016); P Baer "Equity, Greenhouse Gas Emissions, and Global Common Resources" in SH Schneider et al (eds) *Climate Change Policy, A Survey* (2002) 393 394.

stacle"[583] as it has provided a legal shield[584] for those developing countries currently producing some of the highest levels of greenhouse gas emissions in the world, such as China. In addition to relying on the principle of common but differentiated responsibility, many developing countries which are heavily reliant on fossil fuels and consequently emit significant greenhouse gases still rely on PSNR as a shield against outside criticism in regard to their usage of national natural resources and corresponding emission of greenhouse gases.[585] Solving this regulatory conundrum might not lie in challenging the meaning of the equity principle of common but differentiated responsibility, but instead might lie in finding a way of distinguishing between developing countries according to greenhouse gas emissions within the international legal climate change regime in a way that would hold all emitters proportionally responsible. This is a challenging task as the question of fair allocation of emission allowances between developed and developing countries must be solved on a legal, political and economic level. The problem with common but differentiated responsibility is further explained below by using the contextual example of China and the USA.

An attempt to solve this problem is evident in the Paris Agreement 2015, which is the latest addition to the international legal climate change regime. The Paris Agreement 2015 sets out, *inter alia* in its preamble, that it is guided by principles such as the principle of common but differentiated responsibility. Article 2(2) of the Paris Agreement further stipulates that the agreement will be implemented to reflect the principle of common but differentiated responsibility.[586] Article 4(3) of the Agreement stipulates that each party's NDC will reflect common but differentiated respon-

583 W Scholtz "Equity as the basis for a future international climate change agreement: between pragmatic panacea and idealistic impediment. The optimisation of the CBDR principle via realism" (2009) 42(2) The Comparative and International Law Journal of Southern Africa 166 167.

584 D Badrinarayana "Global Warming: A Second Coming for International Law?" (2010) 85 Washington Law Review 253 253.

585 JE Hickey "Some legal Impacts of the Emerging International Climate Change Regime on Energy Prices" (2013) 4 *Global Business Law Review* 1 7.

586 Article 2(2) of the Paris Agreement reads as follows: "This Agreement will be implemented to reflect equity and the principle of common but differentiated responsibilities and respective capabilities, in the light of different national circumstances.".

sibilities.[587] Even though the Paris Agreement is still clearly implementing the principle of common but differentiated responsibility, it has managed to address the abovementioned criticism against the Kyoto Protocol by including a system of NDCs. The Agreement is discussed in greater detail further down in Chapter 3.

A noteworthy addition to the discussion of common but differentiated responsibility is the consideration of the Oslo Principles on Global Climate Change (Oslo Principles) formulated in 2015. The approach of the Oslo Principles to the principle of common but differentiated responsibility is contained in principle 14.[588] In this regard the committee which drafted the Oslo Principles decided that the matter should not be overcomplicated, and gave special attention to (or differentiated) countries that it deemed "vulnerable".[589] Furthermore, the committee employed a "per capita-approach" as a point of departure, which entails that each human being on earth is entitled to emit the same amount of greenhouse gases.[590] This approach is useful as it provides certainty; it is easily calculated; it allows for indirect accommodation of the level of development of a country; it allows for broad accommodation of historical greenhouse gas emissions; and, finally, it allows countries with modest greenhouse gas emissions to increase their emissions (in accordance with principles 14 and 15) until they reach the permissible level.[591] This being said, even the commit-

587 Article 4(3) of the Paris Agreement reads as follows: "Each Party's successive nationally determined contribution will represent a progression beyond the Party's then current nationally determined contribution and reflect its highest possible ambition, reflecting its common but differentiated responsibilities and respective capabilities, in the light of different national circumstances.".

588 Principle 14 of the Oslo Principles reads as follows: "The obligations of States are common but differentiated".

589 A Benjamin et al "Oslo Principles Commentary" (15-7-2015) (10) *Yale Global Justice Program* <http://globaljustice.macmillan.yale.edu/sites/default/files/files/Oslo%20Principles%20Commentary.pdf > (accessed 11-05-2015).
In this regard see principles 8, 9, 10, 13, 14, 15, 16, 17, 18, 19b, 20,21 and 23 [A Benjamin et al "Oslo Principles Commentary" (15-7-2015) (11) *Yale Global Justice Program* <http://globaljustice.macmillan.yale.edu/sites/default/files/files/Oslo%20Principles%20Commentary.pdf> (accessed 11-05-2015).].

590 A Benjamin et al "Oslo Principles Commentary" (15-7-2015) (11) *Yale Global Justice Program* <http://globaljustice.macmillan.yale.edu/sites/default/files/files/Oslo%20Principles%20Commentary.pdf> (accessed 11-05-2015).

591 Principle 14 of the Oslo Principles reads as follows: "The obligations of States are common but differentiated."

tee concedes that it is still possible to find a more equitable approach to the allocation of greenhouse gas emission limitations.[592] The Oslo Principles are discussed further in Chapter 4 of the research project.

3 7 2 2 2 The problem with the principle of common but differentiated responsibility as illustrated by the example of China and the United States of America

China and the USA, the world's two largest greenhouse gas emitters, provide a practical example of the failure of the principle of common but dif-

---

Principle 15 of the Oslo Principles reads as follows: "Least developed countries do not have a legal obligation to reduce GHG emissions at their own expense. They are subject only to the duties set out in Principles 7, 8, and 9."
Principle 7 of the Oslo Principles reads as follows: "All States and enterprises must reduce their GHG emissions to the extent that they can achieve such reduction without relevant additional cost. Relevant measures include switching off power-consuming equipment when not in use; eliminating excessive power consumption where possible, including for heating, cooling and lighting; promoting, to the maximum extent possible, measures that will reduce the need for consuming energy, such as improved insulation of buildings and improved efficiency of energy-consuming devices; elimination of broad fossil-fuel subsidies, including tax exemptions for certain industries, such as air transportation."
Principle 8 of the Oslo Principles reads as follows: "States and enterprises must refrain from starting new activities that cause excessive GHG emissions, including, for example, erecting or expanding coal-fired power plants, without taking countervailing measures, unless the relevant activities can be shown to be indispensable in light of prevailing circumstances, as might be the case, in particular, in the least developed countries. If the new activities are shown to be indispensable, a least developed country is obligated to opt for less GHG-emitting new activities only if and to the extent that developed countries or other entities provide the relevant least developed country with the additional means to meet this obligation."
Principle 9 of the Oslo Principles reads as follows: "Developed and developing countries, as well as enterprises, must take available GHG- reduction measures that entail costs if the costs will be offset through future savings or financial gains. Least developed countries and local enterprises in least developed countries have the same obligation to the extent that other entities provide the financial and technical means required without imposing more than a minimal financial burden on the relevant least developed countries or enterprises.".

592 A Benjamin et al "Oslo Principles Commentary" (15-7-2015) (10-11) *Yale Global Justice Program* <http://globaljustice.macmillan.yale.edu/sites/default/files/files/Oslo%20Principles%20Commentary.pdf> (accessed 11-05-2015).

ferentiated responsibility within the international legal climate change regime, and specifically within the Kyoto Protocol. Furthermore, a consideration of these two countries in this context demonstrates the regulatory difficulties involved in differentiating between developing and developed countries within the international legal climate change regime. This discussion illustrates the need to devise a method (attempted by the Paris Agreement) of addressing developing countries within the international legal climate change regime which are responsible for significant contributions to worldwide greenhouse gas emissions, while still treating developing countries making minimal contributions in an equitable way.[593]

China is considered a non-Annex I country[594] in terms of the Kyoto Protocol, and the USA is considered an Annex I country in terms of the Protocol (though it did not ratify the Protocol). China and the USA are the biggest producers of greenhouse gas emissions in the world and together they produce more than 40% of all carbon dioxide emissions and about 35% of all greenhouse gas emissions.[595] Multiple statistics show the annual carbon dioxide emissions of China to amount to about 28% of global emissions; the emissions of the United States to amount to about 16% of global emissions; the emissions of the European Union to amount to about 10–11% of global emissions; the emissions of India to amount to about 6% of global emissions; the emissions of Japan to amount to about 4% of global emissions; and the emissions of Russia to amount to about 6% of global emissions.[596] The negative influence of China and the USA on the

---

593 This chapter contains all updates until June 2016.

594 See UNFCCC website for a list of non-Annex I parties. [UNFCCC "Parties and Observers" (date unavailable) (page numbers unavailable) *UNFCCC* <http://unfccc.int/parties_and_observers/items/2704.php> (accessed 29-04-2013).]
Non-Annex I countries are mostly developing countries (of which certain developing countries such as low-lying coastal areas for example, are considered particularly vulnerable to the impact of climate change). [UNFCCC "Parties and Observers" (date unavailable) (page numbers unavailable) *UNFCCC* <http://unfccc.int/parties_and_observers/items/2704.php> (accessed 29-04-2013).].

595 Think Global Green "China Overview" (2008) (page numbers unavailable) *Think Global Green* <http://www.thinkglobalgreen.org/CHINA.html> (accessed 14-04-2015).

596 United States Environmental Protection Agency "Global Greenhouse Gas Emissions Data" (10-06-2016) (page numbers unavailable) *United States Environmental Protection Agency* <https://www3.epa.gov/climatechange/ghgemissions/global.html> (accessed 04-07-2016); M Gallucci & P Horn "For U.S and China, World's Biggest Climate Polluters, It's Still Business as Usual" (04-02-2013)

effectiveness of the climate change regime cannot therefore be overemphasised.[597] China, a developing country, overtook the USA, a developed country, in the production of greenhouse gas emissions in 2006.[598] Furthermore, while coal has the highest carbon content of all fossil fuels, the two countries combined already hold a massive lead in international coal consumption and both continue to show a massive reliance on coal as a source of energy moving into the future.[599] China relies on coal for approximately 80% of its current electricity generation, while the USA relies on coal for approximately 42% of its current electricity generation.[600]

(page numbers unavailable) *Inside Climate News* <http://insideclimatenews.org/news/20130129/united-states-china-carbon-greenhouse-gas-emissions-renewable-energy-coal-plants-pollution-global-warming-climate> (accessed 14-04-2015).

597 M Gallucci & P Horn "For U.S and China, World's Biggest Climate Polluters, It's Still Business as Usual" (04-02-2013) (page numbers unavailable) *Inside Climate News* <http://insideclimatenews.org/news/20130129/united-states-china-carbon-greenhouse-gas-emissions-renewable-energy-coal-plants-pollution-global-warming-climate> (accessed 14-04-2015).

598 China is the world's biggest producer of carbon dioxide and produced produced 6,200 million tons of carbon dioxide in 2006, which is 8% more carbon dioxide than the USA in 2006. [J Vidal & D Adam "China overtakes US as world's biggest CO2 emitter" (19-07-2007) (page numbers unavailable) *The Guardian* <http://www.theguardian.com/environment/2007/jun/19/china.usnews> (accessed 14-09-2015)].

599 Center for Climate and Energy Solutions "Coal" (date of publication unavailable) (page numbers unavailable) Center for Climate and Energy Solutions <http://www.c2es.org/energy/source/coal> (accessed 10-07-2016); M Gallucci & P Horn "For U.S and China, World's Biggest Climate Polluters, It's Still Business as Usual" (04-02-2013) (page numbers unavailable) Inside Climate News <http://insideclimatenews.org/news/20130129/united-states-china-carbon-greenhouse-gas-emissions-renewable-energy-coal-plants-pollution-global-warming-climate> (accessed 14-04-2015).
Most recent statistics show that China, the USA and India together account for more than 70% of the world's coal consumption. China accounts for 48% of the global coal consumption, and its consumption of coal far outweighs the consumption of coal of India and the USA [U.S Energy Information Administration "International Energy Outlook 2016" (11-05-2016) (page numbers unavailable) U.S Energy Information Administration <http://www.eia.gov/forecasts/ieo/coal.cfm> (accessed: 04-07-2016).].

600 M Gallucci & P Horn "For U.S and China, World's Biggest Climate Polluters, It's Still Business as Usual" (04-02-2013) (page numbers unavailable) *Inside Climate News* <http://insideclimatenews.org/news/20130129/united-states-china-carbon-greenhouse-gas-emissions-renewable-energy-coal-plants-pollution-global-warming-climate> (accessed 14-04-2015).

From these figures, it is clear to see that an urgent plan must be formulated to address the current absence of regulation concerning developing countries, China in particular, within the international climate change regime and in particular the Kyoto Protocol. In addition, it is apparent that a successful international legal climate change regime would need to have the support and participation of the USA.

The Conference of the Parties (COP) in Paris in December 2015 concluded with an indication of political will-power to move towards a low carbon future from China and the USA, who seem to be on board with the system of NDCs of the Paris Agreement. Under the Paris Agreement, China has pledged *inter alia* the following actions by 2030: the county's carbon dioxide emissions will peak by 2030; it will promote the use of non-fossil-fuel-based energy sources significantly; and it will increase the square metres of forest stock significantly.[601] China will also implement emissions trading from 2017.[602] In addition, China pledged to contribute more than $3 billion to developing countries to assist them in meeting their climate change objectives in terms of the Paris Agreement.[603] Furthermore, China and the USA pledged to double their respective clean energy investments and expand cooperation in promoting clean energy.[604] Under the Paris Agreement, the USA pledged to reduce greenhouse gas emissions by between 26 and 28% compared to its emission levels of 2005 by the year 2025.[605] The Obama administration of the USA also pledged

601 J Busby "After Paris: Good Enough Climate Governance" (2016) January *Current History* 1 6; Center for Climate and Energy Studies "Submitted Intended Nationally Determined Contributions (INDCS)" (date of publication unavailable) (page numbers unavailable) *Center for Climate and Energy Studies* <http://www.c2es.org/international/2015-agreement/indcs> (accessed 20-07-2016).

602 S Dröge "The Paris Agreement 2015, Turning Point for the International Climate Regime" (2016) (February) *German Institute for International and Security Affairs* 1 17.

603 J Busby "After Paris: Good Enough Climate Governance" (2016) (January) *Current History* 1 6.

604 Office of the Press Secretary "FACT SHEET: U.S.-China Joint Announcement on Climate Change and Clean Energy Cooperation" (11-11-2014) (page numbers unavailable) *The Whitehouse* <https://www.whitehouse.gov/the-press-office/2014/11/11/fact-sheet-us-china-joint-announcement-climate-change-and-clean-energy-c> (accessed 10-07-2016); J Busby "After Paris: Good Enough Climate Governance" (2016) (January) *Current History* 1 6.

605 S Dröge "The Paris Agreement 2015, Turning Point for the International Climate Regime" (2016) (February) *German Institute for International and Security Af-*

to contribute $800 million by 2020 to assist poorer countries to adapt to the impacts of climate change.[606] It remains to be seen whether the approach of the Paris Agreement to common but differentiated responsibility will improve the effectiveness of the international legal climate change regime.

3 7 2 3 Precautionary principle

As an equity principle of the international legal climate change regime, the precautionary principle entails that, where a substance or an activity represents a threat to the environment, the activity is not allowed despite absence of conclusive scientific evidence that the substance or activity will in fact adversely affect the environment.[607] Simply put, the principle implies that "we should not wait for scientific certainty before acting on a threat"[608] to the environment and conduct should be guided by the saying "better safe than sorry".[609]

Principle 15 of the Rio Declaration sets out the precautionary principle as follows:

> "In order to protect the environment, the precautionary approach shall be widely applied by States according to their capabilities. Where there are threats of serious or irreversible damage, lack of full scientific certainty shall not be used as a reason for postponing cost-effective measures to prevent environmental degradation."

---

fairs 1 17; Center for Climate and Energy Studies "Submitted Intended Nationally Determined Contributions (INDCS)" (date of publication unavailable) (page numbers unavailable) *Center for Climate and Energy Studies* <http://www.c2es.org/international/2015-agreement/indcs> (accessed 20-07-2016).

606 J Busby "After Paris: Good Enough Climate Governance" (2016) (January) *Current History* 1 7.

607 J Cameron & J Abouchar "The Precautionary Principle: A Fundamental Principle of Law and Policy for the Protection of the Global Environment" (1991) XIV(1) *Boston College International & Comparative Law Review* 2.

608 L Heinzerling "Climate Change, Human Health, and the Post-Cautionary Principle" (2008) 96 *The Georgetown Law Journal* 445 452.

609 FB Cross "Paradoxical Perils of the Precautionary Principle" (1996) 53(3) *Washington and Lee Law Review* 851 851.

The principle has, more recently, also been set out in article 3(3) of the UNFCCC, which reads as follows:

> "The Parties should take precautionary measures to anticipate, prevent or minimize the causes of climate change and mitigate its adverse effects. Where there are threats of serious or irreversible damage, lack of full scientific certainty should not be used as a reason for postponing such measures, taking into account that policies and measures to deal with climate change should be cost-effective so as to ensure global benefits at the lowest possible cost. To achieve this, such policies and measures should take into account different socio-economic contexts, be comprehensive, cover all relevant sources, sinks and reservoirs of greenhouse gases and adaptation, and comprise all economic sectors. Efforts to address climate change may be carried out cooperatively by interested Parties."[610]

The Paris Agreement vaguely references the principle in its preamble, which identifies "the need for an effective and progressive response to the urgent threat of climate change on the basis of the best available scientific knowledge". Similarly, article 4(1) of the Paris Agreement refers to undertaking rapid reductions of greenhouse gas emissions in accordance with "best available science". These inclusions in the Agreement can be interpreted as meaning that should the best available scientific knowledge be uncertain, the need for an effective response to climate change still exists.

In trying to ascertain the content of the principle, it is useful to consider the approach to the principle by the European Commission. The European Commission informed all member states of the Commission that the principle "may be invoked when a phenomenon, product or process may have a dangerous effect, identified by a scientific and objective evaluation, if this evaluation does not allow the risk to be determined with sufficient certainty".[611] Kiss and Shelton contend that the European Commission applies the precautionary principle as an interim measure.[612] This is clearly not the position taken by the UNFCCC, as article 3(3), as set out above, does not frame the principle as an interim measure.

---

610 Article 3(3), UNFCCC.

611 EUR-Lex "The precautionary principle" (21-09-2015) (page numbers unavailable) *EUR-Lex* <http://eur-lex.europa.eu/legal-content/EN/TXT/?uri=URISERV%3Al32042> (accessed 03-09-2016).

612 A Kiss & D Shelton *International Environmental Law* 3ed (2004) 208. Cameron and Abouchar and refer to the principle as a "guiding principle". [J Cameron & J Abouchar "The Precautionary Principle: A Fundamental Principle of Law and Policy for the Protection of the Global Environment" (12-01-1999) 14(1) *Boston College International and Comparative Law Review* 1 2.].

Furthermore, in trying to ascertain the level of certainty that the principle advocates, the approach of the European Court of Justice is useful. The Court sheds light on the level of certainty in question, specifically in the context of the effect of the environment on human health.[613] The European Court of Justice takes the view that, where it is a question of human health, protective interim measures may be instituted with a very low level of certainty that a threat exists to human health.[614] Furthermore, in the application of the European Commission and the European Court of Justice above, the principle is applied so that human health and environmental concerns are given priority over economic concerns.[615]

Climate change has long been viewed as the ideal environment in which to apply the precautionary principle.[616] However, theoretically, it can be argued that the precautionary period for climate change was sometime during the 1980s and the world is now in a "post-cautionary" period of climate change.[617] It is the view of the research project that there is scientific certainty about climate change, that the threats to the environment and to human health and well-being are clear, and that the world is now in a post-cautionary period of climate change. However, climate change sceptics may argue that there does exist significant scientific uncertainty in regard to climate change. In this case the precautionary principle is to be applied in its simplest form, namely that it imposes "a duty on states to prevent pollution despite a lack of scientific evidence".[618] Regardless of the difference in viewpoints toward climate change, the principle is and continues to be a principle that supports international action in the fight against climate change.

---

613 A few cases were decided in this context. For a brief summary on the relevant case law in this regard, see: A Kiss & D Shelton *International Environmental Law* 3ed (2004) 208-209.

614 A Kiss & D Shelton *International Environmental Law* 3ed (2004) 209.

615 A Kiss & D Shelton *International Environmental Law* 3ed (2004) 208.

616 L Heinzerling "Climate Change, Human Health, and the Post-Cautionary Principle" (2008) 96 *The Georgetown Law Journal* 445 452.

617 L Heinzerling "Climate Change, Human Health, and the Post-Cautionary Principle" (2008) 96 *The Georgetown Law Journal* 445 452.

618 JE Hickey & VR Walker "Refining the precautionary principle in international environmental law" (1994-1995) 14 *Virginia Environmental Law Journal* 423 425.

3 7 2 4 Polluter-pays principle

The polluter-pays principle was originally constructed as an economic principle, and later developed into a legal principle.[619] The principle was adopted by the Organisation for Economic Co-operation and Development (OECD) Council in 1972[620] as a means to allocate the costs incurred by pollution.[621] The OECD recommendations of 1974[622] and 1989[623] further updated the principle. The newest recommendations describe the principle as meaning that the polluter should bear the expenses of pollution prevention and control of pollution.[624] The principle has also been interpreted to mean that the cost of pollution prevention and control should be reflected in the costs of goods and services produced or consumed in regard to the pollution.[625]

The precautionary principle was included in Principle 16 of the Rio Declaration, which reads as follows:

> "National authorities should endeavour to promote the internalization of environmental costs and the use of economic instruments, taking into account the approach that the polluter should, in principle, bear the cost of pollution, with due regard to the public interest and without distorting international trade and investment."

Simply put, the polluter-pays principle means that the party responsible for the pollution will pay the costs of the pollution (the cost refers to the environmental harm as a whole). It has also been interpreted to mean that those who utilise a resource should compensate the owners of the re-

---

619 E Woerdman et al "Emissions Trading and the Polluter-Pays Principle: Do Polluters Pay Under Grandfathering?" (2008) 14 *Rotterdam Institute of Law and Economics Working Paper Series* 11.

620 OECD "Guiding Principles concerning International Economic Aspects of Environmental Policies" (adopted 26 May 1972) [C(72)128].

621 O Vícha "The polluter-pays principle in OECD recommendation and its application in international and EC/EU law" (2011) *Czech Yearbook of Public & Private International Law* 57 58.

622 OECD "Implementation of the Polluter-Pays Principle" (adopted 14 November 1974) [C(74)223].

623 OECD "Application of the Polluter-Pays Principle to Accidental Pollution" (adopted 7 July 1989) [c(89)88].

624 OECD "Application of the Polluter-Pays Principle to Accidental Pollution" (adopted 7 July 1989) [c(89)88].

625 OECD "Application of the Polluter-Pays Principle to Accidental Pollution" (adopted 7 July 1989) [c(89)88].

source, which would mean the resource utilised is the environment and the owner is the public.[626] This creates a strong link between the right of states to dispose freely of, explore and exploit resources in terms of PSNR in accordance with the corresponding state duty to exercise PSNR for national development and the well-being of the people, as well as the duty to conserve and sustainably use natural resources.[627]

Although the UNFCCC does not make direct reference to the principle, the principle can be considered to be applicable to the international legal climate change regime in the context of historical accountability for greenhouse gas emissions, and accordingly those states historically responsible for anthropocentric greenhouse gas emissions causing climate change must now be held responsible.[628] It can also be argued that the inclusion of the principle of common but differentiated responsibility throughout the UNFCCC is an implied recognition of the polluter-pays principle within the Convention.[629]

It is more readily acknowledged that the polluter-pays principle is applicable to the international legal climate change regime in the context of international emissions trading (also referred to as the carbon market) and carbon tax.[630] The carbon market functions on the basis of emission rights which are allocated to member states in terms of the Kyoto Protocol. The carbon market uses a method known as 'grandfathering'. Grandfathering can be explained as the basis upon which emission rights are allocated, and means that states receive emission rights based on their historical emission levels.[631] The Kyoto Protocol allocates emission rights to mem-

626 L Ringius, A Torvanger & A Underdal "Burden Sharing and Fairness principles in International Climate Policy" (2002) 2 *International Environmental Agreements: Politics, Law and Economics* 1 18.

627 The rights and duties of states in discussed in Chapter 4.

628 E Neumayer "In defence of historical accountability for greenhouse gas emissions" (2000) 33(2) *Ecological Economics* 185-192.

629 MR Khan "Polluter-Pays-Principle: The Cardinal Instrument for Addressing Climate Change" (2015) (4) Laws 638 638.

630 M Paterson "Principles of Justice in the Context of Global Climate Change" (2001) (122) Graduate Institute Geneva <http://graduateinstitute.ch/files/live/sites/iheid/files/shared/iheid/800/luterbacher/luterbacher%20chapter%206%20106.pdf> (accessed 09-09-2016).

631 E Woerdman et al "Emissions Trading and the Polluter-Pays Principle: Do Polluters Pay Under Grandfathering?" (2008) 14 *Rotterdam Institute of Law and Economics Working Paper Series* 4 <http://ssrn.com/abstract=1271843> (accessed 23-04-2015).

ber states on the basis of the states' emission levels in 1990. Emission rights are therefore received free of charge.[632] It has been argued that grandfathering is not consistent with the polluter-pays principle, because emission rights are received free of charge.[633] However, it can also be argued that grandfathering is consistent with the polluter-pays principle because historic emissions are used when determining that country's future emission limitations in terms of the Kyoto Protocol, and whether or not that country will have a limitation on its emissions. It falls outside the ambit of this research project to discuss this debate in greater detail, and it is sufficient to simply take note that grandfathering is a component of the discussion of the polluter-pays principle in the context of the international legal climate change regime.

3 7 2 5 Principle of sustainable development

Although it is difficult to provide a single overarching definition for sustainable development, the term is largely understood to contain, at a minimum, a concern for future generations.[634] A particularly well-known and useful definition of sustainable development is found in the Brundtland Report, which reads as follows:

> "Sustainable development is development that meets the needs of the present without compromising the ability of future generations to meet their own needs."[635]

---

632 E Woerdman et al "Emissions Trading and the Polluter-Pays Principle: Do Polluters Pay Under Grandfathering?" (2008) 14 *Rotterdam Institute of Law and Economics Working Paper Series* 13 <http://ssrn.com/abstract=1271843> (accessed 23-04-2015).

633 E Woerdman et al "Emissions Trading and the Polluter-Pays Principle: Do Polluters Pay Under Grandfathering?" (2008) 14 *Rotterdam Institute of Law and Economics Working Paper Series* <http://ssrn.com/abstract=1271843> (accessed 23-04-2015).

634 MA Toman "The Difficulty in Defining Sustainability" in J Darmstadter (ed) *Global Development and the Environment, Perspectives on Sustainability* (1992) 15 16-17.

635 World Commission on Environment and Development *Our Common Future* (1987) 43.

Sustainable development is also often understood as simultaneously promoting environmental, social and economic values.[636] Sustainable development, as defined above in the Brundtland Report, defines sustainable development in terms of needs of the present and needs of the future. According to Blanco and Razzaque, the term 'needs' in this context pertain especially to the needs of the poor.[637] Therefore, it is possible to submit that the principle of sustainable development is intended to promote development, particularly in developing countries where poverty tends to be more widespread. As a peripheral issue, it should be noted that the principle of sustainable development includes the principle of sustainable use, although sustainable use is often differentiated from sustainable development as a separate value to be considered in natural resource management.[638]

As stressed by Kiss, in order to successfully integrate environmental, social and economic policies in a regime such as the international legal climate change regime, transparency and public participation in government decision-making is required.[639] Public participation is a right of people relevant in the context of PSNR (this will be illustrated in Chapter 4) and, accordingly, reinterpreting PSNR so that it is a tool that can be employed by people (by which to hold states accountable for the climate commitments they make in terms of the international legal climate change regime) fits well into Kiss's aforementioned submission. Of all the equity principles on which the international legal climate change regime is founded, explained in Section 3.7.2, sustainable development displays the

---

636 E Blanco & J Razzaque Globalisation and Natural Resources Law: Challenges, Key Issues and Perspectives (2011) 94.

637 E Blanco & J Razzaque Globalisation and Natural Resources Law: Challenges, Key Issues and Perspectives (2011) 93.

638 Article 2 of the Biodiversity Convention defines sustainable use as follows: "the use of components of biological diversity in a way and at a rate that does not lead to the long-term decline of biological diversity, thereby maintaining its potential to meet the needs and aspirations of present and future generations." Furthermore, article 6 of the Biodiversity Convention sets out 'General Measures for Conservation and Sustainable Use"; article 8(c) of the Biodiversity Convention further makes reference to sustainable use when parties "regulate or manage biological resources important for the conservation of biological diversity whether within or outside protected areas"; and finally article 10 sets out "Sustainable Use of Components of Biological Diversity".

639 A Kiss & D Shelton *International Environmental Law* 3ed (2004) 217.

strongest ties to PSNR, perhaps because ultimately sustainable development is, like PSNR, "a notion that conveys both rights and duties".[640]

Principle 2 of the Stockholm Declaration stipulates that "the natural resources of the earth [...] must be safeguarded for the benefit of present and future generations through careful planning or management". Principle 2 of the Stockholm Declaration therefore manages to include in this single principle both PSNR and sustainable development as a principle of intergenerational equity, and in so doing emphasises the historic nexus between PSNR and the foundational principle of equity of the international legal climate change regime (which in turn creates a nexus between PSNR and climate change). With similar effect, principle 5 of the Stockholm Declaration stipulates that "non-renewable resources of the earth must be employed in such a way as to guard against the danger of their future exhaustion and to ensure that benefits from such employment are shared by all mankind".

The Rio Declaration also contains references to sustainable development as an essential principle to consider in the context of environmental management.[641] Principle 4 of the Rio Declaration stipulates that "in order to achieve sustainable development, environmental protection shall constitute an integral part of the development process and cannot be considered in isolation from it". Accordingly, principle 4 of the Rio Declaration can be considered as founded in the state duty of PSNR which promotes the conservation and sustainable use of natural resources.[642]

Inclusions of the principle of sustainable development, specifically in the context of the international legal climate change regime, include article 2 of the UNFCCC, which stipulates that the time frame for addressing cli-

---

640 E B Weiss "In Fairness To Future Generations and Sustainable Development" (1992) 8(1) *American University International Law Review* 19 20.

641 For instance, principle 1, 4 and 8 of the Rio Declaration. Principle 1 of the Rio Declaration reads as follows: "Human beings are at the centre of concerns for sustainable development. They are entitled to a healthy and productive life in harmony with nature." Principle 4 Rio Declaration reads as follows: "In order to achieve sustainable development, environmental protection shall constitute an integral part of the development process and cannot be considered in isolation from it." Principle 8 of the Rio Declaration reads as follows: "To achieve sustainable development and a higher quality of life for all people, States should reduce and eliminate unsustainable patterns of production and consumption and promote appropriate demographic policies.".

642 The state rights and duties in terms of PSNR are discussed in Chapter 4.

mate change must "enable economic development to proceed in a sustainable manner".[643] Article 3(4) of the UNFCCC also applies the principle of sustainable development to the regime, and stipulates that "[p]arties have a right to, and should, promote sustainable development", while taking into account the particular conditions of each party.

In addition, article 2(1) of the Kyoto Protocol sets out examples of policies and measures that Annex 1 countries can implement to promote sustainable development.[644] Furthermore, article 12 of the Protocol, which deals with the clean development mechanism, stipulates in article 12(2) that the purpose of the mechanism is to assist countries not included in Annex I to achieve sustainable development. Article 10 of the Protocol further indicates that achieving sustainable development is an important principle of the Kyoto Protocol.[645] The latest component of the international legal climate change regime to incorporate and mention the princi-

---

643 Article 2, UNFCCC.

644 Examples of policies or measures which can be implemented or elaborated by Annex I countries to help promote sustainable development include the following, according to Article 2(1)(a)(i) to (viii) of the Kyoto Protocol: "(i) Enhancement of energy efficiency in relevant sectors of the national economy; (ii) Protection and enhancement of sinks and reservoirs of greenhouse gases not controlled by the Montreal Protocol, taking into account its commitments under relevant international environmental agreements; promotion of sustainable forest management practices, afforestation and reforestation; (iii) Promotion of sustainable forms of agriculture in light of climate change considerations; (iv) Research on, and promotion, development and increased use of, new and renewable forms of energy, of carbon dioxide sequestration technologies and of advanced and innovative environmentally sound technologies; (v) Progressive reduction or phasing out of market imperfections, fiscal incentives, tax and duty exemptions and subsidies in all greenhouse gas emitting sectors that run counter to the objective of the Convention and application of market instruments; (vi) Encouragement of appropriate reforms in relevant sectors aimed at promoting policies and measures which limit or reduce emissions of greenhouse gases not controlled by the Montreal Protocol; (vii) Measures to limit and/or reduce emissions of greenhouse gases not controlled by the Montreal Protocol in the transport sector; (viii) Limitation and/or reduction of methane emissions through recovery and use in waste management, as well as in the production, transport and distribution of energy".

645 Article 10 of the Kyoto Protocol reads as follows: "All Parties, taking into account their common but differentiated responsibilities and their specific national and regional development priorities, objectives and circumstances, without introducing any new commitments for Parties not included in Annex I, but reaffirming existing commitments under Article 4, paragraph 1, of the Convention, and continuing to advance the implementation of these commitments in order to achieve

ple of sustainable development is the Paris Agreement. Article 2 of the Agreement stipulates *inter alia* that it "aims to strengthen the global response to the threat of climate change, in the context of sustainable development". Article 4 stipulates that *inter alia* the temperature goal set out in article 2 of the Agreement must be achieved "within the context of sustainable development". Furthermore, article 6(1) of the Agreement stipulates *inter alia* that parties implement NDCs in the pursuit of promoting sustainable development; and article 6(4) establishes a mechanism to contribute to the mitigation of emissions and to support sustainable development. In addition, article 7 of the Agreement states that adaptation to climate change must occur "with a view to contributing to sustainable development". Finally, article 10 stipulates *inter alia* that the acceleration of innovation, in the context of technology transfer, is critical for promoting sustainable development.

3 7 3 Conference of the parties

The annual COP to the UNFCCC constitutes an integral component of the international legal climate change regime. The COPs are held in order to examine the obligations, institutional arrangements, and the implementation of and possible amendments to the UNFCCC.[646] In addition to the presence of member states at the COPs, the presence of observers, and environmental and industry representatives at the conferences exert pressure on the success of the process and contribute valuable inputs during the negotiations.[647] The COP is the ultimate decision-making body of the UN-

sustainable development, taking into account Article 4, paragraphs 3, 5 and 7, of the Convention, shall: […].".

646 UNFCCC "Conference of the Parties (COP)" (date of publication unavailable) (page numbers unavailable) UNFCCC <http://unfccc.int/bodies/body/6383.php> (accessed 10-07-2016); D Bodansky & L Rajamani "The Evolution and Governance Architecture of the Climate Change Regime" (2013) (30) SSRN <http://papers.ssrn.com/sol3/papers.cfm?abstract_id=2168859> (accessed 12-05-2015).

647 D Bodansky & L Rajamani "The Evolution and Governance Architecture of the Climate Change Regime" (2013) (30) SSRN <http://papers.ssrn.com/sol3/papers.cfm?abstract_id=2168859> (accessed 12-05-2015).

FCCC.[648] Usually, decisions reached at the conferences do not have legal force, unless the UNFCCC expressly provides for the conference to adopt rules for a particular matter.[649] Since 1995, there have been many COPs to the UNFCCC. The amount of COPs and decisions to date illustrate the scale of the political organisation and legal action within the international community that is required to formulate and maintain the international legal climate change regime. The table below lists COPs held to date.[650]

Table 3.1: Summary of the COPs to the UNFCCC

COP	Date	Location
COP 1[651]	1995	Berlin, Germany

648 UNFCCC "Conference of the Parties (COP)" (date of publication unavailable) (page numbers unavailable) UNFCCC <http://unfccc.int/bodies/body/6383.php> (accessed 10-07-2016).

649 D Bodansky et al "Issues for a 2015 climate agreement" (May 2015) (5) *Center for Climate and Energy Solutions* <http://www.c2es.org/docUploads/issues-for-a-2015-climate-agreement.pdf> (accessed 04-05-2015).

650 Note that this chapter only includes developments up until the end of June 2016.

651 The most prominent outcome of the COP 1 in Berlin led to the adoption of the Berlin Mandate, which authorised inter alia negotiations that would strengthen commitments in terms of the UNFCCC since parties agreed that the UNFCCC commitments are inadequate in certain respects. [D Bodansky "The History of the Global Climate Change Regime" in U Luterbacher & DF Sprinz (eds) International Relations and Global Climate Change 2ed (2001) 23 26; UNFCCC "A Brief Overview of Decisions" (date of publication unavailable) (page numbers unavailable) UNFCCC <http://unfccc.int/documentation/decisions/items/2964.php> (accessed 10-07-2016).] For instance, the commitments in the UNFCCC are vague, which result in a lack of concrete obligations and no specific time frames by which industrialised countries (or Annex 1 countries) should reduce greenhouse gas emissions in terms of the convention. [D Freestone "The UN Framework Convention on Climate Change, the Kyoto Protocol and the Kyoto Mechanisms" in D Freestone & C Streck (eds) Legal Aspects of Implementing the Kyoto Protocol Mechanisms, Making Kyoto Work (2005) 7 3-24.] The Berlin Mandate left developed countries with commitments which were not mirrored by developing countries, a fact which would go on to form the future basis of widespread criticism against the international legal climate change regime and specifically the Kyoto Protocol.

COP	Date	Location
COP 2[652]	1996	Geneva, Switzerland
COP 3[653]	1997	Kyoto, Japan
COP 4[654]	1998	Buenos Aires, Argentina
COP 5[655]	1999	Bonn, Germany
COP 6[656]	2000	6-1: The Hague, Netherlands 6-2: Bonn, Germany
COP 7[657]	2001	Marrakesh, Morocco
COP 8[658]	2002	New Delhi, India
COP 9[659]	2003	Milan, Italy

652 Highlights of COP 2 included the discussion of Quantified Emissions Limitation and Reduction Objectives (QELRO's) and the progression in terms of the Berlin Mandate in preparation for COP 3.

653 Significantly, the Kyoto Protocol was adopted at COP 3, which set legally binding emission targets for the six major greenhouse gases for developed country parties (or Annex 1 countries). These targets were to be reached in the time period spanning from 2008 to 2012. [UNFCCC "Meetings" (date of publication unavailable) UNFCCC (page numbers unavailable) <http://unfccc.int/meetings/items/6240.php> (accessed 16-04-2013).].

654 COP 4 led to the Buenos Aires Plan of Action, which focused on maintaining momentum in regard to the Kyoto Protocol, strengthening the financial mechanism of the UNFCCC, and the development and transfer of technology between parties. [UNFCCC "Meetings" (date of publication unavailable) UNFCCC (page numbers unavailable) <http://unfccc.int/meetings/items/6240.php> (accessed 16-04-2013).].

655 COP 5, held in Bonn in 1999, focused on the adoption of guidelines pertaining to the preparation of national communications of Annex 1 countries in relation to the Kyoto Protocol, capacity building, the transfer of technology and flexible mechanisms. [UNFCCC "Meetings" (date of publication unavailable) UNFCCC (page numbers unavailable) <http://unfccc.int/meetings/items/6240.php> (accessed 16-04-2013).].

656 At COP 6 consensus was reached on the Bonn Agreements. In addition, decisions were made based on the Bonn Agreements and pertaining to capacity building for developing countries and countries with transitioning economies. [UNFCCC "Meetings" (date of publication unavailable) UNFCCC (page numbers unavailable) <http://unfccc.int/meetings/items/6240.php> (accessed 16-04-2013).].

657 The Marrakech Ministerial Declaration (as an input into the World Summit on Sustainable Development in Johannesburg) was adopted at COP 7. In addition, COP 7 also adopted the Marrakesh Accords, which sets out the detailed rules for implementing the Kyoto Protocol. [UNFCCC "Kyoto Protocol" (date of publication unavailable) (page numbers unavailable) *UNFCCC* <http://unfccc.int/kyoto_protocol/items/2830.php> (accessed 29-04-2013).].

COP	Date	Location
COP 10[660]	2004	Buenos Aires, Argentina
COP 11[661]	2005	Montreal, Canada
COP 12[662]	2006	Nairobi, Kenya
COP 13[663]	2007	Bali, Indonesia
COP 14[664]	2008	Poznan, Poland

658 COP 8 led to the Delhi Ministerial Declaration on Climate Change and Sustainable Development, which reiterated the need to build on the outcomes of the World Summit on Sustainable Development in Johannesburg. [UNFCCC "Meetings" (date of publication unavailable) *UNFCCC* (page numbers unavailable) <http://unfccc.int/meetings/items/6240.php> (accessed 16-04-2013).].

659 In 2003, at COP 9, attention was focused on the new emission reporting guidelines based on the good-practice guidance provided by the IPCC, the agreement on the modalities and scope for carbon absorbing forest-management projects in the CDM, as well as the special climate fund and the least developed countries fund. [UNFCCC "Meetings" (date of publication unavailable) *UNFCCC* (page numbers unavailable) <http://unfccc.int/meetings/items/6240.php> (accessed 16-04-2013).].

660 COP 10 was concerned with a variety of issues including the development and transfer of technologies, land use, land-use change and forestry, the UNFCCC's financial mechanism, Annex I national communications, capacity building, the UNFCCC's adaptation and response measures, developed countries and future strategies to address climate change. [UNFCCC "Meetings" (date of publication unavailable) (page numbers unavailable) *UNFCCC* <http://unfccc.int/meetings/items/6240.php> (accessed 16-04-2013).].

661 COP 11 was held in 2005 and was a significant event, since for the first time two conferences ran parallel, namely the Meeting of the Parties to the Kyoto Protocol (CMP) and the COP. [UNFCCC "Meetings" (date of publication unavailable) (page numbers unavailable) *UNFCCC* <http://unfccc.int/meetings/montreal_nov_2005/meeting/6329.php> (accessed 13-05-2013).].

662 COP 12 in Nairobi marked the second Meeting of the Parties to the Kyoto Protocol.

663 At COP 13, the IPCC's Fourth Assessment Report was presented and the Bali Road Map was adopted. The Bali Road Map contains the vital decisions vital in regard to the global climate change agreement between countries, the Bali Action Plan, decisions on deforestation and forest management, decisions on technology for developed countries, establishes the Adaptation Fund Board and reviews the Financial Mechanism. The Bali Action plan was an important part in initiating and emphasising the importance of actions of adaptation within the context of climate change. [UNFCCC "Meetings" (date of publication unavailable) (page numbers unavailable) *UNFCCC* <http://unfccc.int/meetings/bali_dec_2007/meeting/6319.php> (accessed 13-05-2013).].

664 COP 14 launched an Adaptation Fund under the Kyoto Protocol and parties identified differing views on increasing the Adaptation Fund and improving the Clean

COP	Date	Location
COP 15[665]	2009	Copenhagen, Denmark
COP 16[666]	2010	Cancun, Mexico

Development Mechanism which was to be resolved in 2009. [UNFCCC "Meetings" (date of publication unavailable) (page numbers unavailable) *UNFCCC* <http://unfccc.int/meetings/bali_dec_2007/meeting/6319.php> (accessed 13-05-2013).].

665 In 2009, COP 15 was considered a significant event, attended by 115 world-leaders, which "raised climate change policy to the highest level". The outcomes at the conference in Copenhagen included advancing global climate change cooperation (including improvements to the Clean Development Mechanism), contributing to the success of future conferences by focusing options and choices that needed to be made on key issues, and producing the Copenhagen Accord. The Copenhagen Accord referred to some important issues regarding finance, reporting of parties and the United Nations collaborative initiative on the REDD Programme. However, most significantly the Copenhagen Accord expressed the political intention not to allow the maximum global average temperature to increase by more than 2 degrees Celsius above pre-industrial levels. [UNFCCC "Meetings" (date of publication unavailable) (page numbers unavailable) *UNFCCC* <http://unfccc.int/meetings/copenhagen_dec_2009/meeting/6295.php> (accessed 13-05-2013).].

666 COP 16 in Cancun was a noteworthy gathering and has been described as "the most comprehensive and far-reaching international response to climate change the world had ever seen to reduce carbon emissions and build a system which made all countries accountable to each other for those reductions". An important outcome of the Cancun conference was the commitment of parties not to allow the maximum global average temperature to increase by more than 2 degrees Celsius above pre-industrial levels (and possibly even lower this number to 1.5,). The conference also made a technology mechanism available, which would increase innovation and development and the sharing of these technologies. In addition, the Green Climate Fund was established to aid developing countries in the undertaking of new projects, programmes and policies to help fight climate change. The parties also agreed on the Cancun Adaptation Framework, which would ultimately promote effective adaptation to climate change. The Cancun Adaptation Framework was built on the Bali Action Plan, which has been discussed above. [UNFCCC "Meetings" (date of publication unavailable) (page numbers unavailable) *UNFCCC* <http://unfccc.int/meetings/cancun_nov_2010/meeting/6266.php> (accessed 13-05-2013).].

COP	Date	Location
COP 17[667]	2011	Durban, South Africa
COP 18[668]	2012	Doha, Qatar
COP 19[669]	2013	Warsaw, Poland

667 COP 17 has been described as the "turning point" in the climate change negotiations. [UNFCCC "Meetings" (date of publication unavailable) (page numbers unavailable) *UNFCCC* <http://unfccc.int/meetings/poznan_dec_2008/meeting/6314.php> (accessed 29-04-2013). The COP ensured prolonged maintenance of the Kyoto Protocol and set the stage for the negotiations of a successor agreement. [Center for Climate and Energy Solutions "Outcomes of the U.N. Climate Change Conference in Lima" (date of publication unavailable) (page numbers unavailable) *Center for Climate and Energy Solutions* <http://www.c2es.org/international/negotiations/cop-19/summary> (accessed 04-05-2015).] A key outcome of this conference was the agreed second commitment period of the Kyoto Protocol. Governments extended the Kyoto Protocol to 2017 and potentially to 2020. [Council on Foreign Relations "The Climate Change Regime" (19-06-2013) (page numbers unavailable) *Council on Foreign Relations* <http://www.cfr.org/climate-change/global-climate-change-regime/p21831> (accessed 17-04-2015).].

668 An important outcome of COP 18 was the Doha Amendment to the Kyoto Protocol and the establishment of the second commitment period of the Kyoto Protocol. [Center for Climate and Energy Solutions "Outcomes of the U.N. Climate Change Conference in Lima" (date of publication unavailable) (page numbers unavailable) *Center for Climate and Energy Solutions* <http://www.c2es.org/international/negotiations/cop-19/summary> (accessed 04-05-2015).] The Doha Amendment to the Kyoto Protocol includes new commitments for Annex 1 countries who agree to take on commitments in a second commitment period to the Kyoto Protocol (which stretches from 1 January 2013 to 31 December 2020), updates to certain articles of the Kyoto Protocol in accordance with the second commitment period, and a revised list of greenhouse gases to report on included in the second commitment period. Importantly, the composition of parties for the second commitment period is different than the composition of parties for the fist commitment period. [UNFCCC "Kyoto Protocol" (date of publication unavailable) (page numbers unavailable) UNFCCC <http://unfccc.int/kyoto_protocol/items/2830.php> (accessed 29-04-2013).].

COP	Date	Location
COP 20[670]	2014	Lima, Peru
COP 21[671]	2015	Paris, France

669 In 2013, COP 19 led to some key decisions concerning the Green Climate Fund and Long-Term Finance, the Warsaw Framework for REDD-Plus, and the Warsaw International Mechanism for Loss and Damage. [UNFCCC "Meetings" (date of publication unavailable) (page numbers unavailable) UNFCCC <http://unfccc.int/meetings/warsaw_nov_2013/meeting/7649.php> (accessed 17-04-2015).]
In particular, the Warsaw conference served to re-focus the final two years of the Durban Platform negotiations, which was to conclude in 2015 in Paris. A timeline was determined at COP 19 by which parties must submit their INDCs in anticipation of the Paris Agreement. [Center for Climate and Energy Solutions "Outcomes of the U.N. Climate Change Conference in Lima" (date of publication unavailable) (page numbers unavailable) *Center for Climate and Energy Solutions* <http://www.c2es.org/international/negotiations/cop-19/summary> (accessed 04-05-2015).].

670 COP 20 was largely overshadowed in anticipation of COP 21, which was to take place in 2015 in Paris, France. The conference in Lima marked the third year of the four-year negotiations, which was to be concluded in Paris at COP 21. Unfortunately, the differentiation between developing and developed countries remained an unresolved and contentious issue, left to be discussed further in Paris. The Lima Call for Climate Action was adopted at COP 20, in which parties to the conference set out their INDCs in anticipation of the Paris Agreement. Other highlights of the conference included the amount of pledges to the Green Climate Fund (nearly $10 billion) received and the announcement from China and the USA regarding their emission targets post-2020. [Center for Climate and Energy Solutions "Outcomes of the U.N. Climate Change Conference in Lima" (date of publication unavailable) (page numbers unavailable) *Center for Climate and Energy Solutions* <http://www.c2es.org/international/negotiations/cop-20-lima/summary> (accessed 04-05-2015).].

671 COP 21 was hailed a successful conference at which a landmark agreement was reached to speed up and amplify the actions that are needed to combat climate change. [UNFCCC "The Paris Agreement" (date of publication unavailable) (page numbers unavailable) *UNFCCC* <http://bigpicture.unfccc.int/#content-the-paris-agreemen> (accessed 06-07-2016).] The outcome of COP 21 is the Paris Agreement, which builds on the foundation established by the UNFCCC and requires parties to combat climate change by means of NDCs. The Paris Agreement is discussed in Section 3.9.

3 7 4 Response to climate change

3 7 4 1 Mitigation of climate change

3 7 4 1 1 Introduction

It is recognised that the issue of climate change will become more urgent in the future.[672] Mitigation of climate change therefore remains necessary and refers particularly to the human intervention to limit or reduce the sources of greenhouse gas emissions, as well as increase or protect carbon sinks and reservoirs.[673] Both mitigation and adaptation to climate change contribute to satisfying the objective of the UNFCCC, as expressed in article 2 of the UNFCCC:

> "The ultimate objective of this Convention and any related legal instruments that the Conference of the Parties may adopt is to achieve, in accordance with the relevant provisions of the Convention, stabilization of greenhouse gas concentrations in the atmosphere at a level that would prevent dangerous anthropogenic interference with the climate system. Such a level should be achieved within a time frame sufficient to allow ecosystems to adapt naturally to climate change, to ensure that food production is not threatened and to enable economic development to proceed in a sustainable manner."

The effectiveness of the state effort to mitigate climate change is dependent on state capacity, which in turn depends on a variety of factors such as the financial ability of the state; socioeconomic circumstances within the state; environmental circumstances of the state; technology available to the state; and the information that the state possesses.[674] Even though states have to develop individual approaches to mitigate climate change, their effort must contribute to the greater international objectives of the in-

---

672 IPCC [TF Stocker et al (eds)] *Summary for Policymakers in Climate Change 2013: The Physical Science Basis*. (2013) 19; IPCC [O Edenhofer et al (eds)] *Summary for Policymakers in Climate Change 2014: Mitigation of Climate Change* (2014) 1.

673 B Metz "International equity in climate change policy" (2000) 1 *Integrated Assessment* 111 111; IPCC [O Edenhofer et al (eds)] *Summary for Policymakers in Climate Change 2014: Mitigation of Climate Change* (2014) 4 UNFCCC "FOCUS: Mitigation" (2014) (page numbers unavailable) *UNFCCC* <http://unfccc.int/focus/mitigation/items/7169.php#intro> (accessed 06-05-2015).

674 UNFCCC "FOCUS: Mitigation" (2014) (page numbers not available) *UNFCCC* <http://unfccc.int/focus/mitigation/items/7169.php#intro> (accessed 06-05-2015).

ternational legal climate change regime, which means their decisions are also to be informed by principles of equity, as discussed above.[675]

Current mitigation commitments of developed countries (or Annex 1 countries) include emission targets for 2020.[676] Each Annex I country must comply with country-specific targets. Developing countries do not need to comply with country-specific targets in the same way as developed countries. Instead, developing countries should take so-called nationally appropriate mitigation actions (NAMAs), specifically in the context of sustainable development.[677] NAMAs can be defined on a national or an international level and are described, by the United Nations, as any action or policy that contributes to the reduction of greenhouse gases under the umbrella of a national government initiative.[678]

There are many strategies aimed at mitigating climate change. Mitigation strategies range across a wide scope of activities, from the use of new technologies and renewable energy to the alteration of management practices and consumer behaviour.[679] Furthermore, mitigation strategies can be complex, occurring on a large scale (such as the activity of planning a green city) or simple, occurring on a small scale (such as the activity of planning a new stove-top design which will be more energy efficient).[680] The IPCC lists key mitigation strategies according to the following sectors: energy supply, transport, buildings, industry, agriculture,

---

675 IPCC [O Edenhofer et al (eds)] *Summary for Policymakers in Climate Change 2014: Mitigation of Climate Change* (2014) 5.

676 The country-specific emission targets for all Annex 1 parties can be found here: Subsidiary Body for Scientific and Technological Advice "Compilation of economy-wide emission reduction targets to be implemented by Parties included in Annex I to the Convention" (4-15 June 2014) (page numbers unavailable) *UNFCCC* <http://unfccc.int/resource/docs/2014/sbsta/eng/inf06.pdf> (accessed 06-05-2015).

677 UNFCCC "FOCUS: Mitigation – NAMAs, Nationally Appropriate Mitigation Actions" (2014) (page numbers unavailable) *UNFCCC* <http://unfccc.int/focus/mitigation/items/7172.php> (accessed 06-05-2015).

678 UNFCCC "FOCUS: Mitigation – NAMAs, Nationally Appropriate Mitigation Actions" (2014) (page numbers unavailable) *UNFCCC* <http://unfccc.int/focus/mitigation/items/7172.php> (accessed 06-05-2015).

679 UNEP "Climate Change Mitigation" (date of publication unavailable) (page numbers unavailable) UNEP <http://www.unep.org/climatechange/mitigation/> (accessed 07-07-2016).

680 UNEP "Climate Change Mitigation" (date of publication unavailable) (page numbers unavailable) UNEP <http://www.unep.org/climatechange/mitigation/> (accessed 07-07-2016).

waste, and forestry.[681] The IPCC explains the mitigating activities and technologies available to each sector. For example, within the forestry sector the following mitigating practices are mentioned by the IPCC: afforestation; reforestation; forest management; reduced deforestation; harvested wood product management; and use of forestry products for bioenergy to replace fossil fuel use.[682] Within the forestry sector the following mitigating technologies are mentioned by the IPCC: tree species improvement to increase biomass productivity and carbon sequestration; improved remote sensing technologies for analysis of vegetation/soil carbon sequestration potential; and monitoring of land-use change.[683] Within this context, an example of a mitigation strategy of the international legal climate change regime is the protection of carbon sinks and reservoirs, which *inter alia* is discussed below. This is a relevant example in the context of the research project, since it also illustrates that a state's management of natural resources (in this case carbon sinks and reservoirs) in terms of PSNR is closely connected with its strategy for managing the effects of climate change. From this perspective, the idea that reinterpreting and reconfiguring PSNR (as a principle focused on people) can aid the international legal climate change regime seems to be a natural progression in the fight against climate change.

3 7 4 1 2 Example of a mitigation strategy: United Nations Programme on Reducing Emissions from Deforestation and Forest Degradation in Developing Countries

Activities within the forestry sector represents a significant portion of the mitigation efforts to combat climate change.[684] Deforestation contributes

---

681 IPCC "Adaptation and mitigation options" (date of publication unavailable) (page numbers unavailable) IPCC <https://www.ipcc.ch/publications_and_data/ar4/syr/en/spms4.html> (accessed 07-07-2016).

682 IPCC "Adaptation and mitigation options" (date of publication unavailable) (page numbers unavailable) IPCC <https://www.ipcc.ch/publications_and_data/ar4/syr/en/spms4.html> (accessed 07-07-2016).

683 IPCC "Adaptation and mitigation options" (date of publication unavailable) (page numbers unavailable) IPCC <https://www.ipcc.ch/publications_and_data/ar4/syr/en/spms4.html> (accessed 07-07-2016).

684 EB Kasimbazi "Policy and legal dimensions of CDM projects in the forestry sector: implications for climate change mitigation and adaptation in Uganda" in BJ

to the production of greenhouse gases as forests are vital carbon sinks.[685] About one-fifth of greenhouse gas emissions originate from land-use, and tropical deforestation can account for up to 15% of annual worldwide carbon dioxide emissions.[686] Moreover, the Stern Review projects that worldwide deforestation annually contributes more to greenhouse gas emissions than the transport sector.[687] Accordingly deforestation is a vital and cost-effective area to focus on in any climate agreement.[688]

For the purposes of the research project, it should be noted that there are already international frameworks in place that can aid in curbing deforestation. In the United Nations REDD programme, poor countries are given a financial incentive to protect their national forests.[689] The Bali Roadmap of 2007 called for the REDD initiative. The 2009 conference of the UNFCCC parties in Copenhagen produced the Copenhagen Accord, which *inter alia* referred to the REDD programme and established a technical framework[690] in this regard. The conference of the parties in Mexico in 2010 converted the REDD programme into the REDD-plus pro-

---

Richardson et al *Climate Law and Developing Countries, Legal and Policy Challenges for the World Economy* (2009) 285 285-310.

685 Council on Foreign Relations "The Climate Change Regime" (19-06-2013) (page numbers unavailable) *Council on Foreign Relations* <http://www.cfr.org/climate-change/global-climate-change-regime/p21831> (accessed 17-04-2015).
Carbon sinks and reservoirs consist of oceans, as well as forests and soils. The difference between carbon sinks and carbon reservoirs is that sinks withdraw carbon from the atmosphere, while reservoirs store carbon previously removed from the atmosphere. If these sinks and reservoirs are disturbed, it is possible that they release carbon into the atmosphere. [C Wold et al *Climate Change and the Law* (2009) 8-9.].

686 Council on Foreign Relations "The Climate Change Regime" (19-06-2013) (page numbers unavailable) *Council on Foreign Relations* <http://www.cfr.org/climate-change/global-climate-change-regime/p21831> (accessed 17-04-2015).

687 N Stern "Stern Review: The Economics of Climate Change" (2006) (xi) *Centro de Ciência do Sistema Terrestre* |<http://mudancasclimaticas.cptec.inpe.br/~rmclima/pdfs/destaques/sternreview_report_complete.pdf> (accessed 30-04-2015).

688 N Stern "Stern Review: The Economics of Climate Change" (2006) (xi) *Centro de Ciência do Sistema Terrestre* http://mudancasclimaticas.cptec.inpe.br/~rmclima/pdfs/destaques/sternreview_report_complete.pdf> (accessed 30-04-2015).

689 Council on Foreign Relations "The Climate Change Regime" (19-06-2013) (page numbers unavailable) *Council on Foreign Relations* <http://www.cfr.org/climate-change/global-climate-change-regime/p21831> (accessed 17-04-2015).

690 Council on Foreign Relations "The Climate Change Regime" (19-06-2013) (page numbers unavailable) *Council on Foreign Relations* <http://www.cfr.org/climate-change/global-climate-change-regime/p21831> (accessed 17-04-2015).

gramme.[691] This meant that components were added to the programme that addressed the conservation of and the expansion of forest carbon sinks, as well as expanding the programme so that forests in all countries are protected.[692] The REDD-plus framework was further developed at the conference of the parties in Warsaw in 2013, by means of the Warsaw Framework. In addition to developing other aspects of the REDD-plus programme, the developments include the specification that funding can come from many sources, such as public, private, bilateral, multilateral, and also market and non-market sources.[693] The Marrakesh Accord provides rules for accounting under the Kyoto Protocol concerning emissions and removals from land-use, land-use change, and forestry.[694]

It falls outside the scope of the research project to examine the REDD or REDD-plus programme in any greater detail. For the purposes of the research project, it should only be noted that this programme exists as part of the international legal climate change regime and that this programme is a valuable supplement to the Kyoto Protocol, since the Protocol does not provide any incentive for countries to avoid deforestation.[695] Instead, the Protocol allows countries to seek credits or financial support after deforestation has already occurred.

The international legal climate change regime is discussed in the succeeding chapter. For the purposes of the research project, the REDD programme as an example of a mitigation strategy is also useful in illustrating the nexus between PSNR and climate change. The object of the REDD programme, namely forests (in simple terms), also forms part of renewable natural resources as the object of PSNR. Accordingly, the manage-

691 D Bodansky et al "Issues for a 2015 climate agreement" (May 2015) (5) *Centre for Climate and Energy Solutions* <http://www.c2es.org/docUploads/issues-for-a-2015-climate-agreement.pdf> (accessed 04-05-2015).

692 D Bodansky et al "Issues for a 2015 climate agreement" (May 2015) (5) *Centre for Climate and Energy Solutions* <http://www.c2es.org/docUploads/issues-for-a-2015-climate-agreement.pdf> (accessed 04-05-2015).

693 D Bodansky et al "Issues for a 2015 climate agreement" (May 2015) (5) *Centre for Climate and Energy Solutions* <http://www.c2es.org/docUploads/issues-for-a-2015-climate-agreement.pdf> (accessed 04-05-2015).

694 D Bodansky et al "Issues for a 2015 climate agreement" (May 2015) (5) *Centre for Climate and Energy Solutions* <http://www.c2es.org/docUploads/issues-for-a-2015-climate-agreement.pdf> (accessed 04-05-2015).

695 Council on Foreign Relations "The Climate Change Regime" (19-06-2013) (page numbers unavailable) *Council on Foreign Relations* <http://www.cfr.org/climate-change/global-climate-change-regime/p21831> (accessed 17-04-2015).

ment by states of their forests must coincide with the rights and duties of states in terms of PSNR. These rights and duties are examined in Chapter 4 of the research project.

3 7 4 2 Adaptation to climate change

Adaptation is a component of the international legal climate change regime. The Stern Review considers adaptation vital to any climate agreement and contends that adaptation must become integrated within national policy and that richer countries should support poorer countries in their adaptation efforts.[696] As mentioned above, both mitigation of and adaptation to climate change contributes to reaching the objective of the UNFCCC as expressed in article 2 of the Convention.[697]

Adaptation can be defined as "the process of adjustment to actual or expected climate and its effects".[698] Adaptation refers to the adaptation required to withstand the present impacts of climate change, building resistance to climate change, and possibly decreasing future impacts of climate change. Even though adaptation is a worldwide undertaking,[699] it occurs on different levels, namely on an international, national, sub-national and individual level. Consequently, adaptation is an enormous undertaking

696 N Stern "Stern Review: The Economics of Climate Change" (2006) (xi) *Centro de Ciência do Sistema Terrestre* |<http://mudancasclimaticas.cptec.inpe.br/~rmclima/pdfs/destaques/sternreview_report_complete.pdf> (accessed 30-04-2015).

697 Article 2 of the UNFCCC reads as follows: "The ultimate objective of this Convention and any related legal instruments that the Conference of the Parties may adopt is to achieve, in accordance with the relevant provisions of the Convention, stabilization of greenhouse gas concentrations in the atmosphere at a level that would prevent dangerous anthropogenic interference with the climate system. Such a level should be achieved within a time frame sufficient to allow ecosystems to adapt naturally to climate change, to ensure that food production is not threatened and to enable economic development to proceed in a sustainable manner.".

698 IPCC [CB Field et al (eds)] *Summary for policymakers in Climate Change 2014: Impacts, Adaptation, and Vulnerability* (2014) 5.

699 Council on Foreign Relations "The Climate Change Regime" (19-06-2013) (page numbers unavailable) *Council on Foreign Relations* <http://www.cfr.org/climate-change/global-climate-change-regime/p21831> (accessed 17-04-2015).

with costs projected to be in the range of $70 billion to $100 billion per year (in the time frame spanning between 2010 and 2050).[700]

The IPCC lists key adaptation strategies according to the following sectors: water, agriculture, infrastructure, human health, tourism, transport, and energy.[701] For each sector a corresponding adaptation strategy is listed by the IPCC. For example, in the case of water, the adaptation strategy concerns the following: expanded rainwater harvesting, water storage and conservation techniques, water re-use, desalination, water-use, and irrigation efficiency.[702] Similar to what was found when considering mitigation strategies, it is clear when considering the sectors to which adaptation strategies are applied that there is a close connection between climate change management and natural resource management (in terms of

---

700 World Bank "The economics of adaptation to climate change" (2010) (10) *World Bank* <http://siteresources.worldbank.org/EXTCC/Resources/EACC_FinalSynthesisReport0803_2010.pdf> (accessed 08-05-2015).

701 IPCC "Adaptation and mitigation options" (date of publication unavailable) (page numbers unavailable) IPCC <https://www.ipcc.ch/publications_and_data/ar4/syr/en/spms4.html> (accessed 07-07-2016).
In the case of agriculture the adaptation strategy concerns the following: "adjustment of planting dates and crop variety; crop relocation; improved land management, e.g. erosion control and soil protection through tree planting". In the case of infrastructure/settlement the adaptation strategy concerns the following: "relocation; seawalls and storm surge barriers; dune reinforcement; land acquisition and creation of marshlands/wetlands as buffer against sea level rise and flooding; protection of existing natural barriers". In the case of human health the adaptation strategy concerns the following: "heat-health action plans; emergency medical services; improved climate-sensitive disease surveillance and control; safe water and improved sanitation". In the case of tourism the adaptation strategy concerns the following: "diversification of tourism attractions and revenues; shifting ski slopes to higher altitudes and glaciers; artificial snow-making". In the case of transport the adaptation strategy concerns the following: "realignment/relocation; design standards and planning for roads, rail and other infrastructure to cope with warming and drainage". In the case of energy the adaptation strategy concerns the following: "strengthening of overhead transmission and distribution infrastructure; underground cabling for utilities; energy efficiency; use of renewable sources; reduced dependence on single sources of energy". [IPCC "Adaptation and mitigation options" (date of publication unavailable) (page numbers unavailable) IPCC <https://www.ipcc.ch/publications_and_data/ar4/syr/en/spms4.html> (accessed 07-07-2016).].

702 IPCC "Adaptation and mitigation options" (date of publication unavailable) (page numbers unavailable) IPCC <https://www.ipcc.ch/publications_and_data/ar4/syr/en/spms4.html> (accessed 07-07-2016).

PSNR). Accordingly, the management of climate change and the management of natural resources in terms of PSNR are able to support or undermine effective management efforts in these sectors.

Considerations or principles for effective adaptation have been identified by the IPCC.[703] The first consideration involves that adaptation is both place- and context-specific, and that no one-size-fits-all solution can be applied in all countries.[704] In addition, effective adaptation to climate change must take account of vulnerability, exposure, socioeconomic factors and sustainable development.[705] In the context of adaptation, the issue of vulnerability is often considered. The vulnerability of an area (or people in an area) to climate change is determined by the following factors: exposure, physical setting, sensitivity, and the ability to adapt.[706] Vulnerability is not necessarily synonymous with poverty, although poverty can be a factor causing vulnerability.[707] The second consideration for effective adaptation is that adaptation must occur through complementary actions of government and individuals.[708] Thirdly, adaptation requires that vulnerability and exposure to climate change be reduced.[709] Fourthly, it must be recognised that governments will have different interests, sociocultural contexts, and risk perceptions, which come into play in national strategies that plan for and implement adaptation strategies.[710] Fifthly, adaptation can be more successful when there are incentives to adapt.[711] Sixthly,

703 IPCC [CB Field et al (eds)] *Summary for policymakers in Climate Change 2014: Impacts, Adaptation, and Vulnerability* (2014) 25.

704 IPCC [CB Field et al (eds)] *Summary for policymakers in Climate Change 2014: Impacts, Adaptation, and Vulnerability* (2014) 25.

705 IPCC [CB Field et al (eds)] *Summary for policymakers in Climate Change 2014: Impacts, Adaptation, and Vulnerability* (2014) 25.

706 WN Adger et al "Adaptation to climate change in the developing world" (2003) 3(3) *Progress in Development Studies* 179 181.

707 WN Adger et al "Adaptation to climate change in the developing world" (2003) 3(3) *Progress in Development Studies* 179 182.

708 IPCC [CB Field et al (eds)] *Summary for policymakers in Climate Change 2014: Impacts, Adaptation, and Vulnerability* (2014) 25.

709 IPCC [CB Field et al (eds)] *Summary for policymakers in Climate Change 2014: Impacts, Adaptation, and Vulnerability* (2014) 25.

710 IPCC [CB Field et al (eds)] *Summary for policymakers in Climate Change 2014: Impacts, Adaptation, and Vulnerability* (2014) 26.

711 IPCC [CB Field et al (eds)] *Summary for policymakers in Climate Change 2014: Impacts, Adaptation, and Vulnerability* (2014) 26.

there are a multitude of constraints confronting effective adaptation.[712] Examples of such constraints include financial constraints, human resource constraints, aspects of governance (including integration and coordination), uncertainties regarding impacts, differing perceptions of risk involved, competing values, and ineffective monitoring of adaptation.[713] Seventhly, adaptation is dependent on an accurate assessment of costs of adaptation and global needs in terms of adaptation.[714] Lastly, adaptation involves the complexity of the interaction between efforts of mitigation and efforts of adaptation across sectors dealing with water, energy, land-use and biodiversity.[715] In this regard the IPCC proposes "climate resilient pathways", which take into consideration both mitigation and adaptation.[716]

Equity is relevant in determining the framework for adaptation among various countries, as the most vulnerable people often inhabit areas of less-developed states that may lack the capacity to adapt. Take for example, the African continent as a collection of less-developed countries. Africa is often considered the "canary in the coalmine" of climate change as it is expected to be the first continent to bear the brunt of climate change impacts.[717] Despite the efforts made to combat climate change in the future, Africa is nevertheless already likely to endure many climate change impacts.[718] This means that adaptation to climate change is a necessity, and not a choice, for a continent such as Africa. Africa is vulnerable particularly because of its substantial dependence on economic sectors which are sensitive to changes in climate and because of its large numbers

---

712 IPCC [CB Field et al (eds)] *Summary for policymakers in Climate Change 2014: Impacts, Adaptation, and Vulnerability* (2014) 25.

713 IPCC [CB Field et al (eds)] *Summary for policymakers in Climate Change 2014: Impacts, Adaptation, and Vulnerability* (2014) 27-28.

714 IPCC [CB Field et al (eds)] *Summary for policymakers in Climate Change 2014: Impacts, Adaptation, and Vulnerability* (2014) 28.

715 IPCC [CB Field et al (eds)] *Summary for policymakers in Climate Change 2014: Impacts, Adaptation, and Vulnerability* (2014) 28.

716 IPCC [CB Field et al (eds)] *Summary for policymakers in Climate Change 2014: Impacts, Adaptation, and Vulnerability* (2014) 28.

717 M Leroy & F Gebresenbet "Science, Facts and Fears: The Debate on Climate Change and Security" in OC Ruppel et al (eds) *Climate Change: International Law and Global Governance, Volume II: Policy, Diplomacy and Governance in a Changing Environment* (2013) 685 686.

718 O Brown et al "Climate change as the 'new' security threat: implications for Africa" (2007) 83 *International Affairs* 1141 1153.

of human settlements in areas which are at greater risk of being affected by climate change, such as low-lying coastal areas.[719] In addition, Africa lacks the capacity which is needed to adapt effectively to climate change. The continent's shortcomings are lack of financial resources, economies built on natural resources, and issues relating to poor government.[720] Furthermore, Africa is also at risk of climate-induced conflict. It has been established that in areas sensitive to climate stress where a risk of conflict exists, conditions which further limit adaptive abilities are more likely to increase the risk of conflict.[721] However, building the capacity needed to adapt to climate change can have positive effects and decrease the risk of climate-induced conflict.[722] Such capacity-building measures may include: diversifying livelihoods, ensuring availability of and access to natural resources, and making use of conflict resolution institutions.[723] In order to be able to adapt effectively to the impacts of climate change, less-developed countries will require funding, which will largely need to be provided by more developed states.

Financial support for adaptation exists in the following forms under the financial mechanism of UNFCCC: the GEF Trust Fund, the Least Developed Countries Fund (LDCF) managed by the GEF, and the Special Climate Change Fund (SCCF) managed by the GEF.[724] The Green Climate Fund (GCF) was established under article 11 of the Kyoto Protocol.[725] In addition, the Adaptation Fund (AF) exists under the Kyoto Pro-

719 O Brown et al "Climate change as the 'new' security threat: implications for Africa" (2007) 83 *International Affairs* 1141 1149.

720 O Brown et al "Climate change as the 'new' security threat: implications for Africa" (2007) 83 *International Affairs* 1141 1149.

721 O Brown et al "Climate change as the 'new' security threat: implications for Africa" (2007) 83 *International Affairs* 1141 1150.

722 O Brown et al "Climate change as the 'new' security threat: implications for Africa" (2007) 83 *International Affairs* 1141 1151.

723 O Brown et al "Climate change as the 'new' security threat: implications for Africa" (2007) 83 *International Affairs* 1141 1151.

724 UNFCCC "Adaptation" (date of publication unavailable) (page numbers unavailable) UNFCCC <http://unfccc.int/adaptation/items/4159.php> (accessed 29-04-2013).

725 UNFCCC "Adaptation" (date of publication unavailable) (page numbers unavailable) *UNFCCC* <http://unfccc.int/adaptation/items/4159.php> (accessed 29-04-2013).

tocol.[726] The AF specifically funds adaptation projects and programmes in vulnerable developing countries.[727]

3 8 The Kyoto Protocol and the effectiveness of the international legal climate change regime

3 8 1 Introduction

The Kyoto Protocol constitutes an important milestone in the development of the international legal climate change regime. The Kyoto Protocol was formulated on the foundation of the UNFCCC and entered into force on 16 February 2005. The Protocol had however already been adopted in 1997, in response to the realisation by parties that the emission reduction targets, as stipulated in the UNFCCC, were not adequate. Currently, the Kyoto Protocol has 192 parties.[728]

The main function of the Kyoto Protocol is to hold certain countries to legally binding greenhouse gas emission targets. More specifically, the Kyoto Protocol turned what were previously referred to as targets and timetables – now referred to as 'quantified emission limitation and reduction objectives' (QELROs) – into legally binding climate commitments for parties. In this regard, the Kyoto Protocol stipulates the emission reduction commitments of Annex 1 countries.[729] Under the Kyoto Protocol parties must reach emission reduction targets mainly through implementing na-

---

726 UNFCCC "Adaptation" (date of publication unavailable) (page numbers unavailable) UNFCCC <http://unfccc.int/adaptation/items/4159.php> (accessed 29-04-2013).

727 UNFCCC "Adaptation" (date of publication unavailable) (page numbers unavailable) UNFCCC <http://unfccc.int/adaptation/items/4159.php> (accessed 29-04-2013).

728 UNFCCC "Status of Ratification of the Kyoto Protocol" (date of publication unavailable) (page numbers unavailable) UNFCCC <http://unfccc.int/kyoto_protocol/status_of_ratification/items/2613.php> (accessed 04-07-2016).
This chapter only developments up until the end of June 2016.

729 See UNFCCC website for a list of the Annex I countries. [UNFCCC "List of Annex I Parties to the Convention" (date unavailable) (page numbers unavailable) UNFCCC <http://unfccc.int/parties_and_observers/parties/annex_i/items/2774.php> (accessed 05-09-2016).].

tional measures.[730] The Protocol helps parties to reach the national emission reduction goal by instituting three market-based mechanisms, namely: international emissions trading; joint implementation; and the clean development mechanism. These mechanisms are discussed in the preceding section of this chapter. The succeeding paragraphs consider the functionality of the Kyoto Protocol further by taking a brief look at the registry of Kyoto units and the reporting procedure of parties.

It is possible that the Kyoto Protocol will expire before the end of its second commitment period in 2020, should the Paris Agreement come into operation before 2020. Regardless of the uncertainty of the Kyoto Protocol's continued existence, the Protocol is a significant current component of the international legal climate change regime and a developmental landmark of the regime. An important part of the examination of the Protocol, as a constituent of the international legal climate change regime, is the question of whether the regime can be considered effective and whether states comply with climate commitments. This issue is also addressed further below.

3 8 2 Market-based mechanisms of the Kyoto Protocol

3 8 2 1 Introduction to international emissions trading

Natural resources (as defined in their capacity as objects of PSNR) are increasingly being regulated as resources forming part of the global market.[731] In the same way, greenhouse gas emissions form part of the global market. Since carbon dioxide is the most prevalent and important greenhouse gas emission being traded, international emissions trading is also referred to as 'carbon trading'. These emissions are traded in terms of the international legal climate change regime, between participants in the market, where the participants may be parties of the Kyoto Protocol or private

730 UNFCCC "Kyoto Protocol" (date of publication unavailable) (page numbers unavailable) UNFCCC <http://unfccc.int/kyoto_protocol/items/2830.php> (accessed 29-04-2013).

731 J Gilbert "The right to freely dispose of natural resources: Utopia or forgotten right?" (2013) 31(2) Netherlands Quarterly of Human Rights 314 317.

entities.[732] Emissions trading is included as a notable component of the international legal climate change regime, as compliance with the Kyoto Protocol (which is discussed further in a succeeding section) is centred around *inter alia* the credibility of the emissions trading system, and transparency of accounting of Kyoto units.[733]

International emission trading is regulated by article 17 of the Kyoto Protocol.[734] Article 17 provides *inter alia* that Annex B parties are allowed to participate in emissions trading in order to fulfil obligations in terms of article 3 of the Protocol. Furthermore, article 17 submits that this trading will be supplemental to the domestic actions of meeting commitments in terms of the Protocol.

The parties which have committed themselves to limiting their emission targets are listed in Annex B of the Kyoto Protocol.[735] In terms of the Kyoto Protocol these Annex B parties have certain permitted amounts that they are allowed to emit.[736] These permitted amounts are divided into 'as-

732 Foundation for International Environmental Law and Development "Final Report: Options for EC Emission Trading" (22-02-2000) (11) Foundation for International Environmental Law and Development <http://ec.europa.eu/environment/archives/docum/pdf/0087_field.pdf> (accessed 07-07-2016).

733 UNFCCC "Kyoto Protocol, Compliance" (date of publication unavailable) (page numbers unavailable) UNFCCC <http://unfccc.int/kyoto_protocol/compliance/items/3024.php> (accessed: 29-04-2012).

734 Article 17 of the Kyoto Protocol reads as follows: "The Conference of the Parties shall define the relevant principles, modalities, rules and guidelines, in particular for verification, reporting and accountability for emissions trading. The Parties included in Annex B may participate in emissions trading for the purposes of fulfilling their commitments under Article 3. Any such trading shall be supplemental to domestic actions for the purpose of meeting quantified emission limitation and reduction commitments under that Article."
See UNFCCC website for a list of the Annex B countries. UNFCCC "Kyoto Protocol" (date unavialable) (page numbers unavialable) UNFCCC <http://unfccc.int/kyoto_protocol/items/3145.php> (accessed 29-04-2013).

735 See UNFCCC website for a list of the Annex B countries. UNFCCC "Kyoto Protocol" (date unavialable) (page numbers unavialable) UNFCCC <http://unfccc.int/kyoto_protocol/items/3145.php> (accessed 29-04-2013).

736 See UNFCCC website for a list of the Annex B countries and their emission targets. UNFCCC "Kyoto Protocol" (date unavialable) (page numbers unavialable) UNFCCC <http://unfccc.int/kyoto_protocol/items/3145.php> (accessed 29-04-2013).

signed amount units' (AAUs).[737] Article 17 of the Kyoto Protocol allows Annex B parties to sell their allowed, but unused, AAUs to other Annex B countries who may need extra AAUs if they are exceeding their allotted number of AAUs. In this way, carbon dioxide is commoditised. Newcomers to the market buy emission units from other parties (or from a government reserve), while those participants leaving the industry are able to keep their AAUs and sell them.[738] Even if the demand for AAUs increases, the available units remain constant as a result of the so-called emission cap. At this juncture of the trading process, the fact that the regime requires effective monitoring and enforcement mechanisms becomes apparent.[739]

Besides AAUs of greenhouse gas emissions being traded on the international market, the trading of other units also makes up a part of international emissions trading. These units are: removal units (RMUs), emission reduction units (ERUs) and certified emission reductions (CERs).[740] One of these units is equal to one ton of carbon dioxide. RMUs are issued to parties on the basis of land use, land-use change, and forestry (LULUCF) activities in accordance with article 3(3) and 3(4) of the Kyoto Protocol.[741] In other words, RMUs are formed by activities which decrease atmospheric levels of carbon dioxide by absorbing the carbon dioxide from the atmosphere and acting as a carbon sink. ERUs and CERs are discussed below.

---

737 UNFCCC "Kyoto Protocol, Emissions Trading" (date of publication unavailable) (page numbers unavailable) UNFCCC <http://unfccc.int/kyoto_protocol/mechanisms/emissions_trading/items/2731.php> (accessed: 29-04-2012).

738 E Woerdman et al "Emissions Trading and the Polluter-Pays Principle: Do Polluters Pay Under Grandfathering?" (2008) 14 *Rotterdam Institute of Law and Economics Working Paper Series* 6 <http://ssrn.com/abstract=1271843> (accessed 23-04-2015).

739 E Woerdman et al "Emissions Trading and the Polluter-Pays Principle: Do Polluters Pay Under Grandfathering?" (2008) 14 *Rotterdam Institute of Law and Economics Working Paper Series* 6 <http://ssrn.com/abstract=1271843> (accessed 23-04-2015).

740 UNFCCC "Kyoto Protocol, Emissions Trading" (date of publication unavailable) (page numbers unavailable) UNFCCC <http://unfccc.int/kyoto_protocol/mechanisms/emissions_trading/items/2731.php> (accessed: 29-04-2012).

741 UNFCCC "Kyoto Protocol, Emissions Trading" (date of publication unavailable) (page numbers unavailable) UNFCCC <http://unfccc.int/kyoto_protocol/mechanisms/emissions_trading/items/2731.php> (accessed: 29-04-2012).
Important actors in this regard are the UNFF, the FAO and the CPF.

3 8 2 2 Joint implementation mechanism

ERUs are formed by a joint implementation project.[742] To understand the clean development and joint implementation trading mechanisms, it is useful to consider that the Kyoto Protocol distinguishes countries based on whether countries are assigned an emissions cap or not. When joint implementation trading occurs, trading occurs between two countries with emission caps.

Article 6 of the Kyoto Protocol is relevant in regard to joint implementation trading.[743] ERUs are formed by a joint implementation project when an Annex B party establishes an emissions removing project within the borders of an Annex B party. The benefit for the party instituting the project is that it will generate ERUs, which count towards that country's

742 UNFCCC "Kyoto Protocol, Emissions Trading" (date of publication unavailable) (page numbers unavailable) UNFCCC <http://unfccc.int/kyoto_protocol/mechanisms/emissions_trading/items/2731.php> (accessed: 29-04-2012).

743 Article 6(1)-(3) of the Kyoto Protocol reads as follows: (1) "For the purpose of meeting its commitments under Article 3, any Party included in Annex I may transfer to, or acquire from, any other such Party emission reduction units resulting from projects aimed at reducing anthropogenic emissions by sources or enhancing anthropogenic removals by sinks of greenhouse gases in any sector of the economy, provided that: (a) Any such project has the approval of the Parties involved; (b) Any such project provides a reduction in emissions by sources, or an enhancement of removals by sinks, that is additional to any that would otherwise occur; (c) It does not acquire any emission reduction units if it is not in compliance with its obligations under Articles 5 and 7; and (d) The acquisition of emission reduction units shall be supplemental to domestic actions for the purposes of meeting commitments under Article 3"; (2) "The Conference of the Parties serving as the meeting of the Parties to this Protocol may, at its first session or as soon as practicable thereafter, further elaborate guidelines for the implementation of this Article, including for verification and reporting"; (3) "A Party included in Annex I may authorize legal entities to participate, under its responsibility, in actions leading to the generation, transfer or acquisition under this Article of emission reduction units"; (4) "If a question of implementation by a Party included in Annex I of the requirements referred to in this Article is identified in accordance with the relevant provisions of Article 8, transfers and acquisitions of emission reduction units may continue to be made after the question has been identified, provided that any such units may not be used by a Party to meet its commitments under Article 3 until any issue of compliance is resolved.".

emission targets.[744] The benefit for the host party lies in receiving foreign investment and technology transfer.[745] Joint implementation projects are prevalent in sectors relating to hydro and wind energy, landfill gas and biomass energy projects.[746] Joint implementation projects are less prevalent in sectors relating to afforestation, agriculture and coal. The joint implementation project must result in a decrease in emissions in addition to compensating for what would have occurred if there were no joint implementation project instituted, a feature similar to the clean development mechanism.[747]

The fundamental difference between a joint implementation mechanism and a clean development mechanism is that the joint development mechanism is only available to Annex 1 countries and accordingly takes place between Annex 1 countries. There are many procedural issues pertaining to joint implementation projects. For instance, the ERUs of joint implementation projects starting in the year 2000 were only allowed to be issued for a crediting period which started in 2008.[748] Furthermore, the host party must authorise or verify a joint implementation project and a party must authorise participants who wish to be involved in the project.[749] Alternatively, it is possible for the project to be verified by the Joint Implementation Supervisory Committee (JISC), which is held accountable by all parties that have ratified the Kyoto Protocol.[750] It is also possible for

744 UNFCCC "Kyoto Protocol, Emissions Trading" (date of publication unavailable) (page numbers unavailable) UNFCCC <http://unfccc.int/kyoto_protocol/mechanisms/emissions_trading/items/2731.php> (accessed: 29-04-2012).

745 UNFCCC "Kyoto Protocol, Emissions Trading" (date of publication unavailable) (page numbers unavailable) UNFCCC <http://unfccc.int/kyoto_protocol/mechanisms/emissions_trading/items/2731.php> (accessed: 29-04-2012).

746 K Karousakis "Joint Implementation: Current Issues and Emerging Challenges" (2006) (9) OECD <https://www.oecd.org/env/cc/37672335.pdf> (accessed 07-07-2016).

747 UNFCCC "Kyoto Protocol, Emissions Trading" (date of publication unavailable) (page numbers unavailable) UNFCCC <http://unfccc.int/kyoto_protocol/mechanisms/emissions_trading/items/2731.php> (accessed: 29-04-2012).

748 UNFCCC "Kyoto Protocol, Emissions Trading" (date of publication unavailable) (page numbers unavailable) UNFCCC <http://unfccc.int/kyoto_protocol/mechanisms/emissions_trading/items/2731.php> (accessed: 29-04-2012).

749 UNFCCC "Kyoto Protocol, Emissions Trading" (date of publication unavailable) (page numbers unavailable) UNFCCC <http://unfccc.int/kyoto_protocol/mechanisms/emissions_trading/items/2731.php> (accessed: 29-04-2012).

750 UNFCCC "The Kyoto Protocol Mechanisms" (2010) (7) UNFCCC <https://cdm.unfccc.int/about/cdm_kpm.pdf> (accessed 07-07-2016) 7.

countries to pool their respective ERUs in accordance with article 4 of the Kyoto Protocol.[751] The formation of countries pooling their ERUs is referred to as a 'bubble', the most prominent example of a bubble being the European Union.[752]

3 8 2 3 Clean development mechanism

Certified emission reductions (CERs) are formed from a clean development project.[753] Article 12 of the Kyoto Protocol regulates the clean development mechanism (CDM).[754] In short, CERs are formed when an An-

751 Article 4(1) of the Kyoto Protocol is relevant in this regard and reads as follows: "Any Parties included in Annex I that have reached an agreement to fulfil their commitments under Article 3 jointly, shall be deemed to have met those commitments provided that their total combined aggregate anthropogenic carbon dioxide equivalent emissions of the greenhouse gases listed in Annex A do not exceed their assigned amounts calculated pursuant to their quantified emission limitation and reduction commitments inscribed in Annex B and in accordance with the provisions of Article 3. The respective emission level allocated to each of the Parties to the agreement shall be set out in that agreement."
See UNFCCC website for a list of the Annex B countries. [UNFCCC "Kyoto Protocol" (date unavialable) (page numbers unavialable) UNFCCC <http://unfccc.int/kyoto_protocol/items/3145.php> (accessed 29-04-2013).].

752 D Bodansky & L Rajamani "The Evolution and Governance Architecture of the Climate Change Regime" (2013) (24) SSRN <http://papers.ssrn.com/sol3/papers.cfm?abstract_id=2168859> (accessed 12-05-2015).

753 UNFCCC "Kyoto Protocol, Emissions Trading" (date of publication unavailable) (date of publication unavailable) UNFCCC <http://unfccc.int/kyoto_protocol/mechanisms/emissions_trading/items/2731.php> (accessed: 29-04-2012).

754 Article 12(1)-(10) of the Kyoto Protocol reads as follows: (1) "A clean development mechanism is hereby defined"; (2) "The purpose of the clean development mechanism shall be to assist Parties not included in Annex I in achieving sustainable development and in contributing to the ultimate objective of the Convention, and to assist Parties included in Annex I in achieving compliance with their quantified emission limitation and reduction commitments under Article 3"; (3) "Under the clean development mechanism: (a) Parties not included in Annex I will benefit from project activities resulting in certified emission reductions; and (b) Parties included in Annex I may use the certified emission reductions accruing from such project activities to contribute to compliance with part of their quantified emission limitation and reduction commitments under Article 3, as determined by the Conference of the Parties serving as the meeting of the Parties to this Protocol"; (4) "The clean development mechanism shall be subject to the authority and guidance of the Conference of the Parties serving as the meeting of

nex B country implements an emission-reducing project in a developing country. CDM projects occur within a wide array of sectors, such as energy; manufacturing and production; transport; waste handling; agriculture; afforestation and reforestation; and construction.[755] An example of a project is installing solar panels in a rural settlement.[756]

The CDM project must provide emission reductions in addition to compensating for what would have occurred if there were no project instituted, a feature similar to the joint development mechanism. The CDM differs from the joint implementation mechanism since the CDM takes place between developing and industrialised countries, whereas the joint implementation mechanism takes place only between industrialised countries.

---

the Parties to this Protocol and be supervised by an executive board of the clean development mechanism"; (5) "Emission reductions resulting from each project activity shall be certified by operational entities to be designated by the Conference of the Parties serving as the meeting of the Parties to this Protocol, on the basis of: (a) Voluntary participation approved by each Party involved; (b) Real, measurable, and long-term benefits related to the mitigation of climate change; and (c) Reductions in emissions that are additional to any that would occur in the absence of the certified project activity"; (6) "The clean development mechanism shall assist in arranging funding of certified project activities as necessary"; (7) "The Conference of the Parties serving as the meeting of the Parties to this Protocol shall, at its first session, elaborate modalities and procedures with the objective of ensuring transparency, efficiency and accountability through independent auditing and verification of project activities." (8) "The Conference of the Parties serving as the meeting of the Parties to this Protocol shall ensure that a share of the proceeds from certified project activities is used to cover administrative expenses as well as to assist developing country Parties that are particularly vulnerable to the adverse effects of climate change to meet the costs of adaptation"; (9) "Participation under the clean development mechanism, including in activities mentioned in paragraph 3 (a) above and in the acquisition of certified emission reductions, may involve private and/or public entities, and is to be subject to whatever guidance may be provided by the executive board of the clean development mechanism"; (10) "Certified emission reductions obtained during the period from the year 2000 up to the beginning of the first commitment period can be used to assist in achieving compliance in the first commitment period.".

755 UNFCCC "Project Search" (date of publication unavailable) (page numbers unavailable) UNFCCC <http://cdm.unfccc.int/Projects/projsearch.html> (accessed 07-07-2016).

756 UNFCCC "Clean Development Mechanism (CDM)" (date of publication unavailable) (page numbers unavailable) UNFCCC <http://unfccc.int/kyoto_protocol/mechanisms/clean_development_mechanism/items/2718.php> (accessed 07-07-2016).

Procedural issues to instituting the CDM include the project undergoing a demanding process of public registration and issuance.[757] This is followed by approval from the designated national authority, which is overseen by the Clean Development Executive Board, accountable to ratified parties of the Kyoto Protocol.[758] Currently, 7 727 clean development mechanism projects have been registered.[759]

3 8 2 4 Registry of Kyoto units acquired through market-based mechanisms

The transfer and acquisition of units, which can also be referred to as Kyoto units, acquired through the abovementioned three market-based mechanisms (international emissions trading, joint implementation mechanism and CDM) are monitored through registry systems of the Kyoto Protocol. The location of the Kyoto units can be tracked at all times by two types of registries, namely the national registries and the CDM registry.[760] The registries operate through a link established with the international transaction log, which is monitored by the UNFCCC secretariat.[761] The log checks the input with independent transaction logs to ascertain whether recordings made in the CDM registry and national registries are accurate.[762]

At the end of a commitment period, the emissions of Annex B countries are compared with their Kyoto units and the balance is determined. It is

757 UNFCCC "The Kyoto Protocol Mechanisms" (2010) 5 UNFCCC <https://cdm.unfccc.int/about/cdm_kpm.pdf > (accessed 07-07-2016).

758 UNFCCC "Kyoto Protocol, Emissions Trading" (date of publication unavailable) (page numbers unavailable) UNFCCC <http://unfccc.int/kyoto_protocol/mechanisms/emissions_trading/items/2731.php> (accessed: 29-04-2012).

759 UNFCCC "Clean Development Mechanism" (date of publication unavailable) (page numbers unavailable) UNFCCC <http://cdm.unfccc.int/index.html> (accessed 07-07-2016).

760 UNFCCC "Kyoto Protocol, Emissions Trading" (date of publication unavailable) (page numbers unavailable) UNFCCC <http://unfccc.int/kyoto_protocol/mechanisms/emissions_trading/items/2731.php> (accessed: 29-04-2012).

761 UNFCCC "Kyoto Protocol, Emissions Trading" (date of publication unavailable) (page numbers unavailable) UNFCCC <http://unfccc.int/kyoto_protocol/mechanisms/emissions_trading/items/2731.php> (accessed: 29-04-2012).

762 UNFCCC "Kyoto Protocol, Emissions Trading" (date of publication unavailable) (page numbers unavailable) UNFCCC <http://unfccc.int/kyoto_protocol/mechanisms/emissions_trading/items/2731.php> (accessed: 29-04-2012).

important that these countries do adhere to their own emission limitations and therefore there are limits to the Kyoto units that parties may sell. In this regard, a "commitment period reserve" must exist within the national registries of the parties.[763]

3 8 3 Effectiveness of the Kyoto Protocol

3 8 3 1 Overall effectiveness of the Kyoto Protocol and the possible movement towards a people-based solution

The Kyoto Protocol has been hailed as a "grand policy experiment" with failures and successes, teaching valuable lessons to be used in future efforts.[764] The Protocol can be considered effective in certain country-specific instances in the context of compliance with climate commitments. When country-specific targets and results are considered (while excluding land use emissions and sinks), the Protocol has achieved success to varying degrees in countries such as Belgium, Ireland, Germany, Croatia, France, Portugal, the United Kingdom, Monaco, Greece, Sweden, the Czech Republic, Poland, Slovakia, the Russian Federation, Hungary, Estonia, Bulgaria, Latvia, Lithuania, Romania, and the Ukraine.[765] However, the European picture of success is complicated by emissions trading. When emissions trading of European countries are taken into account, countries may reach their Kyoto Protocol targets on paper, but their car-

---

763 UNFCCC "Kyoto Protocol, Emissions Trading" (date of publication unavailable) (page numbers unavailable) UNFCCC <http://unfccc.int/kyoto_protocol/mechanisms/emissions_trading/items/2731.php> accessed: 29-04-2012).
The details of the commitment period reserve can be found at: UNFCCC "Kyoto Protocol, Emissions Trading" (date of publication unavailable) (page numbers unavailable) UNFCCC <http://unfccc.int/kyoto_protocol/mechanisms/emissions_trading/items/2731.php> (accessed: 29-04-2012).

764 PG Harris "Collective Action on Climate Change: The Logic of Regime Failure" (2007) 47 Natural Resources Journal 195; Q Schiermeier "The Kyoto Protocol: Hot Air" (28-11-2012) (page numbers unavailable) Nature <http://www.nature.com/news/the-kyoto-protocol-hot-air-1.11882> (accessed 08-07-2016).

765 D Clark "Has the Kyoto protocol made any difference to carbon emissions?" (26-11-2015) (page numbers unavailable) *The Guardian* <http://www.theguardian.com/environment/blog/2012/nov/26/kyoto-protocol-carbon-emissions> (accessed 03-06-2015).

bon emissions are not necessarily in line with their targets.[766] However, it can be submitted that overall the Kyoto Protocol has had greater success than failure in decreasing greenhouse gas emissions of parties, even when taking emissions trading into account.[767] In other words, the total emissions of parties with targets in terms of the Kyoto Protocol have decreased significantly. In a broader context, the Kyoto Protocol has made a great contribution to the international legal climate change regime by developing how countries measure their greenhouse gas emissions and developing the reporting and accounting systems needed in this context.[768] Accordingly, the Kyoto Protocol can be considered effective if taking into account "new [state] behaviors" [769] that it introduced in the context of monitoring greenhouse gas emissions and increasing awareness of the need to develop clean energy and non-reliance on fossil fuels.

The successes of the Kyoto Protocol are often overshadowed by the increase in greenhouse gas emissions by the rest of the world (and in particular by developing countries such as China). Furthermore, many emissions are due to the production of goods in developing countries which are subsequently exported to the Kyoto parties. When the overall carbon footprint of Kyoto parties are calculated by means of including imports and excluding exports, the effectiveness of the Kyoto Protocol is questionable.[770] The Kyoto Protocol is criticised for not being able to significantly slow down

---

766 D Clark "Has the Kyoto protocol made any difference to carbon emissions?" (26-11-2015) (page numbers unavailable) *The Guardian* <http://www.theguardian.com/environment/blog/2012/nov/26/kyoto-protocol-carbon-emissions> (accessed 03-06-2015).

767 D Clark "Has the Kyoto protocol made any difference to carbon emissions" (26-11-2012) (page numbers unavailable) The Guardian <https://www.theguardian.com/environment/blog/2012/nov/26/kyoto-protocol-carbon-emissions> (accessed 03-06-2015).

768 S Gelis "Kyoto Protocol, 10 years later: Did deal to combat greenhouse emissions work and what of its future?" (14-02-2015) (page numbers unavailable) National Post <http://news.nationalpost.com/news/world/kyoto-protocol-10-years-later-was-the-deal-to-combat-greenhouse-emissions-successful-and-what-of-its-future> (accessed 08-07-2016).

769 R Mitchell "Implementation of the FCCC – Compliance, Effectiveness and Institutional Design" in D Sprinz & U Luterbacher (eds) International Relations and Global Climate Change (2001) 68 69-70.

770 D Clark "Has the Kyoto protocol made any difference to carbon emissions" (26-11-2012) (page numbers unavailable) The Guardian <https://www.theguardian.com/environment/blog/2012/nov/26/kyoto-protocol-carbon-emissions> (accessed 08-07-2016).

overall emissions of greenhouse gases by state parties. In addition, the Protocol has also been criticised for its short-term goals and its lack of effective incentives.[771] Much of the criticism against the Kyoto Protocol is based around the state's role in the international legal regime. In particular, the lack of state compliance within the regime is a great cause of ineffectiveness of the regime. Some criticise states facing borderless environmental crises including climate change, contending that states are unable to deal with such crises.[772] Moreover, this leads to the introduction of non-state actors as participants in the international legal climate change regime, and the expansion of the accustomed position of international law which recognised only states as actors.[773] These developments feed into the idea of increased participation by people within the state (and the implementation of the rights of people in terms of PSNR), in order to influence and possibly improve state compliance within the international legal climate change regime.

The future of the Kyoto Protocol is further destabilised by the newly adopted Paris Agreement, which is discussed in the next section of this chapter.

3 8 3 2 Reporting as a component of an effective international legal climate change regime

Reporting and review of Annex I countries are addressed in articles 5, 7 and 8 of the Kyoto Protocol.[774] These articles also address the national

---

771 PG Harris "Collective Action on Climate Change: The Logic of Regime Failure" (2007) 47 Natural Resources Journal 195 195; S Barrett & R Stavins "Increasing Participation and Compliance in International Climate Change Agreements" (2002) 94 Nota di Lavoro 8.

772 JA Camilleri & J Falk The End of Sovereignty? The Politics of a Shrinking and Fragmenting World (1992) 192.

773 W Scholtz "Greening Permanent Sovereignty through the Common Concern in the Climate Change Regime: Awake Custodial Sovereignty!" in OC Ruppel et al (eds) Climate Change: International Law and Global Governance. Volume II: Policy, Diplomacy and Governance in a Changing Environment (2013) 201 206.

774 UNFCCC "Kyoto Protocol, Reporting" (date of publication unavailable) (page numbers unavailable) UNFCCC <http://unfccc.int/kyoto_protocol/reporting/items/3879.php> (accessed: 29-04-2012).

systems and methods of preparing greenhouse gas inventories.[775] Reporting is a vital part of the international climate change regime and measures the effectiveness of the Kyoto Protocol. There are many reasons why reporting in terms of the Kyoto Protocol must be reliable. Only if the information reported is reliable can the effectiveness of the Kyoto Protocol be properly evaluated.[776] In addition, parties not complying with their commitments will be easily recognisable if reporting is reliable.[777] Furthermore, through reliable reporting, information is shared, which makes it possible for parties to observe and learn from each other.[778]

Developed countries report their greenhouse gas emissions every year, while developing countries report their emissions every six years.[779] Overall, reporting from developed countries is considered to be reliable. Except for CDM projects, reporting from developing countries is often not delivered on time and is not considered as reliable.[780] There are a few reasons why reporting from developing countries is considered to be less reliable. Firstly, it is possible that these countries simply do not have the domestic capacity required to monitor greenhouse gas emissions successfully.[781] Secondly, where the domestic capacity does exist to monitor greenhouse gases successfully, countries may be hesitant to report accurately to avoid

775 UNFCCC "Kyoto Protocol, Reporting" (date of publication unavailable) (page numbers unavailable) UNFCCC <http://unfccc.int/kyoto_protocol/reporting/items/3879.php> (accessed: 29-04-2012).

776 D Bodansky & L Rajamani "The Evolution and Governance Architecture of the Climate Change Regime" (2013) (27) SSRN <http://papers.ssrn.com/sol3/papers.cfm?abstract_id=2168859> (accessed 12-05-2015).

777 D Bodansky & L Rajamani "The Evolution and Governance Architecture of the Climate Change Regime" (2013) (27) SSRN <http://papers.ssrn.com/sol3/papers.cfm?abstract_id=2168859> (accessed 12-05-2015).

778 D Bodansky & L Rajamani "The Evolution and Governance Architecture of the Climate Change Regime" (2013) (27) SSRN <http://papers.ssrn.com/sol3/papers.cfm?abstract_id=2168859> (accessed 12-05-2015).

779 Council on Foreign Relations "The Climate Change Regime" (19-06-2013) (page numbers unavailable) *Council on Foreign Relations* <http://www.cfr.org/climate-change/global-climate-change-regime/p21831> (accessed 17-04-2015).

780 Council on Foreign Relations "The Climate Change Regime" (19-06-2013) (page numbers unavailable) *Council on Foreign Relations* <http://www.cfr.org/climate-change/global-climate-change-regime/p21831> (accessed 17-04-2015).

781 Council on Foreign Relations "The Climate Change Regime" (19-06-2013) (page numbers unavailable) *Council on Foreign Relations* <http://www.cfr.org/climate-change/global-climate-change-regime/p21831> (accessed 17-04-2015).

being required to cap emissions at a later stage of climate negotiations.[782] Thirdly, countries may consider monitoring to be a breach of their state sovereignty, as has reportedly been the case with China in the past.[783]

In order to improve reporting within the international legal climate change regime, some policy suggestions are made by Mitchell, which include *inter alia* making data more straight-forward to collect; improving facilitation of evaluation of data; and establishing a system that encourages third parties such as environmental NGOs also to submit information.[784] The regime should further aim to move to an independent reporting system, instead of a self-reporting system.[785]

As submitted above, the information reported by parties must not only be reliable, but must also comply with the climate commitments they have made under the Kyoto Protocol (and in future under the Paris Agreement).[786] Compliance as a component of effectiveness is discussed below.

---

782 Council on Foreign Relations "The Climate Change Regime" (19-06-2013) (page numbers unavailable) *Council on Foreign Relations* <http://www.cfr.org/climate-change/global-climate-change-regime/p21831> (accessed 17-04-2015).

783 The Economist "A deal in Durban" (17-12-2011) (page numbers unavailable) *The Economist* <http://www.economist.com/node/21541806> (accessed 17-04-2015); Council on Foreign Relations "The Climate Change Regime" (19-06-2013) (page numbers unavailable) *Council on Foreign Relations* <http://www.cfr.org/climate-change/global-climate-change-regime/p21831> (accessed 17-04-2015).

784 R Mitchell "Implementation of the FCCC – Compliance, Effectiveness and Institutional Design" in D Sprinz & U Luterbacher (eds) International Relations and Global Climate Change (2001) 68 71-72.

785 R Mitchell "Implementation of the FCCC – Compliance, Effectiveness and Institutional Design" in D Sprinz & U Luterbacher (eds) International Relations and Global Climate Change (2001) 68 71-72.

786 UNFCCC "Kyoto Protocol, Reporting" (date of publication unavailable) (page numbers unavailable) UNFCCC <http://unfccc.int/kyoto_protocol/reporting/items/3879.php> (accessed: 29-04-2012).

3 8 3 3 State compliance as a component of an effective international legal climate change regime

3 8 3 3 1 Compliance provisions in the Kyoto Protocol and the inadequate repercussions of non-compliance

There are different ways in which to measure the effectiveness of the international legal climate change regime.[787] This research project is concerned with one method in particular, namely the consideration of state compliance specifically in regard to its climate commitments in terms of the Kyoto Protocol and the Paris Agreement. For the Kyoto Protocol to be effective, success in two major areas is required. Firstly, the information reported by parties must be reliable and, secondly, parties must comply with the commitments they have made under the Protocol.[788] The research project is more concerned with the second area, as the implementation of people's rights in terms of PSNR would be able to affect this second area more directly than the first.[789]

According to article 14 of the UNFCCC, non-compliance with the UNFCCC will result in the settlement of disputes by negotiation, arbitration, conciliation or settlement by the International Court of Justice (ICJ).[790] In

787 For an overview on compliance and other methods by which effectiveness is measured (for instance, did the UNFCCC make contributions to environmental conservation that would not have occurred had the UNFCCC not been effected, or did the treaty induce worth-while new state behaviour?), see: R Mitchell "Implementation of the FCCC – Compliance, Effectiveness and Institutional Design" in D Sprinz & U Luterbacher (eds) International Relations and Global Climate Change (2001) 68 68-77.

788 UNFCCC "Kyoto Protocol, Reporting" (date of publication unavailable) (page numbers unavailable) UNFCCC <http://unfccc.int/kyoto_protocol/reporting/items/3879.php> (accessed: 29-04-2012).

789 For a more detailed exposition of reporting within the international legal climate change regime within the context of determining effectiveness of the regime, see R Mitchell "Implementation of the FCCC – Compliance, Effectiveness and Institutional Design" in D Sprinz & U Luterbacher (eds) International Relations and Global Climate Change (2001) 68 74-77.

790 Article 14(1)-(8) of the UNFCCC reads as follows: (1) "In the event of a dispute between any two or more Parties concerning the interpretation or application of the Convention, the Parties concerned shall seek a settlement of the dispute through negotiation or any other peaceful means of their own choice"; (2) "When ratifying, accepting, approving or acceding to the Convention, or at any time thereafter, a Party which is not a regional economic integration organization may

an effort to maximise compliance with the international legal climate change regime, a compliance committee for the Kyoto Protocol has been formed, comprising two branches, namely the facilitative branch and the enforcement branch.[791] The facilitative branch provides assistance to parties in their quest to comply with the Protocol by facilitating financial and technical assistance; providing advice to parties on compliance; and warning parties in early stages of non-compliance with the Protocol.[792] The enforcement branch decides what to do when parties fail to comply with the Protocol, and their decisions in this regard mainly relate to: compliance of

declare in a written instrument submitted to the Depositary that, in respect of any dispute concerning the interpretation or application of the Convention, it recognizes as compulsory ipso facto and without special agreement, in relation to any Party accepting the same obligation: (a) Submission of the dispute to the International Court of Justice; and/or (b) Arbitration in accordance with procedures to be adopted by the Conference of the Parties as soon as practicable, in an annex on arbitration. A Party which is a regional economic integration organization may make a declaration with like effect in relation to arbitration in accordance with the procedures referred to in subparagraph (b) above"; (3) "A declaration made under paragraph 2 above shall remain in force until it expires in accordance with its terms or until three months after written notice of its revocation has been deposited with the Depositary"; (4) "A new declaration, a notice of revocation or the expiry of a declaration shall not in any way affect proceedings pending before the International Court of Justice or the arbitral tribunal, unless the parties to the dispute otherwise agree"; (5) "Subject to the operation of paragraph 2 above, if after twelve months following notification by one Party to another that a dispute exists between them, the Parties concerned have not been able to settle their dispute through the means mentioned in paragraph 1 above, the dispute shall be submitted, at the request of any of the parties to the dispute, to conciliation"; (6) "A conciliation commission shall be created upon the request of one of the parties to the dispute. The commission shall be composed of an equal number of members appointed by each party concerned and a chairman chosen jointly by the members appointed by each party. The commission shall render a recommendatory award, which the parties shall consider in good faith"; (7) "Additional procedures relating to conciliation shall be adopted by the Conference of the Parties, as soon as practicable, in an annex on conciliation"; (8) "The provisions of this Article shall apply to any related legal instrument which the Conference of the Parties may adopt, unless the instrument provides otherwise.".

791 For composition of members, see: UNFCCC "Kyoto Protocol, Compliance" (date of publication unavailable) (page numbers unavailable) UNFCCC <http://unfccc.int/kyoto_protocol/compliance/items/3024.php> (accessed: 29-04-2012).

792 UNFCCC "Kyoto Protocol, Compliance" (date of publication unavailable) (page numbers unavailable) UNFCCC <http://unfccc.int/kyoto_protocol/compliance/items/3024.php> (accessed: 29-04-2012).

an Annex 1 party with its emission targets; methodology and reporting requirements relating to greenhouse gas inventories; and eligibility requirements under the mechanisms.[793] The compliance system in terms of the UNFCCC and Kyoto Protocol is complex and each type of non-compliance will require a different course of action.[794]

There are certain repercussions for non-compliance to the Kyoto Protocol, but these are easily evaded. Countries that do not comply with their legally binding Kyoto target emissions are legally required to subtract their shortfall, as well as a 30% penalty, from their allowed emissions in the next phase of the Protocol.[795] However, this legal penalty has proved meaningless since countries can negotiate future allowed emissions so that their loss is offset, or countries can simply withdraw from the Protocol.

3 8 3 3 2 Increasing compliance with the Kyoto Protocol and the principle of sovereignty over natural resources as a negative incentive

First of all, it should be submitted that effective compliance with the Kyoto Protocol begins with participation in the Protocol (the same applies in the case of the Paris Agreement, as will be discussed in the succeeding section of this chapter). Full participation and compliance in the international legal climate change regime, and therefore the Kyoto Protocol, would ensure an efficient climate change response.[796] In general, the international legal system struggles to ensure international state participation

793 UNFCCC "Kyoto Protocol, Compliance" (date of publication unavailable) (page numbers unavailable) UNFCCC <http://unfccc.int/kyoto_protocol/compliance/items/3024.php> (accessed: 29-04-2012).

794 For a diagram depicting the compliance system of the international legal climate change regime, refer to UNFCCC "Kyoto Protocol, Compliance" (date of publication unavailable) (page numbers unavailable) UNFCCC <https://unfccc.int/files/kyoto_mechanisms/compliance/application/pdf/comp_schematic.pdf> (accessed: 21-07-2016).

795 Council on Foreign Relations "The Climate Change Regime" (19-06-2013) (page numbers unavailable) *Council on Foreign Relations* <http://www.cfr.org/climate-change/global-climate-change-regime/p21831> (accessed 17-04-2015).

796 RO Keohane & K Raustiala "Toward a Post-Kyoto Climate Change Architecture: a Political Analysis" in JA Aldy & RN Stavins (eds) *Post-Kyoto International Climate Policy* (2010) 372 372.

and compliance with international law, which includes international treaties such as the UNFCCC and the Kyoto Protocol.[797]

It is challenging to ensure participation by states when states are required to make sacrifices in order to participate in the regime.[798] In this context, states are required to make a sacrifice, for instance, by reducing greenhouse gas emissions and adjusting plans in regard to economic development. A key problem that the international legal climate change regime must solve is to ensure participation in the regime, and to ensure compliance by states within the regime. Non-compliance with climate commitments can be due to a variety of factors that may be difficult or time-consuming for some states to address, such as underdeveloped financial or administrative capacity, or technological capacity or infrastructure.[799] At this juncture it will be explained how incentives can increase compliance with climate commitments in cases of non-complying states.

Ensuring participation and compliance in the international legal climate change regime can incorporate a wide variety of approaches.[800] In this regard, this research project introduces positive and negative incentives as an approach to increasing state compliance with climate commitments. The reinterpretation and reconfiguration of PSNR to enable people to hold states accountable in terms of PSNR can be construed as a negative incentive for state compliance. An example of a positive incentive to increase

797 S Barrett & R Stavins "Increasing Participation and Compliance in International Climate Change Agreements" (2003) 3 International Environmental Agreements: Politics, Law and Economics 349 349.

798 S Barrett & R Stavins "Increasing Participation and Compliance in International Climate Change Agreements" (2002) 94 *Nota di Lavoro* 15.

799 R Mitchell "Implementation of the FCCC – Compliance, Effectiveness and Institutional Design" in D Sprinz & U Luterbacher (eds) International Relations and Global Climate Change (2001) 68 70.
However, as an admittedly broad generalised statement, these seem to be problems that one would expect to encounter in the same countries where legal national systems may not enable people to hold states accountable for climate commitments and accordingly do not impact the research objectives of this research project.

800 For instance, preclusive strategies, or heightening environmental concern. See: R Mitchell "Implementation of the FCCC – Compliance, Effectiveness and Institutional Design" in D Sprinz & U Luterbacher (eds) International Relations and Global Climate Change (2001) 68 71-72.

state compliance may include financial compensation.[801] In the context of climate change, this would mean that industrialised countries compensate developing countries for their participation in (and compliance with) mitigating climate change.[802] In this way a positive incentive is provided for developing countries to participate in reducing or limiting greenhouse gas emissions. However, it remains to be figured out what may positively incentivise industrialised countries to compensate developing countries, besides ethical considerations such as considerations of historical responsibility for greenhouse gas emissions. Another positive incentive may be co-operation between countries in the context of promoting the mitigation of climate change which is mutually beneficial (albeit for various reasons) to all countries involved. This example would occur mainly in the context of international relations between states.

In order to ensure participation, incentives can also be negative in their nature. Examples of negative incentives may be reciprocal measures, financial penalties and trade restrictions inflicted on parties which do not comply with their climate commitments in terms of the international legal climate change regime.[803] For negative incentives to be effective, the errant state must be "significantly worse off" when it does not comply compared to when it does comply.[804] It is unrealistic to expect that should these types of actions, for example sanctions, actually be implemented, that they would be severe enough to make an impact.[805] In the context of negative incentives, it is also possible for international relations to play a

801 L Engberg-Pedersen "Climate Change negotiations and their implications for international development cooperation" (2011) *Danish Institute for International Studies* 19; S Barrett & R Stavins "Increasing Participation and Compliance in International Climate Change Agreements" (2002) 94 *Nota di Lavoro* 16.

802 S Barrett & R Stavins "Increasing Participation and Compliance in International Climate Change Agreements" (2002) 94 *Nota di Lavoro* 16.

803 L Engberg-Pedersen "Climate Change negotiations and their implications for international development cooperation" (2011) *Danish Institute for International Studies* 19; S Barrett & R Stavins "Increasing Participation and Compliance in International Climate Change Agreements" (2003) 3 *International Environmental Agreements: Politics, Law and Economics* 349 362.

804 S Barrett & R Stavins "Increasing Participation and Compliance in International Climate Change Agreements" (2003) 3 *International Environmental Agreements: Politics, Law and Economics* 349 362-363.

805 R Mitchell "Implementation of the FCCC – Compliance, Effectiveness and Institutional Design" in D Sprinz & U Luterbacher (eds) International Relations and Global Climate Change (2001) 68 72.

role. Overall, the UNFCCC and the Kyoto Protocol do suffer from a lack of incentive mechanisms. The incentive mechanisms employed by the Paris Agreement are discussed in the following section of this chapter.

Regardless of the existing incentives for states to comply with the international legal climate change regime, there is still not an acceptable level of compliance by states with the regime as the grave impacts of climate change continue to loom over present and future generations. There are a myriad of different avenues along which one can set out in order to attempt to increase state compliance in the context of the international legal climate change regime.[806] This research project explores one specific avenue, namely whether PSNR can be implemented as a tool by which to increase state compliance within the international legal climate change regime. This occurs by constructing the rights of people in terms of PSNR as a negative incentive for states to abide by their climate commitments. This action therefore requires, firstly, that a state actually makes a climate commitment within the international legal climate change regime and, secondly, that the state implements a national legal system whereby people of the state are able to hold the state accountable. The idea of state compliance with the international legal climate change regime is ultimately somewhat idealistic, especially when it is submitted that states will comply merely for the sake of the global good or in order to ensure the well-being of people.[807] In this regard, the people-based action in terms of PSNR is a worthy approach, since people are likely to pursue such action if it is for their own well-being (and the well-being of future generations), and this action by people will ultimately contribute to the well-being of the greater public. The rights of people in terms of PSNR as a method or negative incentive to increase state compliance is explored further in Chapter 4.

The approach suggested above whereby people of the state use their rights in terms of PSNR to hold the state accountable is an approach warranted by the gravity of climate change. In its current state, the interna-

806 At this point it should be noted that the scope of the research project does not allow a full exposition of state compliance within the international legal climate change regime. However, it is not necessary to include a full exposition of state compliance in order to achieve the objectives of this research project explained in Chapter 1.

807 See Shelton within the context of common concern and state action: D Shelton "Common Concern of Humanity" (2009) 1 *Iustum Aequum Salutare* 33 39.

tional climate change regime is not able to effectively ensure state compliance with climate commitments. If an already established principle of international law, in this case PSNR, contains the components that are able to harmoniously contribute to solving this problem, why not explore this possibility? Therefore, this research project submits that the gravity of climate change has an impact on PSNR by extracting from PSNR that which will allow the climate regime to function more effectively – in this case it happens to be the implementation and enforcement of the rights of people in terms of PSNR. Chapter 4 will go on to explore the specific rights of people in terms of PSNR and their practical implementation and enforcement.

3 9 The Paris Agreement

3 9 1 Introduction

As submitted above, a landmark agreement was reached in Paris at COP 21 to speed up and amplify the actions that are needed globally to combat climate change.[808] The Paris Agreement was adopted at the end of 2015,opened for signature on 22 April 2016, and will enter into force 30 days after it has been ratified by 55 countries accounting for at least 55% of global greenhouse gas emissions.[809] As at 6 July 2016, the Paris Agreement has 178 signatories and has been ratified by 19 states, accounting for 0,18% of total greenhouse gas emissions.[810] If the criteria are met for the Paris Agreement to come into force before 2020, this will mean that the Kyoto Protocol could be dissolved before the end of its second commitment period commencing in 2020.

---

808 UNFCCC "The Paris Agreement" (date of publication unavailable) (page numbers unavailable) UNFCCC <http://bigpicture.unfccc.int/#content-the-paris-agreemen> (accessed 06-07-2016).

809 UNFCCC "The Paris Agreement" (date of publication unavailable) (page numbers unavailable) UNFCCC <http://bigpicture.unfccc.int/#content-the-paris-agreemen> (accessed 06-07-2016).

810 UNFCCC "Paris Agreement – Status of Ratification" (date of publication unavailable) (page numbers unavailable) UNFCCC <http://unfccc.int/paris_agreement/items/9444.php> (accessed 06-07-2016). This Chapter only contains developments until the end of June 2016.

The Paris Agreement, like the Kyoto Protocol, builds on the foundation established by the UNFCCC. Article 2 of the Agreement sets out the objective of the Agreement, which is to keep the temperature rise below 2 degrees Celsius above pre-industrial levels, and pursues further efforts that would limit the increase in temperature to 1,5 degrees Celsius above pre-industrial levels.[811]

3 9 2 Outline of differentiation between countries and nationally determined contributions in the Paris Agreement

The problem of how to deal with differentiation between developed and developing countries in the nature of the commitments of the Paris Agreement was considered a key issue to be addressed by the Agreement.[812] Overall, the Paris Agreement addresses the principle of common but differentiated responsibility in the UNFCCC by including responsibilities for developing and developed parties, but different approaches throughout the Agreement state that differentiation remains between developed and developing countries.[813] There are many examples throughout the Agreement that illustrate this point, a few of which will be elucidated below. This activity of elucidation serves the additional function of setting out key provisions within the Agreement at the same time.

The preamble of the Agreement recognises the "specific needs and special circumstances of developing country parties, especially those that are particularly vulnerable to the adverse effects of climate change". Article 3 includes the "need to support developing country parties for effective implementation" of the Agreement. Article 4 of the Agreement further illustrates the differentiation between developed and developing countries. Article 4(1) stipulates that peaking of greenhouse gas emissions "will take longer for developing countries". Article 4(3) provides that parties' NDCs will represent "its highest possible ambition", but will reflect the principle

811 Article 2(1)(a), Paris Agreement.

812 D Bodansky et al "Issues for a 2015 climate agreement" (May 2015) (2) *Center for Climate and Energy Solutions* <http://www.c2es.org/docUploads/issues-for-a-2015-climate-agreement.pdf> (accessed 04-05-2015).

813 This approach is explained as so-called braking new ground by: R Bodle et al "The Paris Agreement: Analysis, Assessment and Outlook" (28-01-2016) 19 *Ecologic* <http://ecologic.eu/sites/files/event/2016/ecologic_institute_2016_paris_agreement_assessment_0.pdf> (accessed 09-07-2016).

of common but differentiated responsibility and the capabilities of each country according to its national circumstances. In addition, article 4(4) stipulates that developed countries "should continue taking the lead" in reducing greenhouse gas emissions.

In the context of adaptation, article 7 of the Paris Agreement also illustrates subtly that developed and developing countries will not be treated equally under the Paris Agreement. For instance, article 7(2) specifically stipulates that the needs of developing countries that are particularly vulnerable to climate change be taken into account in the formulation of the international legal climate change regime.[814] In addition, article 7(6) specifically reminds parties to take into account the needs of developing countries.[815] Furthermore, article 7(9) stipulates that each party to the Agreement shall engage in various actions "as appropriate".[816]

Furthermore, in terms of adaptation and mitigation, article 9 of the Agreement stipulates *inter alia* that developed countries shall assist developing countries and take into account the needs and priorities of developing countries. Article 9(1) of the Agreement states that "developed countries shall provide financial resources to assist developing country parties with respect to both mitigation and adaptation" and article 9(3) reiterates that developed countries shall take the lead in this regard.

---

814 Article 7(2)of the Paris Agreement reads as follows: "Parties recognize that adaptation is a global challenge faced by all with local, subnational, national, regional and international dimensions, and that it is a key component of and makes a contribution to the long-term global response to climate change to protect people, livelihoods and ecosystems, taking into account the urgent and immediate needs of those developing country Parties that are particularly vulnerable to the adverse effects of climate change.".

815 Article 7(6) of the Paris Agreement reads as follows: "Parties recognize the importance of support for and international cooperation on adaptation efforts and the importance of taking into account the needs of developing country Parties, especially those that are particularly vulnerable to the adverse effects of climate change.".

816 Article 7(9) of the Paris Agreement reads as follows: "Each Party shall, as appropriate, engage in adaptation planning processes and the implementation of actions, including the development or enhancement of relevant plans, policies and/or contributions, which may include: (a) The implementation of adaptation actions, undertakings and/or efforts;
(b) The process to formulate and implement national adaptation plans;
(c) The assessment of climate change impacts and vulnerability, with a
view to formulating nationally determined prioritized actions, taking into account vulnerable people, places and ecosystems.".

According to the Stern Review, technology cooperation is a vital area that any future climate agreement, such as the Paris Agreement, should focus on, as it is important to develop energy effectiveness and low-carbon technologies within the international legal climate change regime.[817] Article 10 of the Paris Agreement manages to strengthen the technology and capacity-building framework of the international legal climate change regime, as well as the financial flow of the regime in general and the flow to developing countries.[818]

While promoting transparency, the differentiation between developed and developing countries in the Paris Agreement is also apparent. Two examples of this differentiation are included in article 13(12), which states that review "shall pay particular attention to the respective national capabilities and circumstances of developing country Parties", and article 13(14), which specifically reads that support "[s]hall be provided to developing countries for the implementation" of article 13.

Another key issue, which the Paris Agreement was expected to address, concerned the legally binding or non-binding nature of the emission or climate commitments.[819] Accordingly, the Paris Agreement requires parties to combat climate change by means of NDCs for developed and develop-

817 N Stern "Stern Review: The Economics of Climate Change" (2006) (xi) *Centro de Ciência do Sistema Terrestre* <http://mudancasclimaticas.cptec.inpe.br/~rmclima/pdfs/destaques/sternreview_report_complete.pdf> (accessed 30-04-2015).

818 Article 10(5)-10(6) of the Paris Agreement reads as follows: "5. Accelerating, encouraging and enabling innovation is critical for an effective, long-term global response to climate change and promoting economic growth and sustainable development. Such effort shall be, as appropriate, supported, including by the Technology Mechanism and, through financial means, by the Financial Mechanism of the Convention, for collaborative approaches to research and development, and facilitating access to technology, in particular for early stages of the technology cycle, to developing country Parties. 6. Support, including financial support, shall be provided to developing country Parties for the implementation of this Article, including for strengthening cooperative action on technology development and transfer at different stages of the technology cycle, with a view to achieving a balance between support for mitigation and adaptation. The global stocktake referred to in Article 14 shall take into account available information on efforts related to support on technology development and transfer for developing country Parties. ".

819 D Bodansky et al "Issues for a 2015 climate agreement" (May 2015) (2) *Center for Climate and Energy Solutions* <http://www.c2es.org/docUploads/issues-for-a-2015-climate-agreement.pdf> (accessed 04-05-2015).

ing countries.[820] The Paris Agreement does not legally bind parties to certain emission levels like the Kyoto Protocol. Instead, the Paris Agreement provides for parties to determine their own emission levels through a system of NDCs. This structure reflects more of a bottom-up approach, rather than the top-down approach structure adhered to in the Kyoto Protocol. The Paris Agreement differs even further from the Kyoto Protocol in that it applies to developed countries and developing countries, of which both groups are supposed to act in accordance with their NDCs. NDCs are a point of criticism and praise for the Paris Agreement, as will be discussed below in this section. The system of NDCs requires that all parties report regularly on their emissions and efforts to implement NDCs according to the stringent accounting methods of article 6(2), thus contributing to positioning transparency high on the agenda of the Paris Agreement.[821] In addition, these NDCs are subject to review every five years after the global stock-take in 2023, according to article 14(2) of the Agreement. Ultimately, the functionality and effectiveness of NDCs, as employed by the Paris Agreement in future, remain to be seen.

3 9 3 Market-based mechanism of the Paris Agreement

The Paris Agreement also contains provisions relating to a significant part of the current international legal climate change regime, namely international emissions trading, or the carbon market. Article 6 of the Paris Agreement does not create a carbon market, but provides that parties of the Paris Agreement have the ability to create a carbon market should they choose to do so, illustrating also a bottom-up approach to emissions trading.[822] According to article 6, countries are able to develop carbon markets domestically and trade carbon internationally by means of trading "inter-

---

820 UNFCCC "The Paris Agreement" (date of publication unavailable) (page numbers unavailable) UNFCCC <http://bigpicture.unfccc.int/#content-the-paris-agree men> (accessed 06-07-2016).

821 UNFCCC "The Paris Agreement" (date of publication unavailable) (page numbers unavailable) UNFCCC <http://bigpicture.unfccc.int/#content-the-paris-agree men> (accessed 06-07-2016).

822 S Zwick "The Road From Paris: Green Lights, Speed Bumps, And The Future of Carbon Markets" (01-02-2016) (page numbers unavailable) Ecosystems Marketplace <http://www.ecosystemmarketplace.com/articles/green-lights-and-speed-bu mps-on-road-to-markets-under-paris-agreement/> (accessed 08-07-2016).

nationally transferred mitigation outcomes",[823] as long as appropriate and "robust accounting"[824] methods ensure that there is no double-counting[825] and systems are in line with the UNFCCC.[826] This approach differs from the approach followed by the Kyoto Protocol – which followed a system where the United Nations drove the market (in a top-down approach).[827]

Article 6(4) establishes a centralised mechanism under the authority and guidance of the Conference of the Parties, in terms of which parties can trade emissions. Article 6(3) stipulates that participation in the carbon market will be voluntary for all countries. Furthermore, article 6(2) of the Agreement stipulates that countries must promote sustainable development within the carbon market. Accordingly, the mechanism in the Paris Agreement has already been informally labelled as the sustainable development mechanism (SDM).[828]

---

823 Article 6(3) of the Paris Agreement reads as follows: "The use of internationally transferred mitigation outcomes to achieve nationally determined contributions under this Agreement shall be voluntary and authorized by participating Parties.".

824 Article 6(2) of the Paris Agreement reads as follows: "Parties shall, where engaging on a voluntary basis in cooperative approaches that involve the use of internationally transferred mitigation outcomes towards nationally determined contributions, promote sustainable development and ensure environmental integrity and transparency, including in governance, and shall apply robust accounting to ensure, inter alia, the avoidance of double counting, consistent with guidance adopted by the Conference of the Parties serving as the meeting of the Parties to this Agreement.".

825 In accordance with article 6(5) of the Paris Agreement, which reads as follows: "Emission reductions resulting from the mechanism referred to in paragraph 4 of this Article shall not be used to demonstrate achievement of the host Party's nationally determined contribution if used by another Party to demonstrate achievement of its nationally determined contribution.".

826 S Zwick "Building on Paris, Countries Assemble The Carbon Markets Of Tomorrow" (29-01-2016) (page numbers unavailable) Ecosystems Marketplace <http://www.ecosystemmarketplace.com/articles/building-on-paris-countries-assemble-the-carbon-markets-of-tomorrow/> (accessed 08-07-2016).

827 S Zwick "Building on Paris, Countries Assemble The Carbon Markets Of Tomorrow" (29-01-2016) (page numbers unavailable) Ecosystems Marketplace <http://www.ecosystemmarketplace.com/articles/building-on-paris-countries-assemble-the-carbon-markets-of-tomorrow/> (accessed 08-07-2016).

828 R Bodle et al "The Paris Agreement: Analysis, Assessment and Outlook" (28-01-2016) (19) Ecologic <http://ecologic.eu/sites/files/event/2016/ecologic_institute_2016_paris_agreement_assessment_0.pdf> (accessed 09-07-2016); S Zwick "The Road From Paris: Green Lights, Speed Bumps, And The Future of Carbon Markets" (01-02-2016) (page numbers unavailable) Ecosystems Market-

Although previously submitted that it would be important to link new trading schemes to existing mechanisms,[829] the position of the joint development mechanism and the CDM is uncertain in terms of the Paris Agreement, and it is unclear whether these existing Kyoto Protocol mechanisms will interact with the SDM when the Agreement enters into force. According to article 6(7) of the Paris Agreement, upcoming COPs, such as COP 22 in Marrakesh at the end of 2016, will *inter alia* deliver more information on this uncertainty in emissions trading under the Paris Agreement, as referred to in article 6(4) of the Paris Agreement. Article 6(8) provides that non-market approaches are also available to parties in terms of the Paris Agreement.[830]

3 9 4 Future outlook on the effectiveness of the Paris Agreement and the position of the principle of sovereignty over natural resources

As mentioned above, the Paris Agreement is non-binding and the predicted effectiveness of its non-binding nature is subject to debate. While many celebrate the non-binding nature of the Paris Agreement as the reason for increased international participation,[831] others believe that its non-binding

place <http://www.ecosystemmarketplace.com/articles/green-lights-and-speed-bumps-on-road-to-markets-under-paris-agreement/> (accessed 08-07-2016).

829 N Stern "Stern Review: The Economics of Climate Change" (2006) (xi) *Centro de Ciência do Sistema Terrestre* <http://mudancasclimaticas.cptec.inpe.br/~rmclima/pdfs/destaques/sternreview_report_complete.pdf> (accessed 30-04-2015).

830 Article 6(8) of the Paris Agreement reads as follows: "Parties recognize the importance of integrated, holistic and balanced non-market approaches being available to Parties to assist in the implementation of their nationally determined contributions, in the context of sustainable development and poverty eradication, in a coordinated and effective manner, including through, inter alia, mitigation, adaptation, finance, technology transfer and capacity- building, as appropriate. These approaches shall aim to: (a) Promote mitigation and adaptation ambition; (b) Enhance public and private sector participation in the implementation of nationally determined contributions; and (c) Enable opportunities for coordination across instruments and relevant institutional arrangements.".

831 M Greenstone "Surprisingly, a Voluntary Climate Treaty Could Actually Work" (13-02-2015) (page numbers unavailable) The New York Times <http://www.nytimes.com/2015/02/15/upshot/surprisingly-a-voluntary-climate-treaty-could-actually-work.html?abt=0002&abg=0> (accessed 05-05-2015); S Page "No, The Paris Agreement Isn't Binding. Here's Why that Doesn't Matter" (14-12-2015) (page

nature will ultimately render it less effective than the Kyoto Protocol which was a legally binding document and yet still arguably ineffective.[832]

The effectiveness of NDCs is questioned in the context of the formal procedures that would have to be developed to monitor and enforce compliance in respect of these contributions. More specifically, many developing countries are not able to measure their greenhouse gas outputs accurately owing to a shortcoming in their national institutional framework.[833] Furthermore, even if countries are indeed able to monitor greenhouse gas emissions accurately, they may lack the internal capacity to audit their total greenhouse gas emissions, or audited emissions may not always be reliable.[834] In addition, NDCs are more conditional and usually hinge on more determinants than a centrally determined contribution would (as was the case under the Kyoto Protocol).

Arguments for why the non-binding nature of the Paris Agreement can be effective, usually hinge on the significance of the positive and negative incentives for cooperation of parties as exerted by international relations. Incentives have already been discussed under the section dealing with compliance with the Kyoto Protocol above. In this context, it is noteworthy that participation at COP 21 and signature[835] of the Paris Agreement showed that the international community is more convinced than ever that cooperation in mitigating the impacts of climate change are urgent. Accordingly, one can argue that pressure resulting from international rela-

---

numbers unavailable) Climate Progress <http://thinkprogress.org/climate/2015/12/14/3731715/paris-agreement-is-an-actual-agreement/> (accessed 09-07-2016).

832 S Page "No, The Paris Agreement Isn't Binding. Here's Why that Doesn't Matter" (14-12-2015) (page numbers unavailable) Climate Progress <http://thinkprogress.org/climate/2015/12/14/3731715/paris-agreement-is-an-actual-agreement/> (accessed 09-07-2016).

833 Council on Foreign Relations "The Climate Change Regime" (19-06-2013) (page numbers unavailable) *Council on Foreign Relations* <http://www.cfr.org/climate-change/global-climate-change-regime/p21831> (accessed 17-04-2015).

834 Council on Foreign Relations "The Climate Change Regime" (19-06-2013) (page numbers not available) *Council on Foreign Relations* <http://www.cfr.org/climate-change/global-climate-change-regime/p21831> (accessed 17-04-2015).

835 More parties signed the Paris Agreement in one day than any other international agreement in history. [UN "Record: 177 parties signed the Paris Agreement" (date of publication unavailable) (page numbers unavailable) Paris 2015, UN Climate Change Conference, COP 21, CMP 11 <http://www.cop21.gouv.fr/en/a-record-over-160-countries-expected-to-sign-the-paris-agreement-in-new-york-on-22-april-2016/> (accessed 09-07-2016).].

tions may amount to more than it did for the Kyoto Protocol and therefore contribute to effective implementation of the Paris Agreement.

A dependable and steadfast reason why the non-binding nature of the Paris Agreement can work is because the robust approach of the Agreement to transparency and accounting methods means that it is easier for people of a state to hold a state accountable for not adhering to their NDCs.[836] The issue of transparency and accountability is purposefully high on the agenda of the Paris Agreement.[837] This characteristic of the Agreement is extremely useful to serving the objective of this research project, which is to explore the possibility of reinterpreting and reconfiguring PSNR so that it becomes more focused on people and accordingly a vehicle for people by which to hold states accountable for climate commitments made in terms of the international legal climate change regime. The transparent accounting under the Paris Agreement will present a support to this suggested line of action by people. This idea is explored further in Chapter 4 of the research project.

PSNR remains an underlying constituent in the latest addition to the international legal climate change regime. The preamble of the Paris Agreement acknowledges that:

> "climate change is a common concern of humankind [and parties] should when taking action to address climate change, respect, promote and consider their respective obligations on human rights, the right to health, the rights of indigenous peoples, local communities, migrants, children, persons with disabilities and people in vulnerable situations and the right to development, as well as gender equality, empowerment of women and intergenerational equity".

With the inclusion of human rights, right to health, the rights of indigenous people, the right to development and the reference to intergenerational equity, it can be submitted that the recognition and connection of PSNR to the Paris Agreement, although somewhat obscured or hidden, is not absent. Article 7(9)(e) of the Paris Agreement reiterates the common

---

836 Page mentions the link between civil society and state accountability. [S Page "No, The Paris Agreement Isn't Binding. Here's Why that Doesn't Matter" (14-12-2015) (page numbers unavailable) Climate Progress <http://thinkprogress.org/climate/2015/12/14/3731715/paris-agreement-is-an-actual-agreement/> (accessed 09-07-2016).].

837 D Bodansky et al "Issues for a 2015 climate agreement" (May 2015) (3) *Center for Climate and Energy Solutions* < http://www.c2es.org/docUploads/issues-for-a-2015-climate-agreement.pdf> (accessed 04-05-2015).

ground between PSNR and the Paris Agreement by its reference to the principle of sustainable development.[838] Article 8 of the Agreement addresses loss and damage associated with adaptation and mitigation of climate change, and again considers sustainable development to play a key role.[839] Furthermore, the principle of common but differentiated responsibility continues to be applied throughout the Agreement, as was elaborated in Section 3.7.2.2. Although PSNR is not as directly inserted into the Paris Agreement as it is in other instruments of the international legal climate change regime, it should be borne in mind that the UNFCCC, in which PSNR is firmly embedded, does still form the cornerstone upon which the Paris Agreement rests and this fact sufficiently confirms the nexus between PSNR and the Paris Agreement.

Only the future will show if the Paris Agreement can increase the effectiveness of the international legal climate change regime. When considering implications beyond "what is actually written in the text"[840] of the Agreement, the content of the Agreement can be praised to a degree at this point of its existence. At the very least it can be submitted that the Paris Agreement has been successful in establishing a global consensus that a

838 Article 7(9)(e) of the Paris Agreement reads as follows: "[E]ach Party shall, as appropriate, engage in adaptation planning processes and the implementation of actions, including the development or enhancement of relevant plans, policies and/or contributions, which may include: […] [b]uilding the resilience of socioeconomic and ecological systems, including through economic diversification and sustainable management of natural resources.".

839 Article 8(1) of the Paris Agreement reads as follows: "Parties recognize the importance of averting, minimizing and addressing loss and damage associated with the adverse effects of climate change, including extreme weather events and slow onset events, and the role of sustainable development in reducing the risk of loss and damage."
As a side note, Article 8 is criticised for not providing a basis for liability or compensation. [LDC News Service "Final Paris Agreement: A nail in the coffin for justice for LDC, says LDC Watch" (13-12-2015) (page number unavailable) LDC News Service <http://ldcnews.com/final-paris-agreement-nail-coffin-justice-ldcs-says-ldc-watch/> (accessed 06-07-2016); B Olivastri "Reflecting on The Paris Climate Agreement Makes Me Glad, Sad and Mad" (13-12-2016) (page numbers unavailable) Friends of the Earth / Les Ami(e)s de la Terre <http://foecanada.org/en/2015/12/reflecting-on-the-paris-climate-agreement-makes-me-glad-sad-and-mad/> (accessed 06-07-2016).].

840 R Bodle et al "The Paris Agreement: Analysis, Assessment and Outlook" (28-01-2016) 21 Ecologic <http://ecologic.eu/sites/files/event/2016/ecologic_institute_2016_paris_agreement_assessment_0.pdf> (accessed 09-07-2016).

business-as-usual approach is no longer a feasible or an acceptable path to national and global development. In addition, the Agreement sends a very clear message to "all relevant actors" that "transformation begins now".[841] Regardless of possible criticisms or doubts, the Paris Agreement has already achieved three concrete and significant successes. Firstly, the Agreement gained unprecedented global support and participation, measured in terms of signatures of an international agreement. Secondly, the Agreement secured the cooperation of the world's two largest greenhouse gas emitters, namely the USA and China. Thirdly, the Agreement was able to secure emission limitations, albeit in the form of bottom-up, NDCs from all parties including developing countries. These three successes have already overcome three massive stumbling blocks that the Kyoto Protocol could not overcome. It is these successes of the Agreement that allow an optimistic anticipation of what the implementation of the Paris Agreement can contribute to ensuring an effective international legal climate change regime in the future.

Regardless of the level of effectiveness that the Paris Agreement is able to achieve in future, the fact that another component of international environmental law, namely PSNR, is able to increase the effectiveness of the Agreement is not to be disregarded. Accordingly, the possibility in Chapter 4 of discussing the implementation of people's rights in terms of PSNR as a negative incentive to increase states' compliance with their climate commitments is equally applicable to climate commitments in terms of the Kyoto Protocol (emission targets) and the Paris Agreement (NDCs).

3 10 Conclusion

This chapter establishes the nexus between climate change and the principle of sovereignty over natural resources, which forms the basis on which the impacts of climate change on the principle are considered. The chapter sets out the scientific basis of climate change and outlines the environmental, social and economic effects of climate change, and the resulting effects on the well-being of people, all in an effort to illustrate the urgency, scale and gravity of the problem of climate change. This chapter examines

841 R Bodle et al "The Paris Agreement: Analysis, Assessment and Outlook" (28-01-2016) 2 Ecologic <http://ecologic.eu/sites/files/event/2016/ecologic_institute_2016_paris_agreement_assessment_0.pdf> (accessed 09-07-2016).

the international legal climate change regime, by defining the term, outlining the development of the regime, and examining the most important constituents of the regime, namely the UNFCCC, the Kyoto Protocol and the latest addition, the Paris Agreement. This chapter of the research project also addresses mitigation and adaptation as components of the international legal climate change regime. In the process of examining the international legal climate change regime, principles on which the regime is founded, namely equity, the principle of common but differentiated responsibility, the precautionary principle, the polluter-pays principle, and the principle of sustainable development, are also discussed. In addition, and throughout the chapter, the embedding of PSNR within the international legal climate change regime is also considered, adding to the establishment of the nexus between PSNR and climate change.

This chapter therefore addresses the primary research objective of the research project, which is to determine whether climate change has, or warrants an impact on the principle of sovereignty over natural resources. As previously submitted, it is accepted that this impact might not be automatic. It is suggested and illustrated in Chapter 3 that the scale and gravity of climate change are so significant, that they warrant or ought to warrant a reconfiguration or reinterpretation of any principle of international environmental law that is able to contribute to making the international legal climate change regime more effective. This first section of Chapter 3 therefore determines that climate change has an impact on PSNR, or more accurately warrants the reinterpretation and reconfiguration of PSNR.

The secondary research objective of the research project is: If the gravity of climate change does indeed have an impact on PSNR (or warrants an impact on PSNR, as has been found in Chapter 3), how exactly does climate change impact PSNR? In this regard, Chapter 3 also considers the international legal climate change regime and one specific source of ineffectiveness, namely the Kyoto Protocol. In addition to explaining the functionality of the Kyoto Protocol, including targets and timetables, and market-based mechanisms, Chapter 3 also considers the contribution of the Kyoto Protocol to the ineffectiveness of the regime, in particular in regard to the component of state compliance with climate commitments (emission targets) in terms of the Protocol. In this regard, state compliance with climate commitments is identified in this chapter as a key component in ensuring a more effective international legal climate change regime. The discussion in Chapter 3 shows that the international legal climate change regime warrants a reinterpretation and reconfiguration of PSNR, if such a reinterpretation and reconfiguration can present a solution to the ineffectiveness of the regime. It will be illustrated in Chapter 4 that in order to increase state compliance with the regime, a viable solution is the implementation of the rights of people in terms of PSNR.

CHAPTER 4: The reinterpretation and reconfiguration of the principle of sovereignty over natural resources and the expansion of the means to combat climate change

4 1 Introduction

Chapter 2 has concluded *inter alia* that a review of PSNR shows that both states and people are considered to be the subjects of PSNR and that states have been and continue to be viewed as the most prevalent subjects of PSNR. It has been determined in Chapter 3 that the gravity of climate change warrants the reinterpretation and reconfiguration of any principle of international environmental law that is able to contribute to increasing the effectiveness of the international legal climate change regime (one way in which to make the international legal climate change regime more effective is by increasing state compliance with the regime). Based on the nexus between PSNR and climate change, and the potential that PSNR has to contribute to the regime, PSNR was identified in the research project as the principle of international environmental law which, when implemented, could present a negative incentive to increase state compliance with climate commitments.

Chapter 4 explores the reinterpretation and reconfiguration of PSNR and the expansion of the means to combat climate change through this reinterpretation and reconfiguration of the principle. The chapter commences by considering the rights and duties of states in terms of PSNR. The environmental, social and economic effects of climate change also mean that the way in which states manage their natural resources is affected. This means that the rights and duties of states in terms of PSNR, which is ultimately the set of rules by which the state manages its natural resources, are affected by the environmental, social and economic effects of climate change. This leads to one of the consequences of the effects of climate change on PSNR, which is the noticeable shift in the interpretation of and understanding of state duties and state rights in terms of PSNR. The international interpretation and understanding of state duties have shifted owing to international environmental concerns, including specifically the concern about the effects of climate change. This has placed more emphasis on state duties (especially in regard to state duties toward their people) than was formerly the case. Chapter 4, along with Chapter 2, therefore confirms PSNR to be a generally state-centric principle, and the shift observed in the interpretation and understanding in regard to the balance of state rights and duties in terms of PSNR (in the context of international environmental concerns) also points to the principle of PSNR being able to evolve, the evolution of which being affected by the gravity of climate change. Chapter 4 therefore completes the exposition of PSNR, which was commenced in Chapter 2. This consideration of the shift observed in regard to the understanding of state rights and duties also lays the groundwork for another component of Chapter 4, namely the exercise of reinterpreting and reconfiguring people's rights in terms of PSNR as an important contemporary component of PSNR and the international legal climate change regime.

In addition to influencing the interpretation and understanding of state rights and duties in terms of PSNR, the gravity of the effects of climate change also warrants a reinterpretation and reconfiguration of PSNR as a principle focused on people. Chapter 4 will accordingly conduct a reinterpretation and reconfiguration of PSNR to investigate, firstly, what people's rights are in terms of PSNR,[842] and, secondly, what the prospects are concerning litigation implemented by people against states in the context of climate change, based on the infringement of people's rights in terms of PSNR. In so doing, it will be determined whether such litigation could provide a worthwhile avenue to pursue in the attempt to increase state compliance with climate commitments undertaken in terms of the international legal climate change regime, and accordingly increase the regime's effectiveness. This exercise serves the research objective of the research project that examines PSNR as a principle focused on people, capable of supporting the international legal climate change regime by allowing people-based action that can constitute a negative incentive for state compliance with climate commitments. In this regard it should be noted that there are a few initial steps, described in this chapter, on which the aforementioned activities are based. The first step is that the state in question, of which compliance is to be addressed, should make a climate commitment in terms of the Kyoto Protocol or the Paris Agreement. The second step is

842 As discussed in Section 1.5, this research project will not consider the duties of people in terms of PSNR.

that the state in question must implement a national legal system whereby citizens or people of the state are able to hold their state accountable. The third step is to establish whether or not only state nationals may institute an action against the state regarding the state's climate commitments.

4 2 Rights and duties of states in terms of the principle of sovereignty over natural resources and the impact of climate change

4 2 1 Introduction to the rights and duties of states in terms of the principle of sovereignty over natural resources

The set of rights and duties for states emanating from the principle of sovereignty over natural resources is evidenced through the examination of sources, such as treaty law; state practice; international court and tribunal decisions; international instruments and United Nations documentation; work by private and public bodies, such as the IPCC or International Law Association (ILA); and international legal literature. The most authoritative list and examination of rights and duties of states emanating from PSNR has been compiled by Schrijver.[843] Not all the rights and duties in this list are discussed in the research project, since they are not all equally relevant to achieving the objectives of this research project.

This section of the research project explores the rights and duties of states emanating from PSNR, specifically within the context of climate change. Certain of the rights and duties will be discussed in greater detail than others dependent on their particular relevance within this research project. As will become apparent below, some of the rights and duties of states in terms of PSNR, discussed below, are not affected to any significant measure by climate change. This section of Chapter 4 includes, to the extent possible, a consideration of how state rights and duties in terms of PSNR are affected by the gravity of climate change. This consideration will reveal a shift, where certain state duties in terms of PSNR gain momentum and certain rights of states in terms of PSNR become constricted. The inclusion of state rights and duties at this point also serve as a component in the examination of PSNR, which was commenced in Chapter 2,

---

843 N Schrijver *Sovereignty Over Natural Resources, Balancing Rights and Duties* (1997) 260-364.

and a component in the examination of the impact of climate change on PSNR, which was commenced in Chapter 3.

Section 4.2.2 will discuss the impact of climate change on the rights of states in terms of PSNR. As submitted in Chapter 2 of the research project, the development of the rights or claims of states in terms of PSNR was based on the original intention behind the inception of PSNR, which was to ensure *inter alia* that developing or previously colonised states were able to regain effective control over the disposal, exploitation and exploration of natural resources falling within their state territory, in order to control their national economic development.

Section 4.2.3 will discuss the impact of climate change on the duties of states in terms of PSNR. Duties of states are also referred to throughout the relevant sources listed in the research project as *inter alia* responsibilities, obligations, undertakings, commitments, or requirements. This collection of terms is referred to throughout the research project, by using the term 'duties', and these duties are understood as obligations freely entered into by the state or a form of conduct that is prescribed by sources of international law.[844] Schrijver submits that the state duties in terms of PSNR can mostly not be legally enforced in court proceedings, but are often adhered to because of factors such as public opinion or political pressure,[845] which are forms of negative incentives. Furthermore, these duties are important because they form part of a "code of conduct", they direct the expectations of other involved parties, and through state practice these duties can become customary international law.[846]

844 Schrijver *Sovereignty Over Natural Resources, Balancing Rights and Duties* (1997) 307.

845 Schrijver *Sovereignty Over Natural Resources, Balancing Rights and Duties* (1997) 307.

846 Schrijver *Sovereignty Over Natural Resources, Balancing Rights and Duties* (1997) 307.

4 2 2 Rights of states in terms of the principle of sovereignty over natural resources and the impact of climate change

4 2 2 1 Right to dispose freely of, explore, and exploit natural resources

At the core of PSNR lies the sovereign right of states to dispose freely of natural resources. A discussion of the alienability of the state right to dispose freely of natural resources would run parallel with a discussion of the alienability of sovereignty. The alienability of sovereignty is discussed in Section 2.5.2.3.3. CERDS clearly states in article 2(1) that the exercise of sovereignty over natural resources includes the disposal of natural resources.[847] An internationally excepted corollary right hereto is the right of states to explore and exploit their natural resources, as confirmed by the Aminoil award.[848] The right to explore and exploit natural resources in particular can be understood as meaning that states have the right to prospect, develop, use, sell and market their natural resources in accordance with their national laws.[849]

The right to dispose freely of, explore and exploit natural resources is evident throughout many UNGA resolutions. Resolution 1515 (XV) recommends that the right of states to dispose freely of their natural resources should be exercised in conformity with the rights and duties of states under international law.[850] Today, these duties would include the duties of

---

847 CERDS is contained in the UNGA resolution 3281(XXIX) of 12 December 1974.
Article 2(1) of CERDS reads as follows: "Every State has and shall freely exercise full permanent sovereignty, including possession, use and disposal, over all its wealth, natural resources and economic activities.".

848 The Aminoil award deals with *inter alia* the decision by Kuwait to terminate Aminoil's concession for exploration and exploitation in the context of nationalisation, which is deemed lawful by the Tribunal. The Tribunal based its decision on Kuwait's sovereign right to freely explore and exploit its natural resources in terms of sovereignty over natural resources. [Arbitral Tribunal: Award in the Matter of an Arbitration between Kuwait and the American Independent Oil Company (Aminoil) 24 March 1982 *International Legal Materials* (1982) 21(5) 976-1053 1021.].

849 N Schrijver *Sovereignty Over Natural Resources, Balancing Rights and Duties* (1997) 264.

850 UNGA resolution 1515 (XV) of 15 December 1960.
UNGA resolution 3175 (XXVIII) of 17 December 1973 deals with this right in the context of the occupied Arab territories.

states in terms of climate commitments made under the international legal climate change regime. According to the UNGA resolution 523 (VI), underdeveloped countries are singled out, in particular, to have the right to determine freely the use of their natural resources, not only in order to promote their national interests and development, but also to promote the expansion of the world economy.[851] Similarly, UNGA resolution 626 (VII) also encourages underdeveloped countries, in particular, in the "proper use and exploitation" of their natural resources.[852] The prevalence of singling out underdeveloped countries in this context is explained in Section 2.2, wherein the foundation and development of PSNR is discussed. UNGA resolution 1803 (XVII) mentions both exploration and disposal of resources by stating the following:

> "The exploration, development and disposition of such resources, as well as the import of the foreign capital required for these purposes, should be in conformity with the rules and conditions which the peoples and nations freely consider to be necessary or desirable with regard to the authorization, restriction or prohibition of such activities."[853]

UNGA resolution 2158 (XXI) also mentions the state's right to dispose freely of, explore and exploit its natural resources according to its national laws, while emphasising the importance of natural resources as a basis for economic and industrial development of the state.[854] In addition, UNGA resolution 3171 (XXVIII) stipulates that the right of the state to exercise sovereignty over natural resources includes full utilisation by the state of its resources at all stages, from exploration to marketing.[855]

The right of states to dispose freely of, explore and exploit is also contained in international instruments and case law. The right of states over

851 UNGA resolution 523 (VI) of 12 January 1952.

852 UNGA resolution 626 (VII) of 21 December 1952.
As a caveat, this resolution maintains that the exercise of this right by member states must be in line with "the need for maintaining the flow of capital in conditions of security, mutual confidence and economic co-operation" between states.

853 UNGA resolution 1803 (XVII) of 14 December 1962.

854 UNGA resolution 2158 (XXI) of 25 November 1966.

855 UNGA resolution 3171 (XXVIII) of 17 December 1973.
This resolution also reiterates that permanent sovereignty over natural resources refers to all resources falling within the jurisdiction of the state, whether they are found on land or in the sea. The precise scope of the term 'natural resources' as a component of PSNR and as a term employed in the research project is examined in Section 2.4.2.

their natural resources is contained in treaties such as the Convention on the Continental Shelf 1958 (Convention on the Continental Shelf), of which article 2 expressly provides that the coastal state has exclusive sovereign rights to explore and exploit its natural resources.[856] In article 56 and article 77 of the Convention on the Law of the Sea, a similar right of the state is mentioned to explore, exploit, conserve and manage its natural resources and to utilise its resources for other economic activities such as the production of energy.[857] The Biodiversity Convention recognises sovereign rights of states over their natural resources.[858] The Energy Charter Treaty includes this right of states, but adds, in particular, that states have the right to decide which of their geographical areas will be explored and developed as energy resources.[859]

---

856 Article 2(1)-(3) of the Convention on the Continental Shelf reads as follows: "(1) The coastal State exercises over the continental shelf sovereign rights for the purpose of exploring it and exploiting its natural resources. (2) The rights referred to in paragraph 1 of this article are exclusive in the sense that if the coastal State does not explore the continental shelf or exploit its natural resources, no one may undertake these activities, or make a claim to the continental shelf, without the express consent of the coastal State. (3) The rights of the coastal State over the continental shelf do not depend on occupation, effective or notional, or on any express proclamation.".

857 Article 56(1)(a) of the Convention on the Law of the Sea reads as follows: "In the exclusive economic zone, the coastal State has:
sovereign rights for the purpose of exploring and exploiting, conserving and managing the natural resources, whether living or non-living, of the waters superjacent to the seabed and of the seabed and its subsoil, and with regard to other activities for the economic exploitation and exploration of the zone, such as the production of energy from the water, currents and winds".
Article 77(1)-(2) of the Convention on the Law of the Sea reads as follows: "(1) The coastal State exercises over the continental shelf sovereign rights for the purpose of exploring it and exploiting its natural resources. (2) The rights referred to in paragraph 1 are exclusive in the sense that if the coastal State does not explore the continental shelf or exploit its natural resources, no one may undertake these activities without the express consent of the coastal State.".

858 Article 15(1) of the Biodiversity Convention reads as follows: "Recognizing the sovereign rights of States over their natural resources, the authority to determine access to genetic resources rests with the national governments and is subject to national legislation.".

859 Article 18(3) of the Energy Charter Treaty reads as follows: "Each state continues to hold in particular the rights to decide the geographical areas within its Area to be made available for exploration and development of its energy resources, the optimalization of their recovery and the rate at which they may be

The right of states over natural resources is also confirmed to varying degrees in arbitration decisions, including for example the Liamco award, the Texaco award and the Aminoil award.[860] In the Liamco award, while confirming the state right over natural resources, the tribunal also stresses the right to contractual agreements and compensation obligations.[861] The Texaco award provides more indirectly that the state's right to exercise national sovereignty is an expression of the territorial sovereignty of the state, which gives the state the prerogative to choose and build its economic system.[862] The Aminoil decision confirms the right of states over natural resources by stressing a national constitutional provision of Kuwait, namely, that "[a]ll the natural wealth and resources are the property of the state".[863]

It is clear from the UNGA resolutions, the international instruments and arbitration awards that the right of states to natural resources, including the right of states to dispose of, explore and exploit natural resources, is well established. This right afforded to states by PSNR is so ingrained in

depleted or otherwise exploited, to specify and enjoy any taxes, royalties or other financial payments payable by virtue of such exploration and exploitation, and to regulate the environmental and safety aspects of such exploration, development and reclamation within its Area, and to participate in such exploration and exploitation, *inter alia*, through direct participation by the government or through state enterprises.".

860 Arbitral Tribunal: Award in the Dispute between Libyan American Oil Company (Liamco) and the Government of the Libyan Arab Republic relating to Petroleum Concessions 12 April 1977 *International Legal Materials* (1981) 20(1) 1-87 53; Arbitral Tribunal: Award on the Merits in Dispute between Texaco Overseas Petroleum Company/California Asiatic Oil Company (Texaco) and the Government of the Libyan Arab Republic 19 January 1977 *International Legal Materials* (1978) 17(1) 1-37 21; Arbitral Tribunal: Award in the Matter of an Arbitration between Kuwait and the American Independent Oil Company (Aminoil) 24 March 1982 *International Legal Materials* (1982) 21(5) 976-1053 992.

861 Arbitral Tribunal: Award in the Dispute between Libyan American Oil Company (Liamco) and the Government of the Libyan Arab Republic relating to Petroleum Concessions 12 April 1977 *International Legal Materials* (1981) 20(1) 1-87 53.

862 Arbitral Tribunal: Award on the Merits in Dispute between Texaco Overseas Petroleum Company/California Asiatic Oil Company and the Government of the Libyan Arab Republic 19 January 1977 *International Legal Materials* (1978) 17(1) 1-37 21.

863 Arbitral Tribunal: Award in the Matter of an Arbitration between Kuwait and the American Independent Oil Company (Aminoil) 24 March 1982 *International Legal Materials* (1982) 21(5) 976-1053 992.

international law, that the possibility of limiting this right as a result of climate change considerations is unlikely. In fact, it could be argued that the impacts of climate change are intensifying opportunities for the exploitation of natural resources, especially in regions such as the Arctic where melting ice makes exploration less inaccessible and more feasible.[864] Accordingly, the gravity of climate change may be adding momentum to this state right, making it unlikely that the exercise of or reliance on this right by states will decrease. However, another perspective is possible, namely that climate change and the climate commitments that states make in terms of the international legal climate change regime are affecting the state's disposal and exploitation of its natural resources since a business-as-usual approach to disposal and exploitation of natural resources will not allow states to meet their climate commitments.

4 2 2 2 Right to exercise effective control over natural resources and the right to receive appropriate compensation

For the purposes of the research project, it is adequate merely to take note of the right of states to exercise effective control over their natural resources and that this right is connected to their right to dispose freely of, explore and exploit natural resources, more so than their right to full or appropriate compensation. This is because both the right to effective control over natural resources and the right to receive appropriate compensation is

---

864 German Federal Ministry of Education and Research "Rapid Climate Change in the Arctic, Polar Research as a Global Responsibility" (2015) (5) *Bundesministerium für Bildung und Forschung* <https://www.bmbf.de/pub/Rapid_Climate_Change_in_the_Artic.pdf> (accessed 07-08-2016); E Rosenthal "Race is On as Ice Melt Reveals Arctic Treasures" (18-09-2012) (page numbers unavailable) *The New York Times* <http://www.nytimes.com/2012/09/19/science/earth/arctic-resources-exposed-by-warming-set-off-competition.html?_r=0> (accessed 07-08-2016); D Cambou & S Smis "Permanent sovereignty over natural resources from a human rights perspective: natural resources exploitation and indigenous peoples' rights in the Arctic" (2013) 22(1) *Michigan State International Law Review* 347 348;
Also take note of the Antarctic Treaty System in this context and see in particular Glazewski: J Glazewski "National sovereignty v the global common good" (22-04-2010) (page numbers unavailable) *UCT* <https://www.uct.ac.za/downloads/uct.ac.za/news/media/oped/Jan_Glazewsk_22_04_10.pdf> (accessed 07-08-2016).

most relevant in instances relating to non-self governing or occupied states or territories.[865] In these instances, the occupying state should pay full compensation to the rightful government for exploiting, depleting or damaging the natural resources in a foreign occupied territory.[866] Besides the reference to the term 'full compensation' in this context, the term 'appropriate compensation' is often mentioned in this regard, more specifically in instances relating to nationalisation, expropriation and requisitioning.[867] It remains controversial and uncertain what can be considered as appropriate compensation, as a common set of international principles regulating what would constitute appropriate compensation in certain instances does not exist.[868] Most opinions seem to diverge on the understanding that appropriate compensation must be fair, just, full, adequate, appropriate,

865 In this regard, see UNGA resolution 3171 (XXVIII) of 17 December 1973 as an example of a resolution supporting states and people striving to regain effective control over their natural resources which may still be under colonial or racial domination or occupation.

866 N Schrijver *Sovereignty Over Natural Resources, Balancing Rights and Duties* (1997) 269.
Article 16(1) of CERDS, as contained in UNGA resolution 3281 (XXIX) of 12 December 1974, sets out compensation within a similar context and also includes a reference to the term 'full compensation'. Article 16(1) of CERDS reads as follows: "It is the right and duty of all States [...] to eliminate colonialism, apartheid, racial discrimination, neo-colonialism and all forms of foreign aggression, occupation and domination, and the economic and social consequences thereof, as a prerequisite for development. States which practise such coercive policies are economically responsible to the countries, territories and peoples affected for the restitution and full compensation for the exploitation and depletion of, and damages to, the natural and all other resources of those countries, territories and peoples".

867 UNGA resolution 1803 (XVII) of 14 December 1962 in particular makes reference to the term 'appropriate compensation' in cases of nationalisation, expropriation or requisitioning. Article 2(c) of CERDS, as contained in the UNGA resolution 3281 (XXIX) of 12 December 1974, also makes reference to the term 'appropriate compensation' in the context of nationalisation, expropriation or the transfer of ownership of foreign property. Furthermore CERDS stipulates in article 2(c) that appropriate compensation shall be determined by the domestic laws and tribunals of the nationalising state unless otherwise agreed.

868 M Sornarajah "The International Law on Foreign Investment" 3ed (2010) 412; E Penrose et al "Nationalisation of Foreign-owned Property for a Public Purpose: An Economic Perspective on Appropriate Compensation" (1992) 55(3) *The Modern Law Review* 351 352.

prompt and effective, but, again, the interpretation of these terms varies widely.[869]

This brief consideration of the right of states to effective control and appropriate compensation is included as it forms part of the complete understanding of state rights afforded by PSNR, and, accordingly, a complete understanding of PSNR. However, it falls outside the scope of the research project to consider this state right in any greater detail. In addition, the impact of climate change on this state right is negligible, especially in light of the research objectives of the research project expressed in Section 1.4.

4 2 2 3 Right to use natural resources for national development

The right to use natural resources for national development is closely connected to the right to dispose freely of, explore and exploit natural resources. This is evidenced by UNGA resolution 626 (VII), which refers to the term 'economic development' in mentioning that states exercise their right to use and exploit their natural resources "for their own progress and development".[870] Furthermore, UNGA resolution 2158 (XXI) recognises natural resources as the basis of national development for developing countries, and the resolution reaffirms that the right to sovereignty over natural resources must be employed in the interest of national development.[871] Similarly, in UNGA resolution 1803 (XVII) the term 'development' is used often, but most notably in the context where the right of states (and people) to natural resources must be exercised in the interest of their national development and the well-being of people of the state.[872] In this context, article 7 of CERDS[873] stipulates that each state has the right and the responsibility to promote *inter alia* the economic development of its people, and in addition to ensure that its people participate fully in the process of development and in the benefits of development. The right of

---

869 E Penrose et al "Nationalisation of Foreign-owned Property for a Public Purpose: An Economic Perspective on Appropriate Compensation" (1992) 55(3) *The Modern Law Review* 351 352; JA Hofbauer *The Principle of Permanent Sovereignty over Natural Resources and Its Modern Implications* LLM thesis, University of Iceland (2009) 22.

870 UNGA resolution 626 (VII) of 21 December 1952.

871 UNGA resolution 2158 (XXI) of 25 November 1966.

872 UNGA resolution 1803 (XVII) of 14 December 1962.

873 As contained in UNGA resolution 3281 (XXIX) of 12 December 1974.

states to use natural resources for national development, specifically a state's "economic and social progress", is also evidenced in the Aminoil award.[874]

The state right to use natural resources for national development is frequently employed by developing countries to strengthen their view that developmental freedom is important, and resource exploitation leading to development is relied upon, even in light of global environmental concerns such as climate change. Evidence of this general attitude was seen early on in the international sphere in the adjustment of principle 21 of the Stockholm Declaration by principle 2 of the Rio Declaration.[875] More recently, this state attitude was showcased in the various COPs to the UNFCCC, when developing countries refused to take on emissions limitations in terms of the Kyoto Protocol, employing the reason that they needed to prioritise national development. Even though the preamble of the UNFCCC makes reference to the "developmental policies" of states by including the provision that states have "the sovereign right to exploit their own resources pursuant to their own environmental and developmental policies", the Paris COP negotiations seem to mark a significant step away from the abovementioned state attitude to prioritise national development, as evidenced through the system of NDCs employed by the Paris Agreement that hold both developing and developed states accountable for emissions

874 Arbitral Tribunal: Award in the Matter of an Arbitration between Kuwait and the American Independent Oil Company (Aminoil) 24 March 1982 *International Legal Materials* (1982) 21(5) 976-1053 1023.

875 Schrijver *Sovereignty Over Natural Resources, Balancing Rights and Duties* (1997) 270; JA Hofbauer *The Principle of Permanent Sovereignty over Natural Resources and Its Modern Implications* LLM thesis, University of Iceland (2009) 17.
Principle 21 of the Stockholm Declaration reads as follows: "States have, in accordance with the Charter of the United Nations and the principle of international law, the sovereign right to exploit their own resources pursuant to their own environmental policies, and the responsibility to ensure that activities within their jurisdiction or control do not cause damage to the environment of other States or of areas beyond the limits of national jurisdiction."
Principle 2 of the Rio Declaration reads as follows: "States have, in accordance with the Charter of the United Nations and the principles of international law, the sovereign right to exploit their own resources pursuant to their own environmental *and developmental* policies, and the responsibility to ensure that activities within their jurisdiction or control do not cause damage to the environment of other States or of areas beyond the limits of national jurisdiction" (emphasis added).

contributions under the international legal climate change regime (discussed in Section 3.9). In this sense the appreciation by states of the effects of climate change had an impact on the right to use natural resources for national development. This submission is supported further by the existence of the corresponding state duty to exercise PSNR not only for national development but also for the well-being of the people, as will be discussed in Section 4.2.3.1.

4 2 2 4 Right to manage natural resources according to national environmental policy

As mentioned above, developing countries have previously greatly relied on the idea that developmental freedom is of great importance even in light of environmental concerns. States are still considered to have a right to pursue their own environmental policies in the management of their natural resources in terms of PSNR, and the tension that has been evident between national development and environmental concerns since before the formulation of the Stockholm Declaration remains evident today.[876] This tension was partly reconciled by the formulation of certain state duties in terms of PSNR, which will be explored in Section 4.2.3.

Noteworthy treaties and conventions that encapsulate the state's right to manage its natural resources according to its national environmental policy include the Vienna Convention for the Protection of the Ozone Layer, which stipulates this right in exact terms.[877] The United Nations Convention to Combat Desertification 1994 (UNCCD) also states this right in ex-

---

876 Schrijver *Sovereignty Over Natural Resources, Balancing Rights and Duties* (1997) 274.

877 The preamble of the Vienna Convention for the Protection of the Ozone Layer reads as follows (after recalling *inter alia* principle 21 of the Stockholm Declaration): "States have, in accordance with the Charter of the United Nations and the principles of international law, the sovereign right to exploit their own resources pursuant to their own *environmental policies*, and the responsibility to ensure that activities within their jurisdiction or control do not cause damage to the environment of other States or of areas beyond the limits of national jurisdiction" (emphasis added).

act terms and adds the reference to environmental management according to "national policies".[878]

The right is further included in many regional treaties, conventions and declarations and especially clearly in documents of areas which were particularly harshly subjected to previous foreign interference into domestic affairs. Examples of further inclusion of the right include the African Convention on the Conservation of Nature and Natural Resources, which contains the right in exact terms.[879] The Treaty for Amazonian Cooperation 1978 (Treaty for Amazonian Cooperation) provides that the utilisation of natural resources shall not be subject to any restrictions other than those imposed by international law, and it can be contended that this means that these states have the right to manage their natural resources pursuant to their national environmental policies.[880] The inclusion of the Stockholm Declaration within the preamble of the Convention on the Conservation of European Wildlife and Natural Habitats 1979 (Bern Convention) means that states have "the sovereign right to exploit their own resources pur-

878 The preamble of the UNCCD reads as follows (after reaffirming principle 2 of the Rio Declaration): "States have, in accordance with the Charter of the United Nations and the principles of international law, the sovereign right to exploit their own resources pursuant to their own environmental and developmental policies, and the responsibility to ensure that activities within their jurisdiction or control do not cause damage to the environment of other States or of areas beyond the limits of national jurisdiction". In addition, article 5(1)(a) of the UNFCCC again refers in particular to assistance by various means, but in accordance with "national policies".

879 The preamble of the African Convention on the Conservation of Nature and Natural Resources reads as follows: "States have, in accordance with the Charter of the United Nations and the principles of international law, a sovereign right to exploit their own resources pursuant to their environmental and developmental policies, and the responsibility to ensure that activities within their jurisdiction or control do not cause damage to the environment of other States or of areas beyond the limits of national jurisdiction".

880 Schrijver *Sovereignty Over Natural Resources, Balancing Rights and Duties* (1997) 275.
Article iv of the Treaty of Amazonian Co-operation reads as follows: "The Contracting Parties declare that the exclusive use and utilization of natural resources within their respective territories is a right inherent in the sovereignty of each state and that the exercise of this right shall not be subject to any restrictions other than those arising from International Law.".

suant to their own environmental policies".[881] The preamble of the United Nations Economic Commission for Europe (UNECE) Convention on Long-Range Transboundary Air Pollution 1979 (Convention on Long-Range Transboundary Air Pollution) also includes the Stockholm Declaration and, accordingly, the right of states to manage natural resources according to national environmental policy.[882] The Association of Southeast Asian Nations (ASEAN) Agreement on the Conservation of Nature and Natural Resources 1985 (Agreement on the Conservation of Nature and Natural Resources) is understood to include the right to manage natural resources according to national environmental policy owing to the various references throughout the Agreement to "national laws".[883]

Article 192 and article 193 of the Convention on the Law of the Sea illustrates the abovementioned conflict between developmental freedom and

881 The preamble of the Bern Convention reads as follows: "Bearing in mind the widespread requests for common action made by governments or by international bodies, in particular the requests expressed by the [Stockholm Declaration]". Principle 21 of the Stockholm Declaration reads as follows: "States have [...] the sovereign right to exploit their own resources pursuant to their own environmental policies". Furthermore, article 3(1) of the Convention makes reference to 'national policies' as it states "[contracting parties] shall take steps to promote national policies for the conservation of wild flora, wild fauna and natural habitats".

882 The preamble of the Convention on Long-Range Transboundary Air Pollution reads as follows: "Considering the pertinent provisions of the [Stockholm Declaration], and in particular principle 21, which expresses the common conviction that States have, in accordance with the Charter of the United Nations and the principles of international law, the sovereign right to exploit their own resources pursuant to their own environmental policies, and the responsibility to ensure that activities within their jurisdiction or control do not cause damage to the environment of other States or of areas beyond the limits of national jurisdiction".

883 For example, article 1(1)-(2) reads as follows: "(1) The Contracting Parties, within the framework of their respective *national laws*, undertake [...] the measures necessary to maintain essential ecological process and life-support systems, to preserve genetic diversity, and to ensure the sustainable utilization of harvested natural resources under their jurisdiction [...] with a view to attaining the goal of sustainable development. (2) To this end they shall develop *national conservation strategies*, and shall co-ordinate such strategies within the framework of a conservation strategy for the Region" (emphasis added). Furthermore, article 2(1) reads as follows: "The Contracting Parties shall take all necessary measures, within the framework of their respective *national laws*, to ensure that conservation and management of natural resources are treated as an integral part of development planning at all stages and at all levels" (emphasis added).

environmental concern particularly well.[884] Article 192 and 193 both bestow a duty on states to "protect and preserve the marine environment". Article 193 at the same time provides that "states have the sovereign right to exploit their natural resources pursuant to their environmental policies", although this right is to be exercised in accordance with their duty to "protect and preserve the environment".Surely this type of conflict must be interpreted to mean that the state's environmental policies must be in line with its duty to protect and preserve the environment. In other words, the state has the right to pursue its own national environmental policies, free from the influence of any other state, with the caveat that these national environmental policies are in line with the protection and preservation of the environment. The reason why the term 'own environmental policies' is stressed within this right seems to be to ensure that environmental concerns are not misused as a reason to interfere in the national affairs of a state.[885]

The UNFCCC also includes this right in its preamble.[886] However, as was stipulated in the case of the state right to use natural resources for national development, the state right to use natural resources according to its own environmental policy is impacted by climate change. Even though the preamble of the UNFCCC makes reference to the "environmental policies" of states by including the provision that states have "the sovereign right to exploit their own resources pursuant to their own environmental and developmental policies",the Paris COP negotiations seem to take a significant step away from the abovementioned state attitude to prioritise national development over environmental concerns, as evidenced through

---

884 Article 192 of the Convention on the Law of the Sea reads as follows: "States have the obligation to protect and preserve the marine environment."
Article 193 of the Convention on the Law of the Sea reads as follows: "States have the sovereign right to exploit their natural resources pursuant to their environmental policies and in accordance with their duty to protect and preserve the marine environment.".

885 Schrijver *Sovereignty Over Natural Resources, Balancing Rights and Duties* (1997) 275.

886 The preamble of the UNFCCC reads as follows: "States have, in accordance with the Charter of the United Nations and the principles of international law, the sovereign right to exploit their own resources pursuant to their own *environmental and developmental policies*, and the responsibility to ensure that activities within their jurisdiction or control do not cause damage to the environment of other States or of areas beyond the limits of national jurisdiction" (emphasis added).

the system of NDCs implemented (discussed in Section 3.9). In light of the shift that has occurred in the international legal climate change regime, it is possible to argue that the urgency, scale and gravity of the problem of climate change have made an impact on this established tension between international environmental concerns and state-specific objectives, and have caused an increase in support for environmental concerns. Thus state duties will outweigh, in particular, this state right to manage natural resources according to state environmental policies, as state policies are increasingly being influenced by the global environmental policy aimed at mitigating the effects of climate change.

4 2 2 5 Right to an equitable share in benefits of transboundary natural resources

As explained in Chapter 2, natural resources occurring within the determined national territory of more than one state are referred to as shared or transboundary natural resources.[887] There exists an overlap in terminology describing natural resources in the context of this state right. The terms such as 'shared', 'transboundary' and 'international' often employed in this context are not always clearly distinguished.

One of the few instances where the right of states to an equitable share in benefits of transboundary natural resources was addressed in a UNGA resolution, was in resolution 3281 (XXIX),[888] which includes CERDS. Article 3 of CERDS confers upon states the right to be informed and consulted by neighbouring states in the case where natural resources are exploited by two or more countries, in order to prohibit damaging legitimate interests of states and in order to contribute to achieving the optimum use of the natural resources. In addition to information and prior consultation, the Action Plan adopted by the United Nations Water Conference (UNWC) stipulates that state policies should allow for equitable utilisation of the

887 K Odendahl "Conservation and Utilization of Natural Resources and Common Spaces" (2002) (page numbers unavailable) *Encyclopedia of Life Support Systems* <http://www.eolss.net/Sample-Chapters/C04/E4-21-06.pdf> (accessed 27-04-2012).

888 UNGA resolution 3281 (XXIX) of 12 December 1974.

shared resource.[889] The UNGA resolution 34/186[890] stipulates that the principles contained in the Action Plan of the UNWC should be viewed as guiding principles in the formulation of bilateral or multilateral conventions concerning natural resources shared by two or more states. Examples of the incorporation of the state right to an equitable share in transboundary natural resources are frequently included in conventions, the most relevant of which are mentioned below.

In the preamble of the Convention on the Conservation of Migratory Species of Wild Animals 1979 (Bonn Convention), which has the main objective to protect migratory species as a shared or transboundary natural resource, it is provided that "man holds the resources of the earth for future generations and has an obligation to ensure that this legacy is conserved and, where utilised, is used wisely". In addition, article 19 of the ASEAN Agreement on the Conservation of Nature and Natural Resources concerns shared resources and article 19(1) refers particularly to harmonious utilisation of transboundary resources.[891] An example of this right within the context of desertification and transboundary resources is found in the UNCCD, in which article 11 provides that sub-regional programmes shall establish mechanisms especially concerned with the transboundary problems associated with desertification and/or drought. Transboundary water resources, in their various forms,[892] are also a vital shared natural resource to which this right of states pertains. Besides the no-harm rule and principle of equitable utilisation, there are a few other legal norms binding states in the context of transboundary water management. These norms include the duty of states to cooperate and negotiate in good faith

---

889 UN "Report of the United Nations Water Conference" (14-25 March 1977) (53) *UN* <http://www.ircwash.org/sites/default/files/71UN77-161.6.pdf> (accessed 06-09-2016).

890 UNGA resolution 34/186 of 18 December 1979 entitled "Co-operation in the field of the environment concerning natural resources shared by two or more states".

891 Article 19(1) of the ASEAN Agreement on the Conservation of Nature and Natural Resources reads as follows: "Contracting Parties that share natural resources shall cooperate concerning their conservation and harmonious utilization, taking into account the sovereignty, rights and interests of the Contracting Parties concerned in accordance with generally accepted principles of international law.".

892 Transboundary water resources appear in the form of groundwater resources or surface water resources. [N Kliot "Institutions for management of transboundary water resources: their nature, characteristics and shortcomings" (2001) 3 *Water Policy* 229 230.].

and the duty of prior consultation.[893] A full exposition of this extensive governing legal framework surrounding transboundary water and other transboundary resources falls outside the ambit of the research project.[894]

Another inclusion of the state right to an equitable share in benefits of transboundary natural resources is found in the Convention on the Law of the Sea,[895] which provides for sharing the resources of the sea. These re-

893 N Kliot et al "Institutions for management of transboundary water resources: their nature, characteristics and shortcomings" (2001) 3 *Water Policy* 229 233.

894 However it should be noted that an understanding of the term "equitable utilisation" may lead to a better understanding the right of a state to an equitable share in the benefits of transboundary resources. In the context of freshwater transboundary resources, equitable utilisation can be considered as a rule of customary international law. [JW Dellapenna "The customary international law of transboundary fresh waters" (2001) 1(3/4) *International Journal of Global Environmental Issues* 264 264, 276.] The formulation of the so-called law of equitable utilisation is most notably contained in the Helsinki Rules and the Convention on the Law of Non-navigational Uses of International Watercourses. In an effort to supplement the formulation of the law contained in the Helsinki Rules, which has been criticised as vague and outdated, the United Nations set out to codify the law in a treaty, namely the Convention on the Law of Non-navigational Uses of International Watercourses, which was adopted by the UNGA in 1997, by resolution 51/229 of 21 May 1997. A contemporary understanding of equitable utilisation includes the understanding that the interests of all relevant states must be taken into account in the course of allocating and using international watercourses. [N Kliot et al "Institutions for management of transboundary water resources: their nature, characteristics and shortcomings" (2001) 3 *Water Policy* 229 229-255.] For discussions on water as a transboundary resource, and the interplay between sovereignty and international water law, see: T Tvedt, O McIntyre & TK Woldetsadik (eds) *A History of Water. Series III, Volume 2: Sovereignty and International Water Law* (2015).

895 For example, article 142 of the Convention on the Law of the Sea, concerning the rights and legitimate interests of coastal States, is relevant. Article 142(1)-(3) reads as follows: "(1) Activities in the Area, with respect to resource deposits in the Area which lie across limits of national jurisdiction, shall be conducted with due regard to the rights and legitimate interests of any coastal State across whose jurisdiction such deposits lie. (2) Consultations, including a system of prior notification, shall be maintained with the State concerned, with a view to avoiding infringement of such rights and interests. In cases where activities in the Area may result in the exploitation of resources lying within national jurisdiction, the prior consent of the coastal State concerned shall be required. (3) Neither this Part nor any rights granted or exercised pursuant thereto shall affect the rights of coastal States to take such measures consistent with the relevant provisions of Part XII as may be necessary to prevent, mitigate or eliminate grave and imminent danger to their coastline, or related interests from pollution or threat thereof

sources can also pertain to resources which are deposited across the limits of national jurisdiction and are as such subject to the principle of the common heritage of humankind.[896]

While the state-based right system in terms of PSNR proved useful during the time of inception of the principle as a tool by which to combat the consequences of colonialism, this system now proves to be a hindrance when it comes to the effective natural resource management of shared or transboundary natural resources.[897] This observation is well-illustrated by the global environmental concern about climate change. The atmosphere, forming the main object of the international legal climate change regime, is considered to fall partly within the jurisdiction of states, and is accordingly a shared or transboundary natural resource.[898] At the same time, the atmosphere is also subject to the principle of the common concern of humankind.[899] In addition, another shared or transboundary natural resource, which forms a significant component of the international legal climate change regime, is forests.[900] Since state parties have to take into consideration the effects of their disposal of natural resources such as the atmosphere and forests (as shared natural resources), the gravity of climate change impacts the contemporary understanding of the state right to an equitable share in the benefits of transboundary natural resources. Put differently, the international legal climate change regime regulates the usage by

---

or from other hazardous occurrences resulting from or caused by any activities in the Area.".

896 Schrijver *Sovereignty Over Natural Resources, Balancing Rights and Duties* (1997) 277.

897 A Mancilla "Shared Sovereignty over Migratory Natural Resources" (2016) 22 *Res Publica* 21 34-35.

898 J Vogler *The Global Commons, Environmental and Technological Governance* 2ed (2000) 6, 122; E Blanco & J Razzaque *Globalisation and Natural Resources Law: Challenges, Key Issues and Perspectives* (2011) 88-89.

899 FL Kirgis "Standing to Challenge Human Endeavors That Could Change the Climate" (1990) 84 *The American Journal of International Law* 525 530; E Blanco & J Razzaque *Globalisation and Natural Resources Law: Challenges, Key Issues and Perspectives* (2011) 91; J Vogler *The Global Commons, Environmental and Technological Governance* 2ed (2000) 122-123; UNGA resolution 43/53 of 6 December 1988; Preamble of the UNFCCC; D Shelton "Common Concern of Humanity" (2009) 1 *Iustum Aequum Salutare* 33 37.

900 E Blanco & J Razzaque *Globalisation and Natural Resources Law: Challenges, Key Issues and Perspectives* (2011) 87. Also see Section 3.7.4.1.2, discussing REDD as an example of a mitigation strategy in terms of the international legal climate change regime.

states of a shared natural resource and a common concern of humankind, which is in this instance primarily the atmosphere. It is therefore possible to submit that this state right in terms of PSNR is affected by a global environmental concern, namely climate change, and the state's disposal of natural resources should correspond with this contemporary environmental challenge.[901] In this regard, state duties in terms of PSNR are set to gain importance. These duties are discussed in the Section 4.2.3.

4 2 2 6 Rights pertaining to foreign investment

4 2 2 6 1 Introduction to and delineation of rights (and duties) pertaining to foreign investment

Certain state rights in terms of PSNR concern rights in the specialised field of foreign investment. These rights pertain to rights to regulate and exercise authority over foreign investment; rights in regard to the taking of property (including expropriation and nationalisation of foreign investment); and possible rights in regard to the settlement of disputes in this context. In order to complete the examination of PSNR, the aspects of PSNR – or, more specifically, rights of states in terms of PSNR that contribute to characterising the principle as an evolutionary concept capable of being affected by climate change – have been focused on above. A discussion of the rights not contributing to the aforementioned objective will not be discussed in great detail, but will be included in this chapter in order to complete a general understanding of PSNR, as commenced in Chapter 2. These rights include the rights pertaining to foreign investment. The duties of states in regard to foreign investment will also not be examined in the research project, and these duties are mentioned only in the current section as conditions to the right pertaining to the taking of foreign property. Therefore, Sections 4.2.2.1 to 4.2.2.5 consider the state rights in terms of PSNR that complete the exposition of PSNR commenced in Chapter 2. These sections also illustrate the effect that climate change has on the understanding of PSNR by indicating the effect that climate change has on the understanding of the specific state right. An impact that climate change has on PSNR, is to affect a shift in the general understanding and

901 A Mancilla "Shared Sovereignty over Migratory Natural Resources" (2016) 22 *Res Publica* 21 22, 33-34.

approach of legal sources to certain state rights and state duties in terms of PSNR, that indicate a tendency to attach increased importance to state duties in terms of PSNR. Another significant impact that climate change has on PSNR, is to warrant a reinterpretation and reconfiguration of people's rights in terms of PSNR. Considering these aforementioned impacts of climate change considered in the research project, it is not necessary to consider the rights of states in terms of PSNR in any greater detail than what is done in Section 4.2.2, and the state rights pertaining to foreign investment discussed in Section 4.2.2.6 do not need to be considered in any greater detail than what is done below.

4 2 2 6 2 Rights pertaining to the regulation and exercise of authority over foreign investment

Foreign investment can be defined as "the transfer of tangible or intangible assets from one country to another for the purpose of their use in that country to generate wealth under the total or partial control of the owner of the assets".[902] The right of states to regulate foreign investment in terms of PSNR extends to aspects such as the admission of foreign investment; the suspension of admission of foreign investment; and the exercise of authority over foreign investment (meaning the exercise of authority over the activities of foreign investors).[903] Article 12(c)(i) of the Final Act of the United Nations Conference on Trade and Employment 1948 (Havana Charter) pertains to the regulation of foreign investment and the exercise of the state's right to exercise authority over foreign investment, and pro-

---

902 M Sornarajah *The International Law on Foreign Investment* 3ed (2010) 8.
For another definition of the term 'foreign direct investment', see: UNCTAD "Definitions of FDI" (date unavailable) (page numbers unavailable) *UNCTAD* <http://unctad.org/en/Pages/DIAE/Definitions-of-FDI.aspx> (accessed 08-08-2016). For the definition of the term 'transnational companies' in this context, see: UNCTAD "Transnational corporations (TNC)" (date unavailable) (page numbers unavailable) *UNCTAD* <http://unctad.org/en/Pages/DIAE/Transnational-corporations-(TNC).aspx> (accessed 07-10-2015).

903 N Schrijver "Permanent sovereignty over natural resources versus the common heritage of mankind, complementary or contradictory principles of international economic law?" in P de Waart et al *International Law and Development* (1988) 87 90; Schrijver *Sovereignty Over Natural Resources, Balancing Rights and Duties* (1997) 278-285.

vides that states may institute safeguards to ensure that foreign investment will not infringe on national policies or the internal affairs of the state. In addition, article 12(c)(ii) of the Charter also provides that the state can determine the terms upon which it will allow future foreign investment. The right to exercise authority over foreign investment is also mentioned in the International Chamber of Commerce (ICC) Guidelines, as the ICC Guidelines state that the host state should do the following: "make known to prospective investors its rules, regulations, policies, relevant official bodies, and general conditions that it wishes to apply to incoming direct private investment in a timely and transparent manner.[904] Other prevalent principles of foreign investment for states and enterprises include *inter alia* the obligation of the foreign investor to comply with the rules, regulations and policies of the host state and the home state, and the obligation of the foreign investor to respect the sovereignty of the host state.[905] The rights of states pertaining to foreign investment are subject to international law, according to principle 5(4) of the Declaration on the Progressive Development of Principles of Public International Law relating to a New International Economic Order 1986 (Seoul Declaration),[906] and the suspension of admission of foreign investment is also particularly dependent on

---

904 ICC "2012 ICC Guidelines for International Investment" (04-05-2012) (7) *ICC* <http://www.iccwbo.org/Advocacy-Codes-and-Rules/Document-centre/2012/2012-ICC-Guidelines-for-International-Investment/> (accessed 02-10-2015).

905 Schrijver *Sovereignty Over Natural Resources, Balancing Rights and Duties* (1997) 279.
For more information in this regard, see the ICC Guidelines for International Investment of 1972, which was revised in 2012 and reaffirmed the fundamental investment principles for investors and governments [ICC "2012 ICC Guidelines for International Investment" (04-05-2012) (page numbers unavailable) *ICC* <http://www.iccwbo.org/Advocacy-Codes-and-Rules/Document-centre/2012/2012-ICC-Guidelines-for-International-Investment/> (accessed 02-10-2015)] and the OECD Guidelines for Multinational Enterprises, which was first adopted in 1976 and consequently revised, which also sets out recommendations for conduct in the context of foreign investment [OECD "OECD Guidelines for Multinational Enterprises" (date of publication unavailable) (page numbers unavailable) *OECD* <http://www.oecd.org/corporate/mne/oecdguidelinesformultinationalenterprises.htm> (accessed 02-10-2015)].

906 Principle 5(4) of the Seoul Declaration reads as follows: "States have a right to regulate, exercise authority, legislate and impose taxes in respect of natural resources enjoyed and economic activities exercised and wealth held in their own territories by foreign interests subject only to any applicable requirements of international law. Except as expressly agreed by treaty or contract or provided by

treaty provisions and contractual agreements, as illustrated in the Liamco award and the Aminoil award.[907]

Other instruments to be consulted while trying to determine the content of the rights of states pertaining to foreign investment in terms of PSNR include the UNGA resolutions 1803 (XVII),[908] 2158 (XXI)[909], 3281 (XXIX),[910] and 3201 (S-VI).[911] UNGA resolution 1803 (XVII) mentions a few issues pertinent in the context of foreign investment. It states that the "import of foreign capital" in the context of exploring, developing and disposing of natural resources must conform with "the rules and conditions which the peoples of the nations freely consider to be necessary or desirable with regard to the authorization, restriction or prohibition of such activities". Therefore this provision in the resolution is not only relevant to the admission of foreign investment but also to the regulation of and the authorisation of foreign investment. Furthermore, when such activity is authorised, the resolution provides that it is governed by the terms of authorisation, by national law and by international law. The stipulations of UNGA resolution 2158 (XXI) differ from resolution 1803 (XVII) as it stipulates that "the exploitation of natural resources in each country shall always be conducted in accordance with its national laws and regulations." Similarly to resolution 2158 (XXI), UNGA resolution 3281 (XXIX), in article 2 of CERDS, provides that the state has the right to "regulate and exercise authority over" foreign investment in its jurisdiction "in accordance with its laws and regulations" and "no state shall be compelled to provide preferential treatment to foreign investment." Article 2 of CERDS therefore supports the exercise of authority over and the regulation of foreign

---

domestic law, no State is required to give preferential treatment to any foreign investment.".

907 Arbitral Tribunal: Award in the dispute between Libyan American Oil Company (Liamco) and the Government of the Libyan Arab Republic 12 April 1977 *International Legal Materials* (1981) 20(1) 1-87 53; Arbitral Tribunal: Award in the Matter of an Arbitration between Kuwait and the American Independent Oil Company (Aminoil) 24 March 1982 *International Legal Materials* (1982) 21(5) p 976-1053 1021-1022.
In regard to the admission of foreign investment, article 12(c)(ii) of the Havana Charter, for example, also provides that the state can determine the possibility and extent as well as the terms on which it will allow foreign investment.

908 UNGA resolution 1803 (XVII) of 14 December 1962.

909 UNGA resolution 2158 (XXI) of 25 November 1966.

910 UNGA resolution 3281 (XXIX) of 12 December 1974, containing CERDS.

911 UNGA resolution 3201 (S-VI) of 1 May 1974.

investment. UNGA resolution 3201 (S-VI) stipulates that NIEO be founded on respect for many principles, including "regulation and supervision of the activities of transnational corporations by taking measures in the interest of the national economies of the countries where such transnational corporations operate on the basis of the full sovereignty of those countries".

Article 2(3) of ICESCR provides that "[d]eveloping countries, with due regard to human rights and their national economy, may determine to what extent they would guarantee the economic rights recognized in the present Covenant to non- nationals" and Schrijver contends that this provision may give states the right to treat their own nationals differently than foreign investors.[912] There are some questions surrounding the treatment of foreign nationals in the context of foreign investment.[913] Most international instruments that deal with foreign investment and the accompanying treatment of foreign nationals stress that conduct in this regard should conform to international law. For instance, article 10(1) of the Energy Charter Treaty states that "in no case shall such [i]nvestments be accorded treatment less favourable than required by international law, including treaty obligations." Article 18(1) of the Energy Charter Treaty reaffirms recognition of state sovereignty over energy resources as well as the accompanying state rights over natural resources, and stipulates that these rights must be exercised "in accordance with subject to the rules of international law". In article 258(a) of the Lomé IV Convention 1995 (Lomé Convention),[914]

912 Schrijver *Sovereignty Over Natural Resources, Balancing Rights and Duties* (1997) 280.

913 Schrijver *Sovereignty Over Natural Resources, Balancing Rights and Duties* (1997) 280.
Note in this context the relatively newly emerged topic of 'Business and Human Rights'. In this regard, see the United Nations Guiding Principles on Business and Human Rights at: Business & Human Rights Resources Centre "UN Guiding Principles" (2011) (page numbers unavailable) *Business & Human Rights Resources Centre* <https://business-humanrights.org/en/un-guiding-principles> (accessed 11-08-2016). There is also a treaty being discussed in the context of business and human rights. [Business & Human Rights Resources Centre "Binding treaty" (2011) (page numbers unavailable) *Business & Human Rights Resources Centre* < https://business-humanrights.org/en/binding-treaty (accessed 11-08-2016).].

914 The Lomé Convention is an international agreement between the EU (the European Community at the time of formulation) and the ACP that aims to serve the development of the ACP. Four such agreements have been formulated: Lomé I

it is stipulated that foreign investment shall be promoted in accordance with state laws and regulations that are appropriate – which indicates that states will not be able to regulate foreign investment in a way that is inconsistent with international law.

4 2 2 6 3 Rights pertaining to the taking of property and the payment of compensation

The lawful 'taking of property' in the context of foreign investment existed even before the adoption of resolutions relating to PSNR by the UNGA.[915] However, today, this right to take foreign property is recognised as a right of PSNR.[916] The term 'taking of property' is a general and inclusive term and is accordingly employed in the heading of this section. Frequently employed and interchangeably used terms in a contemporary context are 'nationalisation' and 'expropriation'. Nationalisation can be defined as a taking by the state of foreign property, accompanied by a transfer of the corresponding economic activity, in an effort to halt economic domination of a section of the economy or of the economy and/or to institute social reform of the public sector.[917] Nationalisation should be understood as "across-the-board takings" intent on restricting economic domi-

---

was signed in 1975; Lomé II was signed in 1979; Lomé III was signed in 1984; and Lomé IV which spans for 10 years form 1990 to 1995 was also revised in 1995. [EC "Development and Relations with African, Caribbean and Pacific States" (date of publication unavailable) (page numbers unavailable) *EC* <http://ec.europa.eu/development/body/cotonou/lome_history_en.htm> (accessed 07-10-2015).].

915 Schrijver *Sovereignty Over Natural Resources, Balancing Rights and Duties* (1997) 285.

916 In the Amoco award, the Tribunal found that the right of states to nationalise is a "fundamental attribute of state sovereignty" and is "unanimously accepted, even by States which reject the principle of permanent sovereignty over natural resources, considered by a majority of States as the legal foundation of such a right". [Iran-United States Claims Tribunal: Partial Award in Amoco International Finance Corporation (Amoco) v Islamic Republic of Iran 14 July 1987 *International Legal Materials* (1988) 27(5) 1314-1405 1344,1358.].

917 M Sornarajah *The International Law on Foreign Investment* 3ed (2010) 365; Schrijver *Sovereignty Over Natural Resources, Balancing Rights and Duties* (1997) 285-286.

nation of a sector of the economy or of the economy.[918] Nationalisation in the context of foreign investment typically occurs today, paired with political unrest within a state.[919] Expropriation, on the other hand, means that the state targets a specific business in general terms, and this represents the more typical form of unilateral state interference today in comparison to nationalisation.[920]

The right pertaining to the taking of property is referred to in treaties, case law, and UNGA resolutions. Article 13 of the Energy Charter Treaty refers to the state right pertaining to the taking of property. Article 12(1)(c)(iii) of the Havana Charter does not explicitly mention this right, but is interpreted by Schrijver to bestow the right to nationalisation on states.[921] The Texaco award,[922] Liamco award,[923] the Aminoil award,[924] and the

918 M Sornarajah *The International Law on Foreign Investment* 3ed (2010) 365-366.

919 M Sornarajah *The International Law on Foreign Investment* 3ed (2010) 366-367.

920 Schrijver *Sovereignty Over Natural Resources, Balancing Rights and Duties* (1997) 285-286; M Sornarajah *The International Law on Foreign Investment* 3ed (2010) 366-367.

921 Schrijver *Sovereignty Over Natural Resources, Balancing Rights and Duties* (1997) 287.
Article 12(1)(c)(iii) of the Havana Charter reads as follows: "The Members recognize that: […] without prejudice to existing international agreements to which Members are parties, a Member has the right: (i) to take any appropriate safeguards necessary to ensure that foreign investment is not used as a basis for interference in its internal affairs or national policies".

922 In the Texaco award, the Tribunal contends "the right of a State to nationalize is unquestionable today. […] The exercise of the national sovereignty to nationalize is regarded as the expression of the State's territorial sovereignty." [Arbitral Tribunal: Award on the merits in dispute between Texaco Overseas Petroleum Company/California Asiatic Oil Company and the Government of the Libyan Arab Republic 19 January 1977 *International Legal Materials* (1978) 17(1) 1-37 21.].

923 In the Liamco award the Tribunal provides that "[t]he right of a State to nationalize its wealth and natural resources is sovereign, subject to the obligation of indemnification for premature termination of concession agreements." [Arbitral Tribunal: Award in the Dispute between Libyan American Oil Company (Liamco) and the Government of the Libyan Arab Republic relating to Petroleum Concessions 12 April 1977 *International Legal Materials* (1981) 20(1) 1-87 120.].

924 In the Aminoil award the Tribunal found that "a nationalization of Aminoil was in itself lawful." [Arbitral Tribunal: Award in the Matter of an Arbitration between Kuwait and the American Independent Oil Company (Aminoil) 24 March 1982 *International Legal Materials* (1982) 21(5) 976-1053 1025.].

Amoco award[925] all recognise the right of a state to nationalise. UNGA resolutions, such as resolutions 1803 (XVII),[926] 3171 (XXVIII),[927] 3201 (S-VI),[928] and 3281 (XXIX)[929] also recognise this state right.

Although the right to take property was described as an inalienable or an unconditional right in some UNGA resolutions,[930] the right is under-

---

925 In the Amoco award, the Tribunal found that the right of states to nationalise is a "fundamental attribute of state sovereignty", "an important tool of economic policy by many countries" and "cannot easily be considered as surrendered". [Iran-United States Claims Tribunal: Partial award in Amoco International Finance Corporation v Islamic Republic of Iran 14 July 1987 *International Legal Materials* (1988) 27(5) 1314-1405 1358.].

926 UNGA resolution 1803 (XVII) of 14 December 1962 contains a provision which refers to "nationalization, expropriation or requisitioning", and specifically the grounds of nationalisation, the compensation in cases of nationalisation and the issue of jurisdiction in cases of nationalisation. UNGA resolution 1803 (XVII) of 14 December 1962 reads as follows: "Nationalization, expropriation or requisitioning shall be based on grounds or reasons of public utility, security or the national interest which are recognized as overriding purely individual or private interests, both domestic and foreign. In such cases the owner shall be paid appropriate compensation, in accordance with the rules in force in the State taking such measures in the exercise of its sovereignty and in accordance with international law.".

927 UNGA resolution 3171 (XXVIII) of 17 December 1973 affirms the right of nationalisation and refers to issues of compensation and jurisdiction in this context. The resolution provides that: "each State is entitled to determine the amount of possible compensation and the mode of payment, and that any disputes which might arise should be settled in accordance with national legislation of each State carrying out such measures".

928 Article 4(e) of the NIEO Declaration, contained in UNGA resolution 3201 (S-VI) of 1 May 1974, expressly describes the right to nationalise as an expression of sovereignty over natural resources and an inalienable right.

929 Article 2(c) of CERDS, contained in UNGA resolution 3281 (XXIX) of 12 December 1974, refers to nationalisation, expropriation or the transfer of foreign property, and refers to issues of compensation and jurisdiction. Article 2(c) of CERDS reads as follows: "To nationalize, expropriate or transfer ownership of foreign property, in which case appropriate compensation should be paid by the State adopting such measures, taking into account its relevant laws and regulations and all circumstances that the State considers pertinent. In any case where the question of compensation gives rise to a controversy, it shall be settled under the domestic law of the nationalizing State and by its tribunals, unless it is freely and mutually agreed by all States concerned that other peaceful means be sought on the basis of the sovereign equality of States and in accordance with the principle of free choice of means.".

930 See for example, UNGA resolution 3201 (S-VI) of 1 May 1974.

stood today as subject to a number of conditions. These conditions pertain mostly to issues of public purpose,[931] public interest,[932] or public utility;[933] due process;[934] the safeguarding of natural resources;[935] non-dis-

---

931 According to the ICC, the host state should only expropriate for a public purpose. [ICC "2012 ICC Guidelines for International Investment" (04-05-2012) (8) *ICC* <http://www.iccwbo.org/Advocacy-Codes-and-Rules/Document-centre/2012/2012-ICC-Guidelines-for-International-Investment/> (accessed 02-10-2015).]
The Texaco award provides that "under international law a State has the right to take property within its jurisdiction for public purposes". [Arbitral Tribunal: Award on the Merits in Dispute between Texaco Overseas Petroleum Company/ California Asiatic Oil Company (Texaco) and the Government of the Libyan Arab Republic 19 January 1977 *International Legal Materials* (1978) 17(1) 1-37 22.] Conversely, the Tribunal in the Liamco award found that "the public utility principle is not a necessary requisite for the legality of a nationalisation". [Arbitral Tribunal: Award in the dispute between Libyan American Oil Company (Liamco) and the Government of the Libyan Arab Republic relating to Petroleum Concessions 12 April 1977 *International Legal Materials* (1981) 20(1) 1-87 58.].

932 Article 3 of the Draft OECD Convention on the Protection of Foreign Property reads as follows: "No Party shall take any measures depriving, directly or indirectly, of his property a national of another Party unless […] [t]he measures are taken in the public interest and under due process of law".
UNGA resolution 1803 (XVII) of 14 December 1962 provides that "[n]ationalization, expropriation or requisitioning shall be based on […] national interest".
Article 13(1)(a) of the Energy Charter Treaty provides that nationalisation be "for a purpose which is in the public interest".

933 UNGA resolution 1803 (XVII) of 14 December 1962 reads as follows: "Nationalization, expropriation or requisitioning shall be based on grounds or reasons of public utility".

934 Article 3 of the Draft OECD Convention on the Protection of Foreign Property 1967 provides *inter alia* that: "No Party shall take any measures depriving, directly or indirectly, of his property a national of another Party unless the […] measures are taken […] under due process of law".
Article 13(1)(c) of the Energy Charter Treaty provides that nationalization be "carried out under due process of law".

935 UNGA resolution 3171 (XXVIII) of 17 December 1973 and resolution 3201 (S-VI) of 1 May 1974 affirms that nationalisation may be carried out by states "in order to safeguard their natural resources".

crimination,[936] payment of compensation[937] (mainly described as "prompt, adequate, effective" or appropriate);[938] and also, albeit less frequently, the

936 Article 3 of the Draft OECD Convention on the Protection of Foreign Property reads as follows: "No Party shall take any measures depriving, directly or indirectly, of his property a national of another Party unless the [...] measures are not discriminatory or contrary to any undertaking which the former Party may have given".
The Liamco Tribunal submits "it is clear and undisputed that non-discrimination is a requisite for the validity of a lawful nationalisation". [Arbitral Tribunal: Award in the Dispute Between Libyan American Oil Company (Liamco) and the Government of the Libyan Arab Republic relating to Petroleum Concessions 12 April 1977 *International Legal Materials* (1981) 20(1) 1-87 58-59.]
Article 13(1)(b) of the Energy Charter Treaty provides that nationalisation be "not discriminatory".

937 Article 3 of the Draft OECD Convention on the Protection of Foreign Property reads as follows: "No Party shall take any measures depriving, directly or indirectly, of his property a national of another Party unless the [...] measures are accompanied by provision for the payment of just compensation. Such compensation shall represent the genuine value of the property affected, shall be paid without undue delay, and shall be transferable to the extent necessary to make it effective for the national entitled thereto."
Article 13(1)(d) of the Energy Charter Treaty also provides that expropriation be "accompanied by the payment of prompt, adequate and effective compensation."
It should be noted that this condition is not considered absolute, especially when considering the inclusion of the term 'possible' in UNGA resolution 3171 (XXVIII) of 17 December 1973, which provides that "each State is entitled to determine the amount of possible compensation and the mode of payment, and that any disputes which might arise should be settled in accordance with the national legislation of each State carrying out such measures." By submitting that the condition is not absolute, there is still support for the contention that the nationalising state may have to pay partial compensation (as opposed to full compensation). [Schrijver *Sovereignty Over Natural Resources, Balancing Rights and Duties* (1997) 292-294, 297.].

938 The Hull Rule required that a state pay "prompt, adequate and effective" compensation when a state exercised the right to the taking of property, but the Hull Rule is no longer a rule of customary international law (if it ever was a rule of customary international law). [AT Guzman "Why LDCs Sign Treaties That Hurt Them: Explaining the Popularity of Bilateral Investment Treaties" (1997) 38 Virginia Journal of International Law 639 641; M Bungenberg "Evolution of Investment Law Protection as Part of a General System of National Resources Sovereignty (and Management)?" in M Bungenberg & S Hobe (eds) Permanent Sovereignty over Natural Resources (2015) 125 127.].

right to be implemented as a measure of last resort.[939] In addition to the aforementioned conditions, principle 5(5) of the Seoul Declaration subjects the right to international law, applicable treaty provisions, and relevant contractual stipulations.[940] It can be inferred from these conditions that states are clearly not entirely at liberty to determine the reason and condition for the taking of property, and the approach of courts causes an extent of uncertainty in this regard.[941] These conditions can also be considered as duties or obligations of states in terms of PSNR when exercising the right to take foreign property.[942]

The question of compensation should not be linked to the lawfulness of the taking of property in any contemporary consideration concerning the right of taking of property.[943] In regard to the amount of compensation that may become payable in this context, it should be noted that there is no significant evidence that a state implementing nationalisation is allowed not to pay compensation, or is unilaterally able to determine the amount of compensation (whether it be partial or full compensation), or the mode of payment.[944] Contemporary compensation is leaning towards "partial and

---

939 According to the ICC, the host state "[s]hould have recourse to expropriation of property of a foreign investor only for a public purpose in accordance with due process and as a last resort." [ICC "2012 ICC Guidelines for International Investment" (04-05-2012) (8) *ICC* <http://www.iccwbo.org/Advocacy-Codes-and-Rules/Document-centre/2012/2012-ICC-Guidelines-for-International-Investment/> (accessed 02-10-2015).].

940 Principle 5(5) of the Seoul Declaration reads as follows: "A State may nationalize, expropriate, exercise eminent domain over or otherwise transfer property, or rights in property, within its territory and jurisdiction subject to the principle of international law requiring a public purpose and non-discrimination; to appropriate compensation as required by international law, and to any applicable treaty, and without prejudice to legal effects flowing from any contractual undertaking.".

941 Schrijver *Sovereignty Over Natural Resources, Balancing Rights and Duties* (1997) 292.

942 Since these duties do not contribute to achieving the objectives of the research project, these duties will not be discussed separately in the following section, which discusses state duties in terms of PSNR. It should suffice to note that these conditions to the right can be construed as state obligations. [Schrijver *Sovereignty Over Natural Resources, Balancing Rights and Duties* (1997) 344-364.].

943 M Sornarajah *The International Law on Foreign Investment* 3ed (2010) 364.

944 Schrijver *Sovereignty Over Natural Resources, Balancing Rights and Duties* (1997) 293-294, 297.

negotiated compensation arrangements",[945] which differ from case to case and can depend on circumstances such as the legitimate expectations of the foreign investor and the reasonableness of the rate of return.[946] Profit foregone can also affect the amount of compensation.[947] It has also been submitted that compensation should take into account the many specific economic aspects of the state exercising the taking of property, such as for example a possible socioeconomic necessity for the taking.[948] State practice in this context has shown that the amount of compensation in this context is ultimately a trade-off between the amount of compensation and other received benefits such as credit facilities, service contracts, management fees and tax concessions.[949] It falls outside the ambit of the research project to discuss compensation in any greater detail.

4 2 2 6 4 Possible right pertaining to the settlement of disputes on the basis of national law

In order to complete the examination of PSNR, it should be noted that it is debatable whether PSNR gives rise to the right of states to settle disputes on the basis of national law. While UNGA resolutions can be said to substantiate the existence of the right to some extent, treaty law proves less

945 Schrijver *Sovereignty Over Natural Resources, Balancing Rights and Duties* (1997) 295-296.

946 Arbitral Tribunal: Award in the Matter of an Arbitration between Kuwait and the American Independent Oil Company (Aminoil) 24 March 1982 *International Legal Materials* (1982) 21(5) 976-1053 1037.

947 In the Amoco award and the AAPL award, it was determined that compensation would not include profit foregone. [Iran-United States Claims Tribunal: Partial Award in Amoco International Finance Corporation (Amoco) v Islamic Republic of Iran 14 July 1987 *International Legal Materials* (1988) 27(5) 1314-1405 1366; ICSID Arbitral Tribunal: Final Award in Asian Agricultural Products Ltd. (AAPL) v. Republic of Sri Lanka 21 June 1990 *International Legal Materials* (1991) 30(3) 624-625.].

948 Schrijver *Sovereignty Over Natural Resources, Balancing Rights and Duties* (1997) 297.

949 Schrijver *Sovereignty Over Natural Resources, Balancing Rights and Duties* (1997) 296.

useful, and some arbitral awards delivered in this context confuse the matter.[950]

UNGA resolutions 1803 (XVII)[951] and 3171 (XXVIII)[952] affirm that in the case where questions of compensation or mode of payment lead to a dispute, the measures within the national jurisdiction of the state implementing the measures shall be exhausted. UNGA resolution 1803 (XVII)[953] provides that, where there is an agreement between states or parties concerned, only then shall the dispute be settled through arbitration or international adjudication; or, according to resolution 3281 (XXIX),[954] "unless it is freely and mutually agreed by all states concerned that other peaceful means be sought on the basis of the sovereign equality of states and in accordance with the principle of free choice of means", the dispute must be settled in accordance with national law. Resolution 31/98[955] adopted the 1976 United Nations Commission on International Trade Law (UNCITRAL) Arbitration Rules, which stipulate the following in article 33(1):

> "[t]he arbitral tribunal shall apply the law designated by the parties as applicable to the substance of the dispute (meaning that parties may have designated the applicable law as national law). Failing such designation by the parties, the arbitral tribunal shall apply the law determined by the conflict of laws rules which it considers applicable."[956]

---

950 Note in this context the Calvo Clause, encompassing the claim that "investment regulation in general and the taking of foreign property in particular are matters of domestic jurisdiction." In other words, the Calvo clause stipulates *inter alia* that state sovereignty entails foreigners and their property are subject to the law and jurisdiction of the state in which they invest. [P Peters & N Schrijver "Latin America and International Regulation of foreign Investment: Changing Perceptions" (1991) *Working Paper Series No. 113* 2; Schrijver *Sovereignty Over Natural Resources, Balancing Rights and Duties* (1997) 178.].

951 UNGA resolution 1803 (XVII) of 14 December 1962.

952 UNGA resolution 3171 (XXVIII) of 17 December 1973.

953 UNGA resolution 1803 (XVII) of 14 December 1962.

954 UNGA resolution 3281 (XXIX) of 12 December 1974, containing CERDS, of which article 2 is relevant.

955 UNGA resolution 31/98 of 15 December 1976 entitled "Arbitration Rules of the United Nations Commission on International Trade Law".

956 This approach is similar to the approach of the ICSID, which stipulates the following in article 42(1) of the ICSID Convention: "The Tribunal shall decide a dispute in accordance with such rules of law as may be agreed by the parties. In the absence of such agreement, the Tribunal shall apply the law of the Contract-

Treaty law appears more vague in designating the right to settle disputes on the basis of national law. For instance, article 26(6) of the Energy Charter Treaty does not mention national law and provides that the issue in dispute be decided "in accordance with [the] Treaty and applicable rules and principles of international law".

The differing approaches to the settlement of disputes on the basis of national law is illustrated by arbitral awards, such as the Texaco award in which it was decided to apply national and international law,[957] and the Aminoil award in which it was decided to apply mainly international law.[958] In the AGIP Company award the tribunal applied, in accordance with the agreement between the parties, the national law of the Congo and

ing State party to the dispute (including its rules on the conflict of laws) and such rules of international law as may be applicable.".

957 The Tribunal decides to apply "the principles of Libyan law and [...] the principles of international law". [Arbitral Tribunal: Award on the merits in dispute between Texaco Overseas Petroleum Company/California Asiatic Oil Company (Texaco) and the Government of the Libyan Arab Republic 19 January 1977 *International Legal Materials* (1978) 17(1) 1-37 36.] The applicable law clause reads as follows: "This concession shall be governed by and interpreted in accordance with the principles of the law of Libya common to the principles of international law and in the absence of such common principles, then by and in accordance with the general principles of law, including such of those principles as may have been applied by international tribunals." [Arbitral Tribunal: Award on the merits in dispute between Texaco Overseas Petroleum Company/California Asiatic Oil Company (Texaco) and the Government of the Libyan Arab Republic 19 January 1977 *International Legal Materials* (1978) 17(1) 1-37 11.].

958 The Tribunal decides to apply "principally international law which is also an integral part of the law of Kuwait" which means that in effect national law of Kuwait was also applied. [Arbitral Tribunal: Award in the Matter of an Arbitration between Kuwait and the American Independent Oil Company (Aminoil) 24 March 1982 *International Legal Materials* (1982) 21(5) 976-1053 1032.] The applicable law clause reads as follows: "[T]he agreements between them shall be given effect, and must be interpreted and applied, in conformity with principles common to the laws of Kuwait and of the State of New York, USA, and in the absence of such common principles, then in conformity with the principles of law normally recognized by civilized states in general, including those which have been applied by international tribunals." [Arbitral Tribunal: Award in the Matter of an Arbitration between Kuwait and the American Independent Oil Company (Aminoil) 24 March 1982 *International Legal Materials* (1982) 21(5) 976-1053 1000.].

applied the principles of international law in a supplementary manner.[959] In the Asian Agricultural Products Ltd. (AAPL) award the parties did not pre-select the applicable law and it was less clear that national law should be applied. The tribunal decided to apply the Sri Lanka/United Kingdom (UK) Bilateral Investment Treaty (BIT) as the main source of law, and also considered rules and principles of international law not specifically included in the treaty.[960] A dissenting opinion in the tribunal finds that national law, namely Sri Lankan law, should be applied as the main source of law.[961]

It falls outside the scope of the research project to find the definitive answer to the undoubted existence of the right to settle disputes on the basis of international law as a right in terms of PSNR. The author submits that there is no uniform right to settle disputes that arise in the context of the taking of foreign property on the basis of national law based on the arbitral awards included above. It is however generally recognised that parties are able to designate the applicable law in case of a dispute, and parties may change this choice of applicable law by mutual consent.[962] Furthermore, it

959 The Tribunal decided to apply, in accordance with article 15 of the agreement between the parties, "Congolese law supplemented by the principles of international law". [ICSID Arbitral Tribunal: Award in the case of AGIP Company (AGIP) v Popular Republic of the Congo 30 November 1979 *International Legal Materials* (1982) 21(4) 726-739 735.].

960 The majority opinion decided that that the Sri-Lanka/UK BIT, customary international law, other international rules and Sri-Lankan law is applicable, and proceeds on the basis that the Sri Lanka/U.K treaty is "the primary source of applicable legal rules". [ICSID Arbitral Tribunal: Final Award in Asian Agricultural Products Ltd. (AAPL) v. Republic of Sri Lanka 21 June 1990 *International Legal Materials* (1991) 30(3) 577-655 577, 597, 630-632].

961 The dissenting opinion of Asante provided that article 42(1) of the ICSID Convention should prevail and the majority is mistaken in not applying Sri-Lankan law as the main source of law, together with applicable international law. [ICSID Arbitral Tribunal: Final Award in Asian Agricultural Products Ltd. (AAPL) v. Republic of Sri Lanka 21 June 1990 *International Legal Materials* (1991) 30(3) 577-655 577, 597, 630-632].

962 Arbitral Tribunal: Award on the Merits in Dispute between Texaco Overseas Petroleum Company/California Asiatic Oil Company (Texaco) and the Government of the Libyan Arab Republic 19 January 1977 *International Legal Materials* (1978) 17(1) 1-37 12, 24.].

is generally recognised that where national law is applicable but not able to solve the dispute, international law must be applied.[963]

In order to complete the examination of PSNR, it should further be noted that it is debatable whether PSNR gives rise in the context of foreign investment to the right of states to choose the means of settlement of a nationalisation dispute. In UNGA resolutions a "local-remedies rule" exists that stipulates that in such a case local remedies must be exhausted before parties can turn to international remedies.[964] General Assembly resolution 1803 (XVII)[965] contains the aforementioned rule and addresses this right in the context of compensation only by stipulating "the national jurisdiction of the State taking such measures shall be exhausted", not leaving room for a choice of the parties. Furthermore, the resolution provides for the possibility of "arbitration or international adjudication" if the parties involved agree to it and only after national measures have been exhausted. General Assembly resolution 3281 (XXIX)[966] also discusses this right in the context of compensation and stipulates in article 2 that the dispute "shall be settled under the domestic law of the nationalizing State and by its tribunals, unless it is freely and mutually agreed by all states concerned that other peaceful means be sought," leaving some room for a possible choice of the parties. General Assembly resolution 3171 (XXVIII)[967] stipulates that "each State is entitled to determine the amount of possible compensation and the mode of payment, and that any disputes which might arise should be settled in accordance with the national legislation of each State carrying out such measures", again not leaving room for the choice of the parties.

963 Schrijver *Sovereignty Over Natural Resources, Balancing Rights and Duties* (1997) 301.
964 Schrijver *Sovereignty Over Natural Resources, Balancing Rights and Duties* (1997) 302.
965 UNGA resolution 1803 (XVII) of 14 December 1962.
966 UNGA resolution 3281 (XXIX) of 12 December 1974.
967 UNGA resolution 3171 (XXVIII) of 17 December 1973.

Since a hearing in local courts can be unfairly prejudicial to a foreign investor, international investment arbitration disposes with the local-remedies rule.[968] In this context, article 26 of the Convention on the Settlement of Investment Disputes between States and Nationals of Other States 1965 (ICSID Convention) provides that this local-remedies rule is not required unless it is stated otherwise.[969] Article X of the ASEAN Agreement for the Promotion and Protection of Investments 1987 (Agreement for the Promotion and Protection of Investments) also does not stipulate that local remedies must be exhausted.[970] Similarly, the North American Free Trade Agreement 1991 (NAFTA), which entered into force in 1994, does not make reference to exhausting local remedies.[971] The Energy Charter Treaty deals with international dispute settlements in article 26(2)(a) and,

---

968 C Schreuer "Calvo's grandchildren: the return of local remedies in investment arbitration" (2005) 1 *The Law and Practice of International Courts and Tribunals* 1 1.

969 C Schreuer "Calvo's grandchildren: the return of local remedies in investment arbitration" (2005) 1 *The Law and Practice of International Courts and Tribunals* 1 1.
Article 26 of the ICSID Convention reads as follows: "Consent of the parties to arbitration under this Convention shall, unless otherwise stated, be deemed consent to such arbitration to the exclusion of any other remedy. A Contracting State may require the exhaustion of local administrative or judicial remedies as a condition of its consent to arbitration under this Convention.".

970 Article X(1)-(2) of the Agreement for the Promotion and Protection of Investments reads as follows: "(1) Any legal dispute arising [...] shall, as far as possible, be settled amicably between the parties to the dispute. (2) If such a dispute cannot thus be settled within six months of its being raised, then either party can elect to submit the dispute for conciliation or arbitration and such election shall be binding on the other party. The dispute may be brought before the [...] [ICSID], [...] [UNCITRAL], the Regional Centre for Arbitration at Kuala Lumpur or any other regional centre for arbitration in ASEAN, whichever body the parties to the dispute mutually agree to appoint for the purposes of Conducting the arbitration.".

971 Chapter 20 of NAFTA sets out the general dispute mechanism, which can be summarised as a three-step approach. Firstly, consultation occurs and the dispute is formally discussed. Secondly, if discussions fail to resolve the dispute, ministers will try to resolve the dispute at a Free Trade Commission meeting. Thirdly, if ministers fail to resolve the dispute, the complainant can request that a panel be established that will make a binding decision on the matter. [Naftanow.org "North American Free Trade Agreement" (09-08-2013) (page numbers unavailable) *Naftanow.org* < http://www.naftanow.org/dispute/default_en.asp> (accessed 26-10-2015).].

although local remedies are mentioned in case of disputes between a contracting party and an investor of another contracting party, the treaty states that the investor party "may choose to submit [the dispute] to the courts or administrative tribunals of the Contracting Party to the dispute", meaning that the investor party does not necessarily have to submit the dispute to these institutions. Regardless of the possible initial and continued involvement of the local court in the dispute, case law[972] and other international instruments[973] indicate that international tribunals do not require local

972 In the Lanco award it was stipulated that the valid consent of parties to international arbitration does not need to be preceded by the exhaustion of local remedies. [ICSID Arbitral Tribunal: Lanco International Inc. v Argentine Republic 8 December 1998 *International Legal Materials* (2001) 40(2) 457-473 469-470]
In the Generation Ukraine award, the tribunal relies on the Lanco award and finds that there is no obligation to exhaust any remedies in the Ukrainian courts before filing its notice of arbitration at the ICSID. [ICSID Arbitral Tribunal: Generation Ukraine Inc v Ukraine 16 September 2003 *International Legal Materials* (2005) 44 404-464 430-431.]
In the Yaung Chi Oo award, the tribunal found that the "non-exhaustion of local remedies cannot exclude the jurisdiction of the tribunal or render the claim inadmissible". [ASEAN Arbitral Tribunal: Yaung Chi Oo Trading Pte Ltd. V Government of the Union of Myanmar 31 March 2003 *International Legal Materials* (2003) 42(3) 540-559 548.].

973 Schrijver *Sovereignty Over Natural Resources, Balancing Rights and Duties* (1997) 303.
The Unified Agreement for the Investment of Arab Capital in the Arab States provides that the court designated in article 28 (the Arab Investment Court until the Arab Court of Justice has been established) will have jurisdiction. According to article 29, the court will have "jurisdiction to settle disputes brought before it by either party to an investment which relate to or arise from application of the provisions of the Agreement." Article 30 stipulates that "[w]here an international Arab agreement setting up an Arab investment or any agreement related to investment within the scope of the League of Arab States stipulates that a matter or dispute should be referred to international arbitration or to an international court, the parties involved may agree to regard it as being within the jurisdiction of the Court." Article 31 stipulates "[t]he Arab investor may have recourse to the courts in the State where the investment is made according to the rules of jurisdiction within such State in the case of matters which fall within the jurisdiction of the Court. However, where the Arab investor brings an action before one authority, he must refrain from so doing before the other." Article 32 provides that, should there be a conflict of jurisdiction between the court and national courts of state parties, the jurisdiction of the Arab Court of Justice will prevail.
Agreement on Promotion, Protection and Guarantee of Investments among Member States of the Organisation of the Islamic Conference gives the investor the

remedies to be exhausted before instituting international arbitration proceedings.[974]

There is support for the claim that the local remedies need not always be exhausted in international arbitration in the context of foreign investment disputes, or at least that states have a free choice of means of settlement of a dispute in the particular context of nationalisation and compensation.[975] However, in disputes between a host state and a foreign investor, it is also possible that parties retain the right to insist that local remedies be exhausted first – unless they tacitly dispense with this right by means of a treaty or otherwise,[976] and this is the cautious viewpoint of the author

---

right to circumvent the national law of the host state. Article 16 reads as follows: "The host state undertakes to allow the investor the right to resort to its national judicial system to complain against a measure adopted by its authorities against him, or to contest the extent of its conformity with the provisions of the regulations and laws in force in its territory, or to complain against the non-adoption by the host state of a certain measure which is in the interest of the investor, and which the state should have adopted, irrespective of whether the complaint is related, or otherwise, to the implementation of the provisions of the Agreement to the relationship between the investor and the host state. Provided that if the investor chooses to raise the complaint before the national courts or before an arbitral tribunal then having done so before one of the two quarters he loses the right of recourse to the other.".

974 C Schreuer "Calvo's grandchildren: the return of local remedies in investment arbitration" (2005) 1 *The Law and Practice of International Courts and Tribunals* 1 2.

975 Schrijver *Sovereignty Over Natural Resources, Balancing Rights and Duties* (1997) 305.

976 In the ELSI case the Chamber finds as follows: "the parties to a treaty can therein either agree that the local remedies rule shall not apply to claims based on alleged breaches of that treaty; or confirm that it shall apply. Yet the Chamber finds itself unable to accept that an important principle of customary international law should be held to have been tacitly dispensed with, in the absence of any words making clear an intention to do so". [ICJ Judgment: Case Concerning Elettronica Sicula S.P.A. (ELSI) (United States v Italy) 20 July 1989 *International Legal Materials* 28(5) (1989) 1109-1164 1125.

Principle 13(1) of the Seoul Declaration provides as follows: "Disputes on questions related to international economic relations have to be settled by peaceful means chosen by the parties concerned, in particular by recourse to international adjudication, international or transnational arbitration, or other international procedures for the settlement of disputes. The principle of local remedies shall be observed, where applicable.".

as well. Examining this point of international arbitration further falls outside the scope of the research project.

The abovementioned rights of states in terms of PSNR have been discussed in order to complete the understanding of the different components of the principle, an exercise which was commenced in Chapter 2. It was indicated how the gravity of climate change has an impact on the understanding of these state rights (except in the case of rights pertaining to foreign investment, as these rights are most negligibly affected by climate change). As anticipated, the majority of the state rights in terms of PSNR have been found largely unaffected by climate change. The exercise below, which examines the state duties in terms of PSNR, is expected to yield more noteworthy results when it is considered what impact of the gravity of climate change has on the contemporary understanding of the state duties in terms of PSNR.

4 2 3 Duties of states in terms of the principle of sovereignty over natural resources and the impact of climate change

4 2 3 1 Duty to exercise sovereignty over natural resources for national development and the well-being of people

Section 3.5.2 explained what is meant by the term 'well-being of people'. As submitted, the research project considers this term as referring to components which affect or have to do with physiology (pertaining to income, food and physical health) and psychology (pertaining to social and mental health) that combine to form the well-being of people. The research project is concerned with how the environmental effects of climate change (leading also to social and economic effects) impact the various stipulated components of physiology and psychology, and ultimately the whole, of the well-being of people. From this perspective the duty of states to exercise PSNR for national development and the well-being of people is considered.

This section of the research project considers the promotion of the well-being of people as a duty of states in exercising PSNR (paired with the state duty of promoting national development). The UNGA resolutions which mention this duty in the most certain terms include resolution 523

(VI),[977] 626 (VII),[978] 1803 (XVII),[979] and 2692 (XXV).[980] Resolution 523 (VI) submits that underdeveloped countries must "utilize such resources in order to be in a better position to further the realization of their plans of economic development in accordance with their national interests." In this case the term 'national interests' can be understood to refer also to the well-being of people. Resolution 626 (VII) simply recommends that member states "exploit their natural wealth and resources wherever deemed desirable by them for their own progress and economic development". In this aforementioned resolution, no direct mention is made of the second part of the duty, namely the well-being of people. The landmark UNGA resolution in regard to PSNR, namely resolution 1803 (XVII), explicitly states that PSNR "must be exercised in the interest of their national development and of the well-being of the people of the state concerned." Resolution 1803 (XVII) therefore links the state right to use natural resources for national development with the state duty to exercise PSNR for national development and the well-being of people. In this context states are trustees or custodians of a country's natural resources, administering and caring for its natural resources for the benefit of its people. Resolution 2692 (XXV) contains and reaffirms the exact phrase of resolution 1803 (XVII), as set out above. Later resolutions refer to this duty in much broader terms, and mostly only in regard to national development, in an effort to impose minimal restrictions on the sovereignty of the state.[981]

This state duty attempts to link the two fundamental pillars or objectives upon which PSNR is founded, namely self-determination of people and economic development of states, with disputable success. At first glance, the state duty to exercise PSNR for national development and the well-being of people seems to be contradictory, since activities that promote national development are more often than not in contradiction with environmentally sustainable development practice, and accordingly contradicts the well-being of people. This is well-illustrated by the unsustainable disposal of natural resources causing climate change, and the result-

977 UNGA resolution 523 (VI) of 12 January 1952.
978 UNGA resolution 626 (VII) of 21 December 1952.
979 UNGA resolution 1803 (XVII) of 14 December 1962.
980 UNGA resolution 2692 (XXV) of 11 December 1970.
981 Schrijver *Sovereignty Over Natural Resources, Balancing Rights and Duties* (1997) 309.

ing impact of climate change on the well-being of people, as discussed above and in Section 3.5.

This state duty has also more recently been encapsulated by treaty law as the state duties in terms of PSNR gain traction in international instruments and become more urgent in light of environmental concerns such as climate change. The state duty to exercise sovereignty over natural resources for national development and the well-being of people is encapsulated in treaty law, albeit indirectly. Common article 1(2) of the International Covenants on Human Rights affords people the right to dispose of natural resources (indirectly promoting the well-being of people).[982] Furthermore, article 25 of the International ICESCR and article 47 of the ICCPR further provide that nothing in these instruments may be interpreted as impairing this inherent right of people (further indirectly promoting the well-being of people).[983] Article 21(1) of the Banjul Charter also determines that the right of people to dispose of natural resources shall be exercised in their exclusive interest, indirectly promoting the well-being of people.[984] With similar effects, the African Convention on the Conservation of Nature and Natural Resources clearly sets out to take into account the social, cultural and economic needs of the people.[985] In this context the preamble of the Treaty for Amazonian Co-operation can also be noted as it states that it is a state responsibility, inherent in state sovereignty, to be

982 Common article 1(2) of International Covenants on Human Rights reads as follows: "All peoples may, for their own ends, freely dispose of their natural wealth and resources without prejudice to any obligations arising out of international economic co-operation, based upon the principle of mutual benefit, and international law. In no case may a people be deprived of its own means of subsistence.".

983 Article 25 of the ICESCR and article 47 of the ICCPR 1966 reads as follows: "Nothing in the present Covenant may be interpreted as impairing the inherent right of all peoples to enjoy and utilize fully and freely their natural wealth and resources.".

984 Article 21(1) of the Banjul Charter reads as follows: "All peoples shall freely dispose of their wealth and natural resources. This right shall be exercised in the exclusive interest of the people. In no case shall a people be deprived of it.".

985 For example, the preamble of the African Convention on the Conservation of Nature and Natural Resources reads as follows: "States are responsible for protecting and conserving their environment and natural resources and for using them in a sustainable manner with the aim to satisfy human needs according to the carrying capacity of the environment"; and article VIII(1)(a) also specifically refers to the "social and economic needs of the people concerned".

simultaneously conscious of socioeconomic development and environmental conservation. Similarly the preamble of the ASEAN Agreement on the Conservation of Nature and Natural Resources also shows that the Agreement is mindful of the interrelationship between socioeconomic development and environmental conservation. Finally, principle 2 of the Earth Charter 2000 (Earth Charter) states that "with the right to own, manage, and use natural resources comes the duty to prevent environmental harm and to protect the rights of people", also indicating the interplay between socioeconomic development and environmental conservation and the state duty towards national development for the well-being of people.[986]

Included in the term 'people' when describing the duty of states to exercise PSNR for national development and the well-being of people is all people or the entire population of a state, and accordingly also a specific group of people, such as indigenous people. Schrijver separated the state duty to respect the interests and rights of indigenous people from the state duty to exercise PSNR for national development and the well-being of people. This makes sense, since indigenous peoples today tend to receive distinguishable attention overall on an international scale. However, for the purposes of this research project, it is not necessary to view or discuss the duty to respect the rights and interests of indigenous people as separate from the duty of states to respect the rights and interests of their entire population, and, accordingly, the state duty to promote the well-being of its people. Within the context of the state duty to exercise PSNR for national development and the well-being of people (including the duty that states have in particular to indigenous people), it should be noted that the impacts of climate change will hit vulnerable communities within the state population, such as indigenous people, particularly hard.[987] Special regard is accorded to indigenous people as a group when discussing the rights emanating from PSNR for people (in Section 4.3.3.2) seeing that these rights are mostly applied in the context of indigenous people.

Section 3.5 discussed the component of well-being of people within the international legal climate change regime, since the well-being of people is affected by the environmental, social and economic impacts of climate change. As submitted in Chapter 3, the international legal climate change

986 Principle 2, Earth Charter.

987 R Tsosie "Indigenous People and Environmental Justice: The impact of climate change" (2007) 78 University of Colorado Law Review 1625 1625, 1628.

regime automatically addresses and promotes human well-being, by addressing the impacts of climate change even though the well-being of people is not a central or main concern of the regime. However, the well-being of people is a central and main concern of the state duty in terms of PSNR, that PSNR is to be exercised for national development and the well-being of people. If, the state duty in terms of PSNR to promote the well-being of people is exercised and the international legal climate change regime objective to mitigate the impacts of climate change is realised (effectively promoting the well-being of people), it can be submitted that the well-being of people is promoted by both PSNR and by the international legal climate change regime when functioning effectively. This means that increasing the effectiveness of the implementation of this state duty in terms of PSNR could increase the effectiveness of the international legal climate change regime and vice versa. It is clear to see from this interaction how the gravity of climate change can contribute to increasing the more effective application of the state duty in terms of PSNR to promote the well-being of people.

4 2 3 2 Duty to cooperate in respect of international development

The duty to cooperate in respect of international development is found in UNGA resolution 523(VI).[988] This resolution stipulates that underdeveloped countries must use their resources in a way that would not only develop their own economy, but would also "further the expansion of the world economy". If interpreted widely, the right to cooperate in respect of international development is also included in UNGA resolutions 837 (IX),[989] 1314 (XIII)[990] and 1514 (XV)[991] as they refer to the duty of "international co-operation". Resolution 837 (IX) stresses "the importance of encouraging international co-operation in the economic development of under-developed countries". Resolution 1314 (XIII) states that "due regard shall be paid to the rights and duties of States under international law and to the importance of encouraging international co-operation in the economic development of under-developed countries". Certain references in

988 UNGA resolution 523 (VI) of 12 January 1952.
989 UNGA resolution 837 (IX) of 14 December 1954.
990 UNGA resolution 1314 (XIII) of 12 December 1958.
991 UNGA resolution 1514 (XV) of 14 December 1960.

CERDS, contained in UNGA resolution 3281(XXIX), have been interpreted as supporting the existence of the state duty to cooperate in respect of international development.[992]

Treaty law contains fewer references to the state duty to cooperate in respect of international development than the UNGA resolutions, and makes reference to the duty in more general terms than the UNGA resolutions. Treaty law frames the references to the duty as objectives of states rather than as duties of states.[993] For instance, article 74 of the United Nations Charter reads that state policy "must be based on the general principle of good-neighbourliness, due account being taken of the interests and well-being of the rest of the world, in social, economic and commercial matters". Article 1(1) of the Havana Charter stipulates that one of its objectives is to contribute to expanding the world economy. The Law of the Sea Convention recognises the following in its preamble, which when widely interpreted can point towards the state duty to cooperate in respect of international development:

> "with due regard for the sovereignty of all States, [it is desirable to establish] a legal order for the seas and oceans which will facilitate international communication, and will promote the peaceful uses of the seas and oceans, [and] the equitable and efficient utilization of their resources".

In its preamble the Energy Charter Protocol on Energy Efficiency and related Environmental Aspects of the Energy Charter Treaty 1994 (Energy Charter Protocol) mentions the need for "intergovernmental co-operation" in the context of energy policy and expresses the desire to undertake "co-operative and coordinated action in the field of energy efficiency and related environmental protection".

These inclusions in treaty law, as submitted, point to general state objectives rather than state duties. One exception seems to be the Seoul Declaration, which, although including only a general reference to state duties

---

992 UNGA resolution 3281 (XXIX) of 12 December 1974, containing CERDS, contains for example, article 4 which provides that states pursue "international trade and other forms of economic co-operation [...] consistent with its international obligations and with the needs of international economic co-operation"; article 5 which provides that states are *inter alia* "to assist in the promotion of sustained growth of the world economy"; and article 13 which provides that all states should "promote international scientific and technological co-operation".

993 Schrijver *Sovereignty Over Natural Resources, Balancing Rights and Duties* (1997) 320.

to cooperate in section 4, at least frames this reference as a duty instead of an obligation or aim.[994] Principle 4(1) of the Seoul Declaration provides as follows:

> "[the] duty to co-operate in economic relations implies the progressive development of this duty in proportion to the growing economic interdependence between States [which] should lead therefore in particular to a reinforced co-operation in the field of […] international protection of the natural environment".

It can be submitted that the importance of the duty to cooperate in respect of international development has continued to grow since the publication of the Seoul Declaration, as the need for reinforced cooperation in the field of international protection of the natural environment has grown in light of contemporary global concerns such as climate change.[995] This supports the notion that climate change has an impact on the interpretation and configuration of state duties in terms of PSNR, in this case particularly on the state duty of cooperation in respect of international development.

The largely politically formulated duty of states in terms of PSNR to cooperate in respect of international development may at first glance seem contrary to the very core of PSNR, since PSNR was in part developed as a response by people for self-determination and a response by developing countries to establish their independence in the context of economic affairs relating to natural resources. It is possible to question this state duty of PSNR, a question that stems from the discussion in Section 2.5.2.2 regarding permanency as an attribute of state sovereignty in PSNR. In examining this state duty, it is possible to argue that sovereignty of a state cannot be limited by the need for development in other states (or throughout the entire world) and, as such, there cannot be a duty of states to cooperate for international development. However, as the discussion in Section 2.5.2.2 shows, there are numerous interpretations of permanency as an attribute of state sovereignty. In this regard, the view of the author is that state rights in terms of PSNR are inalienable, but that a state may enter in-

994 Section 4 of the Seoul Declaration is entitled "The duty to co-operate for global development".

995 L Engberg-Pedersen "Climate Change negotiations and their implications for international development cooperation" (2011) (1-50) *Danish Institute for International Studies* <https://www.ciaonet.org/attachments/18066/uploads> (accessed 10-09-2016).

to arrangements that limit these inalienable rights on the condition that these arrangements are entered into freely and in the best interest and for the well-being of its people. In this way, it is possible that the state duty to cooperate in respect of international development can exist as a state duty in terms of PSNR, but only if states freely take on this duty and exercise it in the best interest of their people.

This duty to cooperate in respect of international development is understood as "taking into account of the interests of others" and practically involves procedures such as sharing information, consultation, negotiation and establishing joint procedures and institutions for these purposes.[996] As climate change will have disastrous consequences for national and consequently international development, the negative incentive and international political pressure surrounding climate change is considered able to convince states to freely take on the state duty in terms of PSNR. In this way climate change has an impact on this state duty, as it can increase the effective application of this duty by states.

4 2 3 3 Duty of conservation and sustainable use of natural resources

The limited natural resource reserves of developing states became a concern during the 1960s, and in response to this concern, the UNGA resolution 1831 (XVII)[997] was adopted.[998] The resolution addresses the concern that the natural resources of developing states in particular were jeopardised by the economic development of developing countries. Whether or not this resolution was a last attempt, springing from the vestiges of colonialism, to exert control over the natural resources of developing countries constitutes an argument falling outside the scope of the research project. In the Stockholm Declaration, the stipulation that the natural resources of the Earth "must be safeguarded for the benefit of present and future generations" is contained in more inclusive terms in principle 2. Chapter III of

996 FX Perrez *Cooperative Sovereignty, From Independence to Interdependence in the Structure of International Environmental Law* (2000) 260.

997 UNGA resolution 1831 (XVII) of 18 December 1962 entitled "Economic development and the conservation of nature".

998 Schrijver *Sovereignty Over Natural Resources, Balancing Rights and Duties* (1997) 324.

CERDS, contained in UNGA resolution 3281 (XXIX),[999] provides in article 30 that the "protection, preservation and enhancement of the environment for the present and future generations is the responsibility of all States". This resolution arguably enhances the duty to conserve the environment to a state responsibility to protect and preserve the environment. Later UNGA resolutions, such as resolution 35/7[1000] and resolution 37/7,[1001] articulate the duty of states to conserve the environment in the context of PSNR in strong terms by including the common provision whereby member states are invited to do the following:

> "in the exercise of their permanent sovereignty over their natural resources, [states are] to conduct their activities in recognition of the supreme importance of protecting natural systems, maintaining the balance and quality of nature and conserving natural resources, in the interests of present and future generations".

UNGA resolution 3326 (XXIX)[1002] supports this formulated state duty with specific reference to PSNR and provides that it is "concerned about the environmental impact of the irrational and wasteful exploitation and consumption of natural resources" and furthermore that this exploitation and consumption constitutes a threat to the exercise of PSNR of states. The articulation of this duty in UNGA resolutions therefore support the notion that the duty of conservation and sustainable use of natural resources goes beyond the established state obligation not to cause transboundary environmental harm to another state.[1003]

---

999 UNGA resolution 3281 (XXIX) of 12 December 1974.

1000 UNGA resolution 35/7 of 30 October 1980.

1001 UNGA resolution 37/7 of 28 October 1982.

1002 UNGA resolution 3326 (XXIX) of 16 December 1974.

1003 The state obligation not to cause transboundary environmental harm to another state is established in the Trail Smelter Arbitration, which provides that: "under the principles of international law […] no State has the right to use or permit the use of its territory in such a manner as to cause injury by fumes in or to the territory of another or the properties or persons therein, when the case is of serious consequence and the injury is established by clear and convincing evidence." [Arbitral Tribunal: Trail Smelter Arbitration (United States v Canada) 16 April 1938 & 11 March 1941 *U.N. Repository of International Arbitration Awards* 1905 1965.]
The decision reached in the Trail Smelter Arbitration was supported by the ICJ Advisory Opinion on the Legality of the Threat or Use of Nuclear Weapons case, in which the ICJ recognises that "the environment is not an abstraction but represents the living space, the quality of life and the very health of human be-

As will be shown by the treaties mentioned below, the state duty of conservation and sustainable use of natural resources is also no longer considered to be an issue of domestic concern. The references to the state duty of conservation and sustainable use of natural resources are amply provided in treaty law, of which the most significant in the context of the research project will be mentioned below. The Bonn Convention contains many references to the conservation of migratory wild animals,[1004] and the preamble states that conservation must occur for the good of mankind, more specifically that "man holds the resources of the earth for future generations and has an obligation to ensure that this legacy is conserved and, where utilized, is used wisely". The Biodiversity Convention recognises the sovereignty of states in its preamble, as well as the need to conserve biological diversity and the sustainable use of the components of biological diversity for the benefit of present and future generations.[1005] The UNFCCC reaffirms the sovereignty of states in its preamble, and provides that it is determined to protect the climate system for present and future generations.[1006] The UNFCCC also includes many references to illustrate the importance of promoting sustainable development throughout the Convention.[1007]

---

ings, including generations unborn. The existence of the general obligation of States to ensure that activities within their jurisdiction and control respect the environment of other States or areas beyond national control is now part of the corpus of international law relating to the environment." [ICJ Advisory Opinion on the Legality of the Threat or Use of Nuclear Weapons 8 July 1996 *ICJ Reports* 226 241-242.].

1004 For example, article II(1) of the Bonn Convention recognizes the "importance of migratory species being conserved" and III(4)(a) provides that parties "shall endeavour to conserve and [...] restore the habitats of the species".

1005 The Brundtland Commission's definition of sustainable development, as included in Section 3.7.2.5, provides that sustainable development means that development meets the needs of present and future generations equally.

1006 The preamble of the UNFCCC also recalls "the provisions of General Assembly resolution 44/228 of 22 December 1989 on the United Nations Conference on Environment and Development, and resolutions 43/53 of 6 December 1988, 44/207 of 22 December 1989, 45/212 of 21 December 1990 and 46/169 of 19 December 1991 on protection of global climate for present and future generations of mankind".

1007 References to sustainable development in the UNFCCC are plentiful. For example: article 2 of the Convention stipulates that economic development should "proceed in a sustainable manner"; article 3(4) provides that parties should promote sustainable development; article 3(5) provides that parties "should cooper-

The UNCCD contains many references that indicate a state duty for parties of conservation and sustainable development. Most notably, article 2(1) of the UNCCD sets out the objective of the Convention "with a view to contributing to the achievement of sustainable development in affected areas", and article 2(2) sets out the objective of the Convention as *inter alia* "conservation and sustainable management of land and water resources". The Convention on the Law of the Sea includes in abundance this state duty as it pertains to all forms of natural resources of the sea. A few pertinent examples include the preamble of the Convention which recognises the desirability that a legal order be established for the equitable and efficient utilisation of resources, for the conservation of living resources, and for the protection and preservation of resources in this context. Article 56 provides that the state has the sovereign right to conserve natural resources within its exclusive economic zone. Article 61 of the Convention on the Law of the Sea also takes note of specific measures required to produce a sustainable yield from the marine environment. Article 119 of the Convention deals with the conservation of the living resources of the high seas and stipulates that states shall take measures in the context of ensuring sustainable yields from the marine environment. Article 145 also considers the conservation of the flora and fauna of the marine environment. The preamble of the International Tropical Timber Agreement 2006 (International Tropical Timber Agreement) recognises that states have "the sovereign right to exploit their own resources pursuant to their own environmental policies and have the responsibility to ensure that activities within their jurisdiction and control do not cause damage to the environment of other States or areas beyond the limits of natural jurisdiction". The objectives of the International Tropical Timber Agreement, specifically included in article 1(f) of the Agreement, promote and support *inter alia* the improvement of forest management and efficiency of wood utilisation and the increase in the capacity to conserve and enhance other

ate to promote a supportive and open international economic system that would lead to sustainable economic growth and development in all Parties"; article 4(2)(a) further highlights the "need to maintain strong and sustainable economic growth"; and article 4(1)(d) provides that all parties shall "[p]romote sustainable management, and promote and cooperate in the conservation and enhancement, as appropriate, of sinks and reservoirs of all greenhouse gases not controlled by the Montreal Protocol, including biomass, forests and oceans as well as other terrestrial, coastal and marine ecosystems.".

forest values in timber producing forests. Article 21(4)(b) of the Agreement contains the provisions relating to a fund for the sustainable management of tropical timber-producing forests, and allocates resources by *inter alia* taking into account conservation programmes in timber-producing forests Article 19(1) of the Energy Charter Treaty provides that in the "pursuit of sustainable development" each party will take into account its environmental obligations under international agreements to which it is a party, and each party shall minimise harmful environmental impacts.

Other treaties including a reference in some form to the state duty of conservation and sustainable use of natural resources include, most notably, the African Convention on the Conservation of Nature and Natural Resources and the Treaty of Amazonian Cooperation. The African Convention on the Conservation of Nature and Natural Resources contains a multitude of inclusions of the state duty to conservation and sustainable use of natural resources applicable to the treaty parties, specifically listing one of its objectives in article II(2) as "to foster the conservation and sustainable use of natural resources". The Treaty of Amazonian Cooperation provides measures for conservation of natural resources and the promotion of sustainable development in the region. Most notably, the preamble provides that "it is necessary to maintain a balance between economic growth and conservation of the environment" and acknowledges "continuing and expanding the joint efforts being made for the ecological conservation of the Amazon region". Furthermore, article 1 stipulates that parties to the treaty's actions produce "conservation and rational utilization of the natural resources of those territories".

It should be noted that, practically, it is possible for the conservation and sustainable development duty of states in terms of PSNR to extend beyond their borders, such as in the instance where a state shares technology that promotes conservation or sustainable development practices with another state.[1008] According to the inclusion of the duty in treaty law, this duty of states has expanded and developed as the global environmental awareness surrounding conservation and sustainable development has mounted. It is feasible to assume that major environmental events such as climate change will push the boundaries of this state duty even further, leaving it well established within the international realm. In this sense, the

1008 N Schrijver *Sovereignty Over Natural Resources, Balancing Rights and Duties* (1997) 332.

gravity of climate change has an impact on the state duty of conservation and sustainable development of natural resources.

4 2 3 4 Duty to share transboundary natural resources equitably

The duty to share transboundary natural resources equitably is established in UNGA-resolutions and treaty law. The UNGA resolution 2995 (XXVII),[1009] entitled "Co-operation between States in the field of the environment", and resolution 3129 (XXIX)[1010] with a similar title, specifically emphasises that "in the exploration, exploitation and development of their natural resources, States must not produce significant harmful effects in zones situated outside their national jurisdiction". Resolution 3281 (XXIX),[1011] provides the following in article 3 of CERDS:

> "[i]n the exploitation of natural resources shared by two or more countries, each state must co-operate on the basis of a system of information and prior consultations in order to achieve optimum use of such resources without causing damage to the legitimate interest of others".

During 1976 and 1978, the Working Group of UNEP drafted principles entitled "Environmental Law Guidelines and Principles on Shared Natural Resources" in response to UNGA resolution 3129 (XXIX). The UNGA used the Working Group's report to formulate resolution 34/186 of 1979,[1012] which contains the following request addressed to states:

> "use the principles as guidelines and recommendations in the formulation of bilateral or multilateral conventions regarding natural resources shared by two or more States, on the basis of the principles of good faith and in the spirit of good neighbourliness and in such a way as to enhance and not to affect adversely development and the interests of all countries, in particular of the developing countries".

The inclusion of the duty by treaty law is seen in article 19(1) of the ASEAN Agreement on the Conservation of Nature and Natural Resources, which concerns shared resources and stipulates that state parties which share natural resources "shall cooperate concerning their conservation and

---

1009 UNGA resolution 2995 (XXVII) of 15 December 1972.
1010 UNGA resolution 3129 (XXVIII) of 13 December 1973, followed by resolution 34/186 of 18 December 1979 with the same title.
1011 UNGA resolution 3281 (XXIX) of 12 December 1974.
1012 UNGA resolution 34/186 of 18 December 1979.

harmonious utilization". Furthermore, article 19(2)(a) of the Agreement determines that the state parties shall also cooperate so as not to prevent another state party from utilising a shared resource. This construction of the Agreement is also relevant in the current functioning of the instruments of the international legal climate change regime, which similarly attempt to have states cooperate as to the management of the atmosphere (and also forests) so that each state is not eventually permanently prevented from utilising this natural resource. Furthermore, the preamble of the International Tropical Timber Agreement recognises that states have "the responsibility to ensure that activities within their jurisdiction and control do not cause damage to the environment of other States or areas beyond the limits of natural jurisdiction".

As also submitted in Section 4.2.2.5 in regard to the state right to an equitable share in transboundary resources, transboundary water resources form a vital shared natural resource to which this duty of states also pertains. As shown by the Lake Lanoux award, this duty of states to share transboundary natural resources involves complex cooperation and the duty to negotiate and consult in good faith.[1013] Another inclusion of the state duty to share transboundary natural resources equitably in regard to transboundary water resources is found in the Convention on the Law of the Sea.[1014] This Convention provides that resources of the sea, and the shar-

1013 N Kliot et al "Institutions for management of transboundary water resources: their nature, characteristics and shortcomings" (2001) 3 *Water Policy* 229 233. The Lake Lanoux Arbitration involved the use of the waters of the lake in which France proposed to carry out certain works involving the use of the waters of the lake. Spain feared that these works would adversely affect its rights and interests The Tribunal determined that there existed a "duty of consultation and of bringing into harmony the respective actions of the two States when general interests are in involved in matters concerning waters". Furthermore, the Tribunal stressed that the "consultations and negotiations between the two States must be genuine". [Arbitral Tribunal "Lake Lanoux Arbitration (Spain v France)" (16 November 1957) (1, 15, 27) *Ecolex* <http://www.ecolex.org/details/court-decision/lake-lanoux-arbitration-france-v-spain-b09cb956-2cb5-479e-ba3a-bbfd4f7b68fc/> (accessed 07-09-2016).].

1014 Article 83 of the Convention on the Law of the Sea deals with the delimitation of the continental shelf between states with opposite or adjacent coasts. In this case, article 83(3) is of particular relevance and reads as follows: "the States concerned, in a spirit of understanding and cooperation, shall make every effort to enter into provisional arrangements of a practical nature and, during this transitional period, not to jeopardize or hamper the reaching of the final agreement. Such arrangements shall be without prejudice to the final delimitation."

ing of these resources, can also pertain to resources which are situated across the limits of national jurisdiction and as such are subject to the principle of the common heritage of humankind.[1015]

There has been a general acceptance for some time that the notion of absolute state sovereignty is being impacted and "gradually replaced"[1016] by the notion that natural resources must be equitably utilised and more commonly managed for the benefit of present and future generations (as discussed in Section 2.5.2.3.3). In practice it means that states are under the obligation today to recognise the rights of other states in terms of PSNR and should at least consult whether the rights of these states might be affected by any action in this context.[1017]

Case law also illustrates this abovementioned position. For example, in the awards regarding the border between Costa Rica and Nicaragua it was

---

Article 142 of the Convention on the Law of the Sea, concerning the rights and legitimate interests of coastal States, is also relevant. Article 142(1)-(3) of the Convention reads as follows: "(1) Activities in the Area, with respect to resource deposits in the Area which lie across limits of national jurisdiction, shall be conducted with due regard to the rights and legitimate interests of any coastal State across whose jurisdiction such deposits lie. (2) Consultations, including a system of prior notification, shall be maintained with the State concerned, with a view to avoiding infringement of such rights and interests. In cases where activities in the Area may result in the exploitation of resources lying within national jurisdiction, the prior consent of the coastal State concerned shall be required. (3) Neither this Part nor any rights granted or exercised pursuant thereto shall affect the rights of coastal States to take such measures consistent with the relevant provisions of Part XII as may be necessary to prevent, mitigate or eliminate grave and imminent danger to their coastline, or related interests from pollution or threat thereof or from other hazardous occurrences resulting from or caused by any activities in the Area.".

1015 Schrijver *Sovereignty Over Natural Resources, Balancing Rights and Duties* (1997) 277.

1016 N Schrijver Sovereignty Over Natural Resources, Balancing Rights and Duties (1997) 338.
Although not specifically related to the context of PSNR, the Island of Palmas Case is relevant in this regard. The Tribunal provides that "[t]erritorial sovereignty, as has already been said, involves the exclusive right to display the activities of a State. This right has as corollary a duty: the obligation to protect within the territory the rights of other States […]" [Arbitral Tribunal: Island of Palmas Case (Netherlands v United States) 4 April 1928 Reports of International Arbitral Awards Vol II 829-871 839.].

1017 N Schrijver Sovereignty Over Natural Resources, Balancing Rights and Duties (1997) 338.

inter alia decided that Costa Rica may not prevent Nicaragua from undertaking work for the improvement of the San Juan river, provided that the work does not damage the territory of Costa Rica and that Costa Rica has the right to indemnification as appropriate.[1018] Accordingly, the state duty of sharing transboundary resources equitably is developing, and occurrences with environmental consequences, such as climate change, have brought into focus the interdependency of states and the need to manage equitably shared natural resources, in particular in the context of the research project, the shared natural resource of the atmosphere.

The duty of states to share transboundary natural resources equitably is the counterpart of the state right to an equitable share in the benefits of transboundary resources, as discussed in Section 4.2.2.5. The state duty to share transboundary natural resources equitably is also connected to the duty of states not to cause transboundary harm, a state duty formulated in the Trail Smelter arbitration,[1019] and consequently confirmed by other sources of international law.[1020] The duty of states not to cause trans-

1018 Arbitral Tribunal: Awards regarding the Border between Costa Rica and Nicaragua (Costa Rica v Nicaragua) 22 March 1888, 30 September 1897, 20 December 1897, 22 March 1898 & 26 July 1899 Reports of International Arbitral Awards Vol XXVIII 189-236 191, 210.]
Furthermore, in the Lake Lanoux Arbitration demonstrates the importance the tribunal attached to the consultation between states, and also the complexity involved in managing transboundary water resources. The tribunal found that there existed a "duty of consultation" and that this duty of consultation "must be genuine". [International Arbitral Tribunal "Lake Lanoux Arbitration (Spain v France)" (16 November 1957) (1, 15, 27) Ecolex <http://www.ecolex.org/details/court-decision/lake-lanoux-arbitration-france-v-spain-b09cb956-2cb5-479e-ba3a-bbfd4f7b68fc/> (accessed 07-09-2016).].

1019 A principle established in the Trail Smelter Arbitration: "under the principles of international law […] no State has the right to use or permit the use of its territory in such a manner as to cause injury by fumes in or to the territory of another or the properties or persons therein, when the case is of serious consequence and the injury is established by clear and convincing evidence." [Arbitral Tribunal: Trail Smelter Arbitration (United States v Canada) (16 April 1938 & 11 March 1941) *U.N. Repository of International Arbitration Awards* 1905-1982 1965.].

1020 The Trail Smelter decision was supported by the ICJ Advisory Opinion on the Legality of the Threat or Use of Nuclear Weapons case: "The Court also recognizes that the environment is not an abstraction but represents the living space, the quality of life and the very health of human beings, including generations unborn. The existence of the general obligation of States to ensure that activities within their jurisdiction and control respect the environment of other States or

boundary harm is well-developed in international law, and it can be put forward that the state duty to share transboundary resources equitably stems from the more well-known state duty not to cause transboundary harm, but is more positively constructed than this state duty. When the state duty is considered in the context of climate change, a more complex exercise than in the Trail Smelter arbitration has to be conducted, as it is not possible to determine the precise damage that one state's greenhouse gas emissions causes to another state. However, it is still possible to argue that, in the context of climate change, states have the same duty not to cause transboundary harm to other states as formulated in the Trail Smelter arbitration,[1021] and in the context of climate change this duty becomes more urgent. This is because the atmosphere is a shared or transboundary natural resource, forming the main component of the international legal climate change regime.[1022] In addition, another shared or transboundary natural resource, which also forms a significant component of an effectively functioning international legal climate change regime, is forests.[1023] Similarly, climate change also has a serious environmental im-

---

areas beyond national control is now part of the corpus of international law relating to the environment." [ICJ Advisory Opinion on the Legality of the Threat or Use of Nuclear Weapons 1996 *ICJ Reports* 226 241-242.]

Principle 21 of the Stockholm Declaration and Article 2 of the Rio Declaration also contains the duty of states not to cause transboundary harm and reads as follows: "States have, in accordance with the Charter of the United Nations and the principles 13 of international law, the sovereign right to exploit their own resources pursuant to their own environmental and developmental policies, and the responsibility to ensure that activities within their jurisdiction or control do not cause damage to the environment of other States or of areas beyond the limits of national jurisdiction".

Article 194 of the Law of the Sea Convention also contains the duty of states not to cause transboundary harm, and provides that "States shall take all measures necessary to ensure that activities under their jurisdiction or control are so conducted as not to cause damage by pollution to other States and their environment, and that pollution arising from incidents or activities under their jurisdiction or control does not spread beyond the areas where they exercise sovereign rights in accordance with this Convention.".

1021 J Dugard *International Law, A South African Perspective* 4ed (2011) 402-404.

1022 J Vogler *The Global Commons, Environmental and Technological Governance* 2ed (2000) 6, 122; E Blanco & J Razzaque *Globalisation and Natural Resources Law: Challenges, Key Issues and Perspectives* (2011) 88-89.

1023 E Blanco & J Razzaque *Globalisation and Natural Resources Law: Challenges, Key Issues and Perspectives* (2011) 87.

pact on the ocean as a shared natural resource (as discussed in Section 3.5.4). Accordingly, the connection between the international legal climate change regime and the state duty to share transboundary natural resources equitably grows, as the problem of climate change becomes more urgent. Therefore, it can be argued that the effects of climate change impact this state duty in terms of PSNR and intensifies the need for the effective and urgent application of this state duty by all states.

4 2 3 5 Duty to respect international law

There are many instances in the context of sovereignty over natural resources where a state duty to respect international law is observed. The duty to respect international law is observed in the UNGA resolution 837 (IX),[1024]which requests the Commission on Human Rights to complete recommendations "concerning international respect for the right of people and nations to self-determination, including recommendations concerning their permanent sovereignty over their natural wealth and resources", while having due regard, in particular, to "the rights and duties of States under international law and to the importance of encouraging international co-operation". Similarly, regard is to be had for "rights and duties of States under international law", according to resolutions 1314 (XIII)[1025] and 837 (IX)[1026], which concerns recommendations regarding international respect for the right of people and nations to self-determination. Resolution 1803 (XVII)[1027] recommends that the right of states to their natural resources be respected, and also declares that "[t]he free and beneficial exercise of the sovereignty of peoples and nations over their natural resources must be furthered by the mutual respect of States based on their sovereign equality." UNGA Resolution 1515 (XV)[1028] recommends that the right of states to their natural resources be respected, but adds that it should be respected "in conformity with the rights and duties of States under international law." The reference to "mutually acceptable contractual

1024 UNGA resolution 837 (IX) of 14 December 1954.
1025 UNGA resolution 1314 (XIII) of 12 December 1958.
1026 UNGA resolution 837 (IX) of 14 December 1954.
1027 UNGA resolution 1803 (XVII) of 14 December 1962.
1028 UNGA resolution 1515 (XV) of 15 December 1960.

practices" in resolution 2158 (XXI)[1029] can also be understood as including international obligations.[1030] Resolution 3281 (XXIX), [1031]containing CERDS, provides in regard to the fundamentals of international economic relations, that relations between states shall be governed by *inter alia* the principles of "fulfillment in good faith of international obligations" and "respect for human rights and international obligations".

The most relevant international instruments that contain the state duty to respect international law are mentioned below. The International Covenants on Human Rights, stipulate in article 1(2), subject to article 25 of the ICESCR and article 47 of the ICCPR,[1032] that people may dispose of their natural resources "without prejudice to any obligations arising out of international economic co-operation, based upon the principle of mutual benefit, and international law." Along similar lines, the Banjul Charter provides in article 21(3) that the right of people to dispose of their natural resources "shall be exercised without prejudice to the obligation of promoting international economic cooperation based on mutual respect, equitable exchange and the principles of international law." The Convention on the Law of the Sea contains a few references to the state duty to respect international law. Most notably, in regard to the rights and duties of the coastal state in the exclusive economic zone, article 56(2) provides that "the coastal state shall have due regard to the rights and duties of other States". Article 300, as a general provision of the Convention, provides that "parties shall fulfill in good faith the obligations assumed under this Convention and shall exercise rights [...] in a manner which would not constitute an abuse of right". The Seoul Declaration also stipulates in principle 5(3) that PSNR is not exempt from "the application of the relevant principles and rules of international law".[1033]

It is concluded that a definite duty for states to respect international law exists in terms of PSNR. More specifically, this duty, as circumscribed in legal sources and international instruments, specifically tends to require of

1029 UNGA resolution 2158 (XXI) of 25 November 1966.

1030 N Schrijver *Sovereignty Over Natural Resources, Balancing Rights and Duties* (1997) 340.

1031 UNGA resolution 3281 (XXIX) of 12 December 1974.

1032 Article 25 of the ICESCR and article 47 of the ICCPR reads as follows: "Nothing in the present Covenant shall be interpreted as impairing the inherent right of all peoples to enjoy and utilize fully and freely their natural wealth and resources.".

1033 Principle 5(3), Seoul Declaration.

states to respect the rights of other states, obey international law and perform their duties in good faith.[1034] The duty is connected to climate change as respect for international law would promote state compliance with the international legal climate change regime that serves to limit or prevent the severe effects of climate change. Accordingly climate change can be said to impact the state duty, as the recognition of the gravity and urgency of the effects of climate change may increase the application of the state duty and vice versa.

4 2 4 Final remarks on the rights and duties of states in terms of sovereignty over natural resources, and the shift in environmental attitudes owing to the effects of climate change

The impact of climate change on the rights and duties of states in terms of PSNR has been discussed above. Although many nuanced points of impact are discussed, some clear conclusions have been reached, which will be summarised below.

The right of states to dispose freely of, explore and exploit natural resources is considered by the author to be affected by the gravity of climate change. This is because climate change, as well as the climate commitments states make in terms of the international legal climate change regime, forces states to re-consider and adapt their disposal and exploitation of natural resources by, for instance, promoting clean energy resources and limiting the utilisation of fossil fuel energy resources.

The state right to use natural resources for national development is also affected by the gravity of climate change, and states find it increasingly difficult to use their developmental status as a reason not to join the global fight against climate change. To illustrate changing attitudes, the NDCs of the Paris Agreement hold developed and developing states to emission limitations; and the excuse of developing countries that they can prioritise national development has become less acceptable in light of the gravity of climate change. The state right to use natural resources for national development pursuant to states' own environmental policy is also affected by the gravity of climate change.

1034 N Schrijver *Sovereignty Over Natural Resources, Balancing Rights and Duties* (1997) 343.

The state right to an equitable share in benefits of transboundary natural resources is significantly challenged by the concern of climate change and the atmosphere as a shared transboundary natural resource. This state right is restricted in light of climate change, since the construction of the atmosphere as a transboundary natural resource to share becomes somewhat convoluted.

The state rights pertaining to foreign investment are included in Section 4.2.2.6 since they form a component in a complete understanding of PSNR. However, the impact of climate change on these rights are found to be negligible.

The duty of states to exercise sovereignty over natural resources for national development and the well-being of the people is impacted to a large extent by the gravity of climate change. As Section 3.5 illustrated, climate change entails significant negative environmental, social, and economic consequences for the well-being of people. It follows that these consequences will promote this state duty and increase its importance as the concern about climate change continues to grow. In the same way as the state right to use natural resources for national development is restricted by climate-related concerns, climate change limits the state duty to exercise PSNR for national development through the incorporation of the duty to exercise PSNR for the well-being of people.

The state duty to cooperate in international development is strengthened by the concern about climate change and the acknowledgement of the disastrous impacts that climate change can have on international development. It is generally argued that climate commitments in terms of the international legal climate change regime support the state duty to cooperate in international development, especially if the international development refers to sustainable international development. With global pressure to conserve the environment and practise sustainable development for the benefit of present and future generations, the duty of states to conserve and use natural resources sustainably is automatically supported. It is argued that climate change has served to heighten this concern and accordingly greatly reinforces the state duty of conservation and sustainable use of natural resources.

The state duty to share transboundary natural resources equitably which developed in reaction to the state right to an equitable share in transboundary natural resources, is growing. In general, a tendency to accept that natural resources need to be more jointly managed by states for the benefit of present and future generations has become more apparent. Environmental concerns, of which climate change has been the most threatening, have underscored the interdependency of states, the interdependency of the ecosystem, and the need to manage in particular the shared resource of the atmosphere jointly and equitably.

The duty to respect international law is a general duty of states. However, the disastrous impacts of climate change have served to emphasise the importance of the state duty to respect international law, and respect for climate commitments in terms of the international legal climate change regime can greatly contribute to increasing state compliance with climate commitments and the international reduction of greenhouse gases.

With the inception of PSNR, states naturally focused on developing their rights in terms of the principle. The rights were crucial in securing domestic control over natural resources, and mainly served to safeguard the state against foreign interests. The duties of states in terms of the principle developed later as a result of and in reaction to the exercise of rights by states in terms of the principle. At one point it was accepted that state duties in terms of PSNR are at least as important as state rights in terms of PSNR.[1035] It is submitted by the author that, since then, these state duties only gained in importance, especially when considered in the context of an environmental crisis such as climate change. At present, state duties in terms of PSNR are no longer only a reaction to state rights. The state duties are in their own right fundamental components of a contemporary application and understanding of PSNR, especially if PSNR is to be employed in the fight against contemporary environmental problems such as climate change. The examination of state rights and duties above indicate that a contemporary understanding of PSNR involves a restriction of state rights and a promotion of state duties in terms. In this context, it is suggested that it is necessary for states to step more fully into their role as "temporary custodians of natural resources charged with managing them for the benefit of all the people in the country", including future generations.[1036]

1035 L Barrera-Hernandez "Sovereignty over natural resources under examination: the Inter-American system for human rights and natural resource allocation" (2006) 12 Ann. Surv. Int'l & Comp L. 43 44.

1036 E Duruigbo "Permanent Sovereignty and Peoples' Ownership of Natural Resources in International Law" 2006 (38) The George Washington International Law Review 33 37.

Although there has been great development in the recognition of the aforementioned necessary shift and a noticeable shift in the role of states in the conservation of the environment, the possibility of effective implementation seems doubtful without the existence of worthwhile incentives urging state implementation of duties in terms of PSNR. It is accepted that international state compliance is a significant hurdle to compliance of the state duties described above. This is the reason why it is also necessary to consider people's rights in terms of PSNR, along with the shift in state rights and duties, when considering the evolution of PSNR. People's rights in terms of PSNR present a means by which to increase state compliance within the international legal climate change regime, as will be discussed in section 4.3 and 4.4.

4 2 5 Increased importance of the common concern as an indicator of the growing importance of the duties of states in terms of the principle of sovereignty over natural resources

It has been submitted in Section 4.2 that state duties in terms of PSNR have gained importance and momentum while state rights have become restricted with the rise of the global concern for the environment and in particular the effects of climate change. The concern for the global interconnected environment is increasing, as evidenced by the Gabčíkovo-Nagymaros Project case in which the court found that environmental law and the effects on the environment of the relevant project are relevant in interpreting the treaty in regard to the project undertaken by Hungary and Slovenia.[1037] The separate opinion of Weeramantry additionally provides as follows:

> "We have entered an era of international law in which international law subserves not only the interests of individual States, but looks beyond them and

---

Also see: W Scholtz "Greening Permanent Sovereignty through the Common Concern in the Climate Change Regime: Awake Custodial Sovereignty!" in OC Ruppel et al (eds) Climate Change: International Law and Global Governance. Volume II: Policy, Diplomacy and Governance in a Changing Environment (2013) 201 201-211.

1037 Case Concerning the Gabčíkovo-Nagymaros Project (Hungary v Slovenia) ICJ Judgment on 25 September 1997, *ICJ Reports 1997* 64-65, 67.

> their parochial concerns to the greater interests of humankind and planetary welfare [...]. International environmental law will need to proceed beyond weighing the rights and obligations of parties within a closed compartment of individual State self-interest, unrelated to the global concerns of humanity as a whole."[1038]

The principle of common concern functions outside the parameters of state interest and forms part of a "global set of values" for humankind, which can be referred to as the "laws of humanity".[1039] Tracing the development of the principle of common concern of humankind, by considering its inclusion in particular international instruments, the exercise illustrates the increased importance attached to state duties in regard to the environment which is to be conserved for present and future generations. The two duties in terms of PSNR which are in line with the goals of the principle of common concern of humankind are the state duty to exercise sovereignty over natural resources for the well-being of people and the state duty of conservation and sustainable use of natural resources.

A brief consideration of the development of the principle of common concern of humankind is included below. Initially, many treaties referred to environmental issues as a "common interest" rather than a "common concern". The International Convention for the Regulation of Whaling 1946 (International Convention for the Regulation of Whaling) makes use of the terms "in the interest of the nations of the world" and "common

---

1038 ICJ Judgment: Separate Opinion of Weeramantry on Case Concerning the Gabčíkovo-Nagymaros Project (Hungary v Slovenia) ICJ Judgment on 25 September 1997 *ICJ Reports 1997* 118.

1039 D Shelton "Common Concern of Humanity" (2009) 1 *Iustum Aequum Salutare* 33 34.
In this regard the Martens Clause is often considered. The Martens Clause forms part of the law of armed conflict. The Martens Clause reads as follows: "Until a more complete code of the laws of war is issued, the High Contracting Parties think it right to declare that in cases not included in the Regulations adopted by them, populations and belligerents remain under the protection and empire of the principles of international law, as they result from the usages established between civilized nations, from the laws of humanity and the requirements of the public conscience." [Preamble of the Convention (II) with Respect to the Laws and Customs of War on Land 1899 and its annex of Regulations concerning the Laws and Customs of War on Land (Hague Convention (II).]
However, Cassese dispels *inter alia* the argument that the Martens clause upgrades the so-called laws of humanity to a source of international law. A Cassese "The Martens Clause: Half a Loaf or Simply Pie in the Sky?" (2000) 11(1) *EJIL* 187-216.

interest" when referring to the safeguarding and level of whale stocks.[1040] It is noteworthy that the convention lifted concern for diminishing whale stocks from a local concern to an international concern to be solved by states, indicating an increase in the global awareness of diminishing natural resources, specifically whale stock supplies.[1041] Similarly, the Tokyo Convention for the High Seas Fisheries of the North Pacific Ocean 1952 (Tokyo Convention) states that it is in the "common interest of mankind" to ensure the sustainability of the North Pacific Ocean's fishery resources,[1042] indicating the awareness of the need to ensure sustainable utilisation of natural resources.

A few years later, the Antarctic Treaty 1959 (Antarctic Treaty) also refers to the "interest of all mankind".[1043] The treaty is considered to represent a major development in the full international recognition of the concept of common concern of humankind.[1044] The Antarctic Treaty is supported by the Convention for the Conservation of Antarctic Marine Living Resources 1980 (Canberra Convention), which explicitly refers to the "interest of all mankind".[1045] In addition, the Protocol on Environmental

---

1040 The preamble of the International Convention for the Regulation of Whaling states: "Recognizing the interest of the nations of the world in safeguarding for future generations the great natural resources represented by the whale stocks; [...] Recognizing that it is in the common interest to achieve the optimum level of whale stocks as rapidly as possible without causing widespread economic and nutritional distress".

1041 A Kiss & D Shelton *International Environmental Law* 3ed (2004) 32.

1042 The preamble of the Tokyo Convention for the High Seas Fisheries of the North Pacific Ocean states: "The Governments of the United States of America, Canada and Japan, whose respective duly accredited representatives have subscribed hereto, Acting as sovereign nations in the light of their rights under the principles of international law and custom to exploit the fishery resources of the high seas, and Believing that it will best serve the common interest of mankind, as well as the interests of the Contracting Parties, to ensure the maximum sustained productivity of the fishery resources of the North Pacific Ocean".

1043 The preamble of the Antarctic Treaty states: "Recognizing that it is in the interest of all mankind that Antarctica shall continue forever to be used exclusively for peaceful purposes and shall not become the scene or object of international discord".

1044 A Kiss & D Shelton *International Environmental Law* 3ed (2004) 32.

1045 The preamble of the Canberra Convention states: "Believing that it is in the interest of all mankind to preserve the waters surrounding the Antarctic continent for peaceful purposes only and to prevent their becoming the scene or object of international discord".

Protection to the Antarctic Treaty 1991 (Madrid Protocol) refers to the "interest of mankind as a whole" when mentioning the regime for the protection of the Antarctic and associated environment.[1046]

The African Convention on the Conservation of Nature and Natural Resources supports both the notions of common interest and common concern and specifically uses the term "common concern of humankind" when referring to the conservation of the global environment.[1047] Furthermore, the preamble of the Convention states that "joint action" is required in regard to conservation, and further provides that natural resources are to be rationally utilised "for the present and future welfare of mankind".[1048] With the inclusion of the phrase "future welfare of mankind", the element of time is added to the concept of common concern,[1049] contributing to the contemporary understanding accorded to the concept that it includes a consideration of both present and future generations. The spirit of the entire Stockholm Declaration reiterates and supports the idea of common interest or common concern. Furthermore, the Bonn Convention,[1050] the Bern Convention[1051] and the World Charter for Nature 1982[1052] expand

---

1046 The preamble of the Protocol on Environmental Protection to the Antarctic Treaty states: "Convinced that the development of a comprehensive regime for the protection of the Antarctic environment and dependent and associated ecosystems is in the interest of mankind as a whole".

1047 The preamble of the African Convention on the Conservation of Nature and Natural Resources states: "that the conservation of the global environment is a common concern of humankind as a whole".

1048 Preamble, African Convention on the Conservation of Nature and Natural Resources.

1049 A Kiss & D Shelton *International Environmental Law* 3ed (2004) 33.

1050 The preamble of the Bonn Convention states: "Recognizing that wild animals in their innumerable forms are an irreplaceable part of the earth's natural system which must be conserved for the good of mankind; Aware that each generation of man holds the resources of the earth for future generations and has an obligation to ensure that this legacy is conserved and, where utilized, is used wisely".

1051 The preamble of the Bern Convention states: "Recognising that wild flora and fauna constitute a natural heritage of aesthetic, scientific, cultural, recreational, economic and intrinsic value that needs to be preserved and handed on to future generations".

1052 UNGA resolution 37/7 of 28 October 1982.
UNGA Resolution 37/7 of 28 October 1982 contains the World Charter for Nature which states: "Conscious of the spirit and terms of its resolutions 35/7 and 36/6, in which it solemnly invited Member States, in the exercise of their permanent sovereignty over their natural resources, to conduct their activities in

the recognition of common interest and concern by also connecting the concepts to the benefit of present and future generations.

More recently, the UNFCCC acknowledges that climate change is a "common concern of humankind",[1053] and the Biodiversity Convention states that conserving biological diversity is a "common concern of humankind".[1054] In addition, both the UNFCCC and the Biodiversity Convention consider the effects of climate change and a loss of biological diversity to be a common concern.[1055] The Convention for the Protection of the Marine Environment of the North-East Atlantic 1992 (Convention for the Protection of the Marine Environment of the North-East Atlantic)[1056] and the UNCCD[1057] also apply the principle of common concern, but with a less international focus.

Through the development of the international instruments mentioned above it is possible to detect an increase in international concern for the

---

recognition of the supreme importance of protecting natural systems, maintaining the balance and quality of nature and conserving natural resources, in the interests of present and future generations".
The draft World Charter for Nature, as found in the UNGA Resolution 37/7 of 28 October 1982, elucidates the growing concern of environmental depredation at the time. UNGA Resolution 37/7 of 28 October 1982 contains the World Charter for Nature which states: "Reaffirming that man must acquire the knowledge to maintain and enhance his ability to use natural resources in a manner which ensures the preservation of the species and ecosystems for the benefit of present and future generations".

1053 The preamble of the UNFCCC states that a "change in the Earth's climate and its adverse effects are a common concern of humankind".

1054 The preamble of the Biodiversity Convention states "the conservation of biological diversity is a common concern of humankind".

1055 D Shelton "Common Concern of Humanity" (2009) 1 *Iustum Aequum Salutare* 33 37.

1056 The preamble of the Convention for the Protection of the Marine Environment of the North-East Atlantic recognises the following: "that concerted action at national, regional and global levels is essential to prevent and eliminate marine pollution and to achieve sustainable management of the maritime area, that is, the management of human activities in such a manner that the marine ecosystem will continue to sustain the legitimate uses of the sea and will continue to meet the needs of present and future generations".

1057 The preamble of the UNCCD reads as follows: "Affirming that human beings in affected or threatened areas are at the centre of concerns to combat desertification and mitigate the effects of drought, Reflecting the urgent concern of the international community, including States and international organizations, about the adverse impacts of desertification and drought".

depletion and unsustainable disposal of natural resources, a global recognition of the interconnected ecosystem, and an awareness that state duties collectively exercised are required in this regard.[1058] This also means that if a common concern poses a problem, the problem cannot be solved by only one state, and if a common concern provides a benefit, the benefit cannot be received by only one state.[1059] As submitted by the ICJ in the Legality of the Threat or Use of Nuclear Weapons case:

> "the environment is not an abstraction but represents a living space, the quality of life and the very health of human beings, including generations unborn. The existence of the general obligation of States to ensure that activities within their jurisdiction and control respect the environment of other States or areas beyond national control is now part of the corpus of international law relating to the environment."[1060]

It can be submitted that the recognition of the common concern of humankind that exists today is a good indication that state duties towards the environment and the well-being of people, including present and future generations, in terms of other principles of international law, such as PSNR, are also gaining momentum. As the concern for the environment grows stronger (not only in regard to the effects of climate change), it will continue to impact the evolution of PSNR, and the understanding of state duties (and rights) in terms of the principle.

4 3 Reinterpreting and reconfiguring the rights of people in terms of the principle of sovereignty over natural resources

4 3 1 Introduction

As determined in Section 2.5.3, people, in addition to states, are considered as the subjects of PSNR. A secondary objective of the research project, as stipulated in 1.4, is to reinterpret and reconfigure PSNR as a

---

1058 D Shelton "Common Concern of Humanity" (2009) 1 *Iustum Aequum Salutare* 33 34.

1059 D Shelton "Common Concern of Humanity" (2009) 1 *Iustum Aequum Salutare* 33 34.

1060 ICJ Advisory Opinion on the Legality of the Threat or Use of Nuclear Weapons 8 July 1996 *ICJ Reports* 226 241-242.

principle focused on people that can contribute to strengthening the international legal climate change regime. In a wider context than pertaining to climate change, it is also submitted that the position of people in the PSNR dialogue needs to change and people ought to be considered as more than "faceless populations" of states.[1061] Furthermore, intervention is required so that the existing provisions of PSNR do not become "meaningless in today's world",[1062] but instead are utilised in advancing people's interests. These two goals are complementary, as illustrated by the objective to reinterpret and reconstruct PSNR so that the principle contributes to the international legal climate change regime.

It was shown in Section 4.2 that there has been a shift in focus between state rights and state duties in terms of PSNR, owing to a growing concern for the interconnectedness of the global environment and the potential devastation that the effects of climate change can cause. This shift means that some state rights are restricted, while state duties typically gain momentum and support due to international environmental concerns such as climate change. This section of the research project considers a consequence of the gravity of climate change on PSNR, namely that a reinterpretation and reconfiguration of the rights of people in terms of PSNR are warranted. The people's rights in terms of PSNR can be used as a basis upon which to hold states accountable for their climate commitments and thereby increase the effectiveness of the international legal climate change regime. In this way the people's rights in terms of PSNR constitutes a negative incentive for states to abide by their climate commitments, not merely prompting compliance with the law but also benefitting the environ-

1061 E Duruigbo "Permanent Sovereignty and Peoples' Ownership of Natural Resources in International Law" 2006 (38) The George Washington International Law Review 33 37.
International environmental law in general focuses on environmental damage rather than "its impact on human beings". [HM Osofsky "Learning from Environmental Justice: A New Model for International Environmental Rights" (2005) 24 Stanford Environmental Law Journal 71 78.].

1062 E Duruigbo "Permanent Sovereignty and Peoples' Ownership of Natural Resources in International Law" 2006 (38) The George Washington International Law Review 33 92.

ment (considered as two key objectives of environmental compliance by Silecchia[1063]).

Within Chapter 1 of the research project, the consistent referral to the 'reinterpretation and reconfiguration of PSNR as a principle focused on people' was explained. As submitted, a reinterpretation or reconfiguration of the principle is proposed in order to accord people prominence as subjects of PSNR, rather than states which have generally been considered the most prominent subjects of PSNR. Furthermore, although people have historically also been considered as the subjects of PSNR, and accordingly held the right to economic self-determination, this right has never taken centre-stage like the rights of states. In this sense, the existing elements of PSNR, namely people's rights (in addition to state duties) should be reinterpreted and reconfigured. In broad terms, it is generally accepted that people have rights and duties in terms of the environment.[1064] However, this research project focuses only on the rights of people in terms of PSNR and not on the duties of people, as submitted in Section 1.5.

In the remaining part of Chapter 4, PSNR will be reinterpreted and reconfigured to investigate, firstly, what the people's rights in terms of PSNR are; and, secondly, what the prospects are for success of litigation based on infringements of people's rights in terms of PSNR in the context of climate change. In so doing, it will be determined if such litigation can provide a worthwhile avenue to pursue in the attempt to increase the state's compliance with climate commitments in terms of the international legal climate change regime and accordingly if the possibility of such litigation could increase the regime's overall effectiveness.

1063 Silecchia LA "Ounces of Prevention and Pounds of Cure: Developing Sound Policies for Environmental Compliance Programs" (1996) VII Fordham Environmental Law Journal 583 595.

1064 For example, the preamble of the Aarhus Convention not only stipulates that "every person has the right to live in an environment adequate to his or her health and well-being" but that every person also has the duty (individually or as part of an association) to "protect and improve the environment for the benefit of present and future generations".

4 3 2 People's rights in terms of the principle of sovereignty over natural resources

4 3 2 1 Introduction to people's rights in terms of the principle of sovereignty over natural resources

As will be shown below, self-determination of people gives rise the people's right of economic self-determination in terms of PSNR. This right is also referred to as the right to natural resource self-determination or simply as the right of people to PSNR. These terms are all used interchangeably throughout the research project. Included in the right to economic self-determination is the right to dispose freely of natural wealth and resources and the right of people not to be deprived of their means of subsistence.[1065] Accordingly, the author refers to rights (in the plural form) of people throughout the research project. Furthermore, this right of people in terms of PSNR can also be claimed indirectly (as will be shown below).

Although PSNR has long been recognised as a right of people,[1066] there has been a noticeable renewed international revival of the general interest in people's rights in terms of PSNR and the notion that people need to participate more actively as rights-owners in the natural resource management of states.[1067] There has been some discussion centred around the top-

---

1065 E Blanco & J Razzaque *Globalisation and Natural Resources Law; Challenges, Key Issues and Perspectives* (2011) 133.

1066 G Abi-Saab "Permanent Sovereignty over Natural Resources and Economic Activities" in M Bedjaoui (ed) *International Law: Achievements and Prospects* (1991) 597 603; JY Qin "Reforming WTO Discipline on Export Duties: Sovereignty over Natural Resources, Economic Development and Environmental Protection" 2012 (4) *Wayne State University Law School Research Paper* <http://papers.ssrn.com/sol3/papers.cfm?abstract_id=2030477> (accessed 16-03-2015).

1067 TT Onifade "Peoples-Based Permanent Sovereignty over Natural Resources: Toward Functional Distributive Justice?" (2015) 16 *Human Rights Review* 343 344.

See for instance: LA Miranda "The Role of International Law in Intrastate Natural Resource Allocation: Sovereignty, Human Rights, and Peoples-Based Development (2012) 45 *Vanderbilt Journal of Transnational Law* 785 785-840; L Barrera-Hernandez "Resources Under Examination: The Inter-American System for Human Rights and Natural Resource Allocation" (2006) 12 *Annual Survey of International & Comparative Law* 43 43-58. HM Haugen "Peoples' right to self-determination and self-governance over natural resources: Possible and desirable?" (2014) 8(1) *Nordic Journal of Applied Ethics* 3 3-21. DP Zongwe

ic referred to as people-based sovereignty over natural resources.[1068] In this regard, people's rights in terms of PSNR are often approached on the basis of common ownership or environmental justice. More established common ownership approaches place people in the position of owners of natural resources, while more recently developed environmental justice approaches rely on environmental justice as a basis for people's rights to natural resources.[1069] The approach of people-based sovereignty over natural resources is based on the pillar of self-determination as opposed to the pillar of PSNR (of which self-determination is a foundational pillar).[1070] Exploring these various approaches to people's rights pertaining to natural resources falls outside the scope of the research project, as the research project is concerned with a construction of people's rights in terms of PSNR that can be used as a basis for instituting action against states for not adhering with climate commitments.

4 3 2 2 Right to economic self-determination

As was submitted in Section 2.2, PSNR is founded on pillars of state sovereignty and self-determination of people. The right to self-determination is a well-established human right belonging to all people.[1071] It is also

---

"The Legal Justifications for a People-Based Approach to the Control of Mineral Resources in the Democratic Republic of the Congo" (2008) 12 *Cornell Law School Inter-University Graduate Student Conference Papers* 1 1-71.

1068 LA Miranda "The Role of International Law in Intrastate Natural Resource Allocation: Sovereignty, Human Rights, and Peoples-Based Development (2012) 45 *Vanderbilt Journal of Transnational Law* 785 785-840; TT Onifade "Peoples-Based Permanent Sovereignty over Natural Resources: Toward Functional Distributive Justice?" (2015) 16 *Human Rights Review* 343 343-368;.

1069 TT Onifade "Peoples-Based Permanent Sovereignty over Natural Resources: Toward Functional Distributive Justice?" (2015) 16 *Human Rights Review* 343 364.

1070 TT Onifade "Peoples-Based Permanent Sovereignty over Natural Resources: Toward Functional Distributive Justice?" (2015) 16 *Human Rights Review* 343 364.

1071 R Pereira & O Gough "Permanent sovereignty over natural resources in the 21st century: natural resource governance and the right to self-determination of indigenous peoples under international law" (2013) 14 *Melbourne Journal of International Law* 1 15; E Blanco & J Razzaque Globalisation and Natural Resources Law; Challenges, Key Issues and Perspectives 2011 133; HM Haugen "Peoples' right to self-determination and self-governance over natural rea significant and complex field of international law. It is accordingly necessary to limit the discussion on self-determination within the research project, as the scope of the research project does not allow for a complete exposition of this right. Economic self-determination must be seen as separate from secessionary self-determination.[1072] Economic self-determination is not concerned with challenging state existence, but with participating in the control over natural resources within state borders.[1073] The research project is concerned only with the branch of self-determination which is referred to as 'economic self-determination' or 'natural resource self-determination'.[1074] According to Farmer, it is natural to conclude that the right to economic self-determination belongs to people, not states.[1075]

The fact that people's right to self-determination includes economic self-determination or, put differently, sovereignty over natural resources is not in dispute. Sovereignty over natural resources has long been accepted

---

sources: Possible and desirable?" (2014) 8(1) *Nordic Journal of Applied Ethics* 3 3-21.

1072 A Farmer "Towards a meaningful rebirth of economic self-determination: Human rights realization in resource-rich countries" 2006 (39) International Law and Politics 417 438, 441.

1073 A Farmer "Towards a meaningful rebirth of economic self-determination: Human rights realization in resource-rich countries" 2006 (39) International Law and Politics 417 441.

1074 D Cambou & S Smis "Permanent sovereignty over natural resources from a human rights perspective: Natural resources exploitation and indigenous peoples' rights in the arctic" (2013) 22(1) Michigan State International Law Review 347 358-359; F Visser "The principle of permanent sovereignty over natural resources and the nationalisation of foreign interests" (1988) XXI Comparative and International Law Journal of South Africa 76 90.

1075 A Farmer "Towards a meaningful rebirth of economic self-determination: Human rights realization in resource-rich countries" 2006 (39) International Law and Politics 417 452.
However, note difficulty existed in the drafting of the ICESCR and ICCPR (in the context in foreign investment), when it came to deciding who the right holders were, namely people or states. [J Gilbert "The right to freely dispose of natural resources: Utopia or forgotten right?" (2013) 31(2) Netherlands Quarterly of Human Rights 314 324.]
Also see discussion in Section 2.5.3 in regard to people as subjects of PSNR and Section 4.3.2.2 in regard to the drafting of the ICESCR and ICCPR.

as a constituent,[1076] a component[1077] and an attribute[1078] of the right to self-determination. The inclusion of the right of people to self-determination and the right of people to economic self-determination, as contained in the most relevant UNGA resolutions and international instruments, will be discussed below.

UNGA resolution 626 (VII)[1079] remembers the "right of peoples freely to use and exploit their natural wealth and resources" and clearly states this right is "inherent in their sovereignty". Resolution 1314 (XIII),[1080] entitled "Recommendations concerning international respect for the right of peoples and nations to self-determination", states clearly that the right of people to self-determination includes "permanent sovereignty over their natural wealth and resources". Resolution 1514 (XV)[1081] affirms this right particularly in the context of people as it contains the Declaration on the Granting of Independence to Colonial Countries and Peoples. The resolution declares that people have the right to self-determination and that "by virtue of that right they freely determine their political status and freely pursue their economic, social and cultural development."[1082] UNGA resolution 1514 (XV)[1083] further stipulates that "peoples may, for their own ends, dispose freely of their natural wealth and resources" without prejudice to any obligations arising out of international economic co-operation" and that the General Assembly is "conscious of the need for [...] respect for the principles of equal rights and self-determination of all peoples, and of universal respect for, and observance of, human rights". The landmark resolution for PSNR, resolution 1803 (XVII),[1084] provides that: "the right

---

1076 RX Perrez "The relationship between 'permanent sovereignty' and the obligation not to cause transboundary environmental damage" 1996 (26) Environmental Law 1187 1190.

1077 G Abi-Saab "Permanent Sovereignty over Natural Resources and Economic Activities" in M Bedjaoui (ed) International Law: Achievements and Prospects (1991) 597 600.

1078 LS Clark "International Law and Natural Resources" 1976-1977 (4) Syracuse Journal of International Law and Commerce 377 378.

1079 UNGA resolution 626 (VII) of 21 December 1952.

1080 UNGA resolution 1314 (XIII) of 12 December 1958.

1081 UNGA resolution 1514 (XV) 14 December 1960.

1082 Article 2, Declaration on the Granting of Independence to Colonial Countries and Peoples 1960 (as adopted by the General Assembly resolution 1514 (XV) 14 December 1960).

1083 UNGA resolution 1514 (XV) of 14 December 1960.

1084 UNGA resolution 1803 (XVII) of 14 December 1962.

of peoples and nations to permanent sovereignty over their natural wealth and resources must be exercised in the interest of their national development and of the well-being of the people of the State concerned".

Article 1(2) of the UN Charter, which sets out the purpose of the United Nations, includes the purpose of developing "friendly relations among nations based on respect for the principle of equal rights and self-determination of peoples". Article 1(1) of the International Covenants on Human Rights mirrors the Declaration on the Granting of Independence to Colonial Countries and Peoples by including a similar phrasing of the right to self-determination, which reads as follows:

> "All peoples have the right of self-determination. By virtue of that right they freely determine their political status and freely pursue their economic, social and cultural development."

Article 1(2) of the International Covenants on Human Rights reads as follows:

> "All peoples may, for their own ends, freely dispose of their natural wealth and resources without prejudice to any obligations arising out of international economic co-operation, based upon the principle of mutual benefit, and international law. In no case may a people be deprived of its own means of subsistence."

Article 1(1) of the International Covenants on Human Rights therefore includes the right to self-determination, while article 1(2) of the International Covenants on Human Rights includes the right to economic self-determination. Article 1(2) also links the right of people to dispose freely of their natural resources with the right of people not to be deprived of their means of subsistence. However, the right to dispose freely with their natural wealth and resources was not initially included in this form in the draft text of the International Covenants on Human Rights. Originally, the right to self-determination was to include an arguably stronger assertion, namely the right to 'permanent sovereignty over their natural wealth and resources'. The original draft of common article 1(3) of the International Covenants on Human Rights read as follows:

> "The right of peoples to self-determination shall also include permanent sovereignty over their natural wealth and resources. In no case may a people

be deprived of its own means of subsistence on the grounds of any rights that may be claimed by other States."[1085]

The inclusion of the phrase 'permanent sovereignty over natural wealth and resources' as part of people's right to self-determination was a very contentious issue during the drafting stage of the International Covenants on Human Rights, and was ultimately not included in the final draft. The inclusion of permanent sovereignty over natural resources as a right of people was seen as contradicting the established international law position that states (most prominently) were to exercise permanent sovereignty over natural resources.[1086] This supports the contention made at various points throughout the research project that PSNR was originally a more state-centric than people-centric principle. It was feared that the inclusion of the term 'permanent sovereignty over natural resources', especially in the context of foreign investment, would authorise "unwarranted" taking of foreign property and would further "subject international agreements and arrangements to unilateral renunciation".[1087]However, the UNGA decided that, even though the right to permanent sovereignty over natural resources was not included in the International Covenants on Human Rights, the right of self-determination included "the simple and elementary principle that a nation or people should be master of its own natural wealth or resources".[1088] Furthermore, the Human Rights Committee confirmed that Article 1(2) of the International Covenants on Human Rights in their current form today can be interpreted so that self-determination confers natu-

---

1085 UNGA "Annotations on the text of the draft International Covenants on Human Rights" (1955) (15) UNGA <http://www2.ohchr.org/english/issues/opinion/articles1920_iccpr/docs/A-2929.pdf> (accessed 12-02-2015).

1086 J Gilbert "The right to freely dispose of natural resources: Utopia or forgotten right?" (2013) 31(2) Netherlands Quarterly of Human Rights 314 322.

1087 UNGA "Annotations on the text of the draft International Covenants on Human Rights" (1955) (15) UNGA <http://www2.ohchr.org/english/issues/opinion/articles1920_iccpr/docs/A-2929.pdf> (accessed 12-02-2015).
Also see: J Gilbert "The right to freely dispose of natural resources: Utopia or forgotten right?" (2013) 31(2) Netherlands Quarterly of Human Rights 314 323-234.

1088 UNGA "Annotations on the text of the draft International Covenants on Human Rights" (1955) (15) UNGA <http://www2.ohchr.org/english/issues/opinion/articles1920_iccpr/docs/A-2929.pdf> (accessed 12-02-2015).

ral resources rights to people and a duty to states to protect these rights.[1089]

The widespread fear that the final text of article 1 of the International Covenants on Human Rights might make it possible for countries to exploit underdeveloped countries under the guise of technical assistance or international cooperation[1090] led to the eventual adoption of article 25 of the ICESCR and article 47 of the ICCPR in the final text of both Covenants. Article 25 of the ICESCR and Article 47 of the ICCPR are also relevant in the context of establishing people's rights to natural resource self-determination and provide that people have an "inherent right" to "enjoy and utilize fully and freely their natural wealth and resources". The ICESCR and the ICCPR therefore both include the people's right to dispose of natural resources as a qualified right in common article 1, and again as an absolute right in articles 25 and 47 respectively. Accordingly the people's right to dispose of natural resources is included twice in both International Covenants on Human Rights.[1091] This inclusion cemented the link between human rights, self-determination and natural resources that has long been presumed.[1092] At present, the existence of the right of people to dispose freely of their natural resources is to a large degree reliant on the development of the International Covenants on Human Rights as discussed above.

1089 Human Rights Committee "Concluding observations of the Human Rights Committee: Canada" (07-04-1999) (page numbers unavailable) CAID <http://caid.ca/UNHRCCan1999.pdf> (accessed 07-09-2016); D Cambou & S Smis "Permanent sovereignty over natural resources from a human rights perspective: Natural resources exploitation and indigenous peoples' rights in the arctic" 2013 22(1) Michigan State International Law Review 347 360.

1090 J Gilbert "The right to freely dispose of natural resources: utopia or forgotten right?" (2013) 31(2) *Netherlands Quarterly of Human Rights* 314 324.

1091 J Gilbert "The right to freely dispose of natural resources: utopia or forgotten right?" (2013) 31(2) *Netherlands Quarterly of Human Rights* 314 324.

1092 RX Perrez "The relationship between 'permanent sovereignty' and the obligation not to cause transboundary environmental damage" (1996) 26 *Environmental Law* 1187 1190; G Abi-Saab "Permanent Sovereignty over Natural Resources and Economic Activities" in M Bedjaoui (ed) *International Law: Achievements and Prospects* (1991) 597 600; Schrijver *Sovereignty Over Natural Resources, Balancing Rights and Duties* (1997) 49-56; LS Clark "International Law and Natural Resources" (1976-1977) 4 *Syracuse Journal of International Law and Commerce* 377 378.

The Declaration on the Right to Development 1986 (Declaration on the Right to Development) clearly states that the right to development and the right to self-determination belong to "all peoples".[1093] In addition, article 2(1) of the Declaration on the Right to Development states that every "human person" should participate in and benefit from the right to development.[1094] Furthermore, article 2(2) makes it clear that "all human beings" have a responsibility in this context, either individually or collectively, which supports the idea of active participation of the community in achieving their right to dispose freely of natural resources,[1095] although people's duties in the context of PSNR is not discussed in the research project as mentioned in Section 1.5. The Declaration also recalls, in its preamble, the right of people to exercise "full and complete sovereignty over all their natural wealth and resources" subject to the provisions of the International Covenants on Human Rights.

In addition, article 21(1) of the Banjul Charter stipulates that:

> "All peoples shall freely dispose of their wealth and natural resources. This right shall be exercised in the exclusive interest of the people. In no case shall a people be deprived of it".[1096]

1093 Article 1(1)-(2) of the Declaration on the Right to Development states the following: "(1) The right to development is an inalienable human right by virtue of which every human person and all peoples are entitled to participate in, contribute to, and enjoy economic, social, cultural and political development, in which all human rights and fundamental freedoms can be fully realized"; (2) "The human right to development also implies the full realization of the right of peoples to self-determination, which includes, subject to the relevant provisions of both International Covenants on Human Rights, the exercise of their inalienable right to full sovereignty over all their natural wealth and resources.".

1094 Article 2(1) of the Declaration on the Right to Development states the following: "The human person is the central subject of development and should be the active participant and beneficiary of the right to development.".

1095 Article 2(2) of the Declaration on the Right to Development states the following: "All human beings have a responsibility for development, individually and collectively, taking into account the need for full respect for their human rights and fundamental freedoms as well as their duties to the community, which alone can ensure the free and complete fulfilment of the human being, and they should therefore promote and protect an appropriate political, social and economic order for development.".

1096 Article 21(2)-21(5) of the Banjul Charter further reads as follows: (2) "In case of spoliation the dispossessed people shall have the right to the lawful recovery of its property as well as to an adequate compensation"; (3) "The free disposal of wealth and natural resources shall be exercised without prejudice to the obli-

Article 21 therefore clearly bestows upon all people a right to dispose freely of their wealth and natural resources, and furthermore stipulates that this right must be exercised by states in the "exclusive interest of the people", a characteristic additional to those mentioned in the International Covenants on Human Rights.

People have already been established as subjects of PSNR in Section 2.5.3, and the content of people's rights in terms of PSNR remains to be determined. Abi-Saab describes PSNR as "essentially a people's right".[1097] Qin also advances PSNR as a right belonging to people and refers specifically to PSNR as a human right.[1098] However, Qin still views the state as a representative of the people, exercising this right in the interest of the people.[1099] A more extreme view, forwarded by Duruigbo, is that the right to PSNR theoretically belongs to people, but since it is a right also belonging to the state, it is unnecessary to refer to this as a right belonging to people.[1100] This is not the viewpoint promoted in the research project. The author submits that PSNR is a right of states and people, and this interpretation means that the state can exercise this right on behalf of people, but people can also exercise this right. This contention is in line with that of Gilbert, where he stipulates that to ensure that people are the

---

gation of promoting international economic cooperation based on mutual respect, equitable exchange and the principles of international law"; (4) "States parties to the present Charter shall individually and collectively exercise the right to free disposal of their wealth and natural resources with a view to strengthening African unity and solidarity"; (5) "States parties to the present Charter shall undertake to eliminate all forms of foreign economic exploitation particularly that practiced by international monopolies so as to enable their peoples to fully benefit from the advantages derived from their national resources".

1097 G Abi-Saab "Permanent Sovereignty over Natural Resources and Economic Activities" in M Bedjaoui (ed) *International Law: Achievements and Prospects* (1991) 597 603.

1098 JY Qin "Reforming WTO Discipline on Export Duties: Sovereignty over Natural Resources, Economic Development and Environmental Protection" 2012 (4) *Wayne State University Law School Research Paper* <http://papers.ssrn.com/sol3/papers.cfm?abstract_id=2030477> (accessed 16-03-2015).

1099 JY Qin "Reforming WTO Discipline on Export Duties: Sovereignty over Natural Resources, Economic Development and Environmental Protection" 2012 (4) *Wayne State University Law School Research Paper* <http://papers.ssrn.com/sol3/papers.cfm?abstract_id=2030477> (accessed 16-03-2015).

1100 E Duruigbo "Permanent Sovereignty and Peoples' Ownership of Natural Resources in International Law" 2006 (38) The *George Washington International Law Review* 33 45, 49.

main beneficiaries of the exploitation of natural resources, they must be designated as right holders of economic self-determination.[1101] It is further submitted by Gilbert that this means that the interest of the people must be the "exclusive driving force" behind all decisions relating to natural resource use.[1102] Section 4.3.3 and 4.4 will show that people can exercise the right of economic self-determination and do not necessarily need the state to exercise it on their behalf. It is in this way that it becomes possible for people to hold the state accountable for climate commitments made in terms of the international legal climate change regime. In fact, where the state is removed as the actor, which is to act on behalf of people, it can help to avoid situations where people's rights are "hijacked" by states and people instead can become the active participants in ensuring that they benefit from the right.[1103] There is ample evidence that where this right is entrusted only to states, people's interests are often neglected. Onifade contends that it is the lack of participation of people within the natural resource sector and the unilateral approach followed by states in terms of PSNR, that have led directly to widespread socioeconomic problems within the natural resource sector, such as the so-called research curse.[1104]

1101 A Farmer "Towards a meaningful rebirth of economic self-determination: Human rights realization in resource-rich countries" (2006) 39 International Law and Politics 417 438, 445.

1102 J Gilbert "The right to freely dispose of natural resources: Utopia or forgotten right?" (2013) 31(2) Netherlands Quarterly of Human Rights 314 325.

1103 E Duruigbo "Permanent Sovereignty and Peoples' Ownership of Natural Resources in International Law" (2006) 38 *The George Washington International Law Review* 33 46, 64.

1104 TT Onifade "Peoples-Based Permanent Sovereignty over Natural Resources: Toward Functional Distributive Justice?" (2015) 16 *Human Rights Review* 343 344.
For an introduction to the so-called resource curse, see JES Fawcett & A Parry *Law and International Resource Conflicts* (1981); RF Mikesell "Explaining the resource curse, with special reference to mineral-exporting countries" (1997) 23 *Resources Policy* 191-199; JD Sachs & AM Warner "Natural Resources and Economic Development, The curse of natural resources" (2001) 45 *European Economic Review* 827-838. For more information to the resource curse within the context of climate change, see: DA Mwiturubani & J van Wyk "Climate Change and Natural Resources Conflicts in Africa" (date unavailable) (page numbers unavailable) *Institute for Security Studies* <https://www.issafrica.org/uploads/Mono170.pdf> (accessed 07-09-2016).

4 3 2 3 Human rights and the right to economic self-determination

As submitted, the research project is concerned with the branch of self-determination known as economic self-determination or natural resource self-determination, and included in the right of economic self-determination is the right to dispose freely of natural wealth and resources and the people's right not to be deprived of their means of subsistence. Economic self-determination cannot be separated from the broader human rights discourse in the context of exploitation of natural resources,[1105] and people's rights in terms of PSNR can be comfortably discussed from a human rights-perspective. However, it should be noted that this research project is not written from a human rights-perspective.

Human rights are generally considered to be capable of being organised into three generations of rights. The first generation of human rights concerns civil and political rights, serving to promote the principle of liberty and protect people against arbitrary state action.[1106] The second generation of human rights concerns economic, social and cultural rights, serving the individual in his capacity as a member of society and serving to create equality among people.[1107] The third generation of human rights focuses

---

1105 A Farmer "Towards a meaningful rebirth of economic self-determination: Human rights realization in resource-rich countries" 2006 (39) International Law and Politics 417 473.
Also see: M Anderson "Human Rights: Approaches to Environmental Protection: An Overview" in Boyle A & M Anderson (eds) Human Rights Approaches to Environmental Protection (1996) 1 1-24.

1106 K Vasak "A 30-Year Struggle, The sustained efforts to give force of law to the Universal Declaration of Human Rights" (1977) The UNESCO Courier 29 29-32; M Doelle "Climate Change and Human Rights: The Role of the International Human Rights in Motivating States to Take Climate Change Seriously" (2004) 1(2) Macquarie Journal of International and Comparative Environmental Law 179 (page numbers unavailable in electronic version); JC Mubangizi "Towards a new approach to the classification of human rights with specific reference to the African context" (2004) African Human Rights Law Journal (2004) 4(1) 93 95.

1107 K Vasak "A 30-Year Struggle, The sustained efforts to give force of law to the Universal Declaration of Human Rights" (1977) The UNESCO Courier 29 29-32; JC Mubangizi "Towards a new approach to the classification of human rights with specific reference to the African context" (2004) African Human Rights Law Journal (2004) 4(1) 93 95; M Doelle "Climate Change and Human Rights: The Role of the International Human Rights in Motivating States to Take Climate Change Seriously" (2004) 1(2) Macquarie Journal of International

on collective, rather that individual rights and are so-called solidarity rights.[1108] First generation rights tend to be more universally accepted, while third generation rights are not.[1109] The right of people to dispose freely of natural resources is considered as a third generation human right, along with other environmental rights, such as the right to a healthy environment.

The status of the right to dispose freely of natural resources is confirmed as a human right in particular by its inclusion in the International Human Rights Covenants 1966,[1110] where it is termed a "qualified" right in article 1 of both Covenants, and an "absolute" right in article 25 of the ICESCR and in article 46 of the ICCPR.[1111] The right of people to dispose freely of natural resources is also included in the Banjul Charter.[1112]

Promoting the right of people to dispose freely of natural resources in terms of PSNR, as is suggested in the research project as an aid to the international legal climate change regime, also contains other advantages. For instance, promoting the right of people to dispose freely of natural resources can encourage greater stability within states.[1113] The reason for this is that the promotion of the human right of people to dispose freely of

---

and Comparative Environmental Law 179 (page numbers unavailable in electronic version).

1108 M Doelle "Climate Change and Human Rights: The Role of the International Human Rights in Motivating States to Take Climate Change Seriously" (2004) 1(2) Macquarie Journal of International and Comparative Environmental Law 179 (page numbers unavailable in electronic version).

1109 M Doelle "Climate Change and Human Rights: The Role of the International Human Rights in Motivating States to Take Climate Change Seriously" (2004) 1(2) Macquarie Journal of International and Comparative Environmental Law 179 (page numbers unavailable in electronic version).

1110 Article 1(2) of the ICCPR and the ICESCR reads as follows: "All peoples may, for their own ends, freely dispose of their natural wealth and resources without prejudice to any obligations arising out of international economic co-operation, based upon the principle of mutual benefit, and international law. In no case may a people be deprived of its own means of subsistence.".

1111 J Gilbert "The right to freely dispose of natural resources: Utopia or forgotten right?" (2013) 31(2) Netherlands Quarterly of Human Rights 314 324.

1112 Article 21(1) of the Banjul Charter reads as follows: "All peoples shall freely dispose of their wealth and natural resources. This right shall be exercised in the exclusive interest of the people. In no case shall a people be deprived of it".

1113 A Farmer "Towards a meaningful rebirth of economic self-determination: Human rights realization in resource-rich countries" (2006) 39 International Law and Politics 417 455.

natural resources will rarely be successfully implemented in isolation and the approach is likely also to serve to advance other important human rights such as the right to life, the right to education, the right to development and the right to a healthy environment.[1114] Accordingly, the improvement of human rights implementation within a state can instil foreign confidence that increases foreign investment in a state. Furthermore, increasing the number of people with a stake in economic stability and growth of the country, as an effect of the implementation of the right to dispose freely of natural resources by people, can also serve to increase stability in foreign investment dealings.[1115] It becomes evident that a snowball effect of development can be caused by promoting the right to economic self-determination, which can in turn lead to even better infrastructure, attracting even more foreign investment.[1116]

It is notable in this regard that the impact of climate change on human rights in general is recognised, as illustrated by the African Union resolution 153, entitled "Resolution on Climate Change and Human Rights and the Need to Study its Impact in Africa".[1117] In this resolution,[1118] the African Commission on Human Rights and Peoples' Rights (ACHPR) considers "the rights of indigenous peoples and communities to their ancestral lands, cultures, lifestyles and a safe satisfactory environment"[1119]

1114 E Blanco & J Razzaque *Globalisation and Natural Resources Law; Challenges, Key Issues and Perspectives* (2011) 132-133; A Farmer "Towards a meaningful rebirth of economic self-determination: Human rights realization in resource-rich countries" (2006) 39 *International Law and Politics* 417 455.

1115 A Farmer "Towards a meaningful rebirth of economic self-determination: Human rights realization in resource-rich countries" (2006) 39 International Law and Politics 417 455.

1116 A Farmer "Towards a meaningful rebirth of economic self-determination: Human rights realization in resource-rich countries" (2006) 39 International Law and Politics 417 455.

1117 OC Ruppel "Climate Change Law Policy Positions in the African Union and Related Developments in Selected African Countries" in OC Ruppel et al (eds) Climate Change: International Law and Global Governance, Volume II: Policy, Diplomacy and Governance in a Changing Environment (2013) 411 418-419.

1118 African Union resolution 153 of 25 November 2009, entitled "Resolution on Climate Change and Human Rights and the Need to Study its Impact in Africa".

1119 OC Ruppel "Climate Change Law Policy Positions in the African Union and Related Developments in Selected African Countries" in OC Ruppel et al (eds) Climate Change: International Law and Global Governance, Volume II: Policy, Diplomacy and Governance in a Changing Environment (2013) 411 418-419.

and furthermore urges the heads of state and government of the African Union:

> "to ensure that human rights standards safeguards, such as the principle of free, prior and informed consent, be included into any adopted legal text on climate change as preventive measures against forced relocation, unfair dispossession of properties, loss of livelihoods and similar human rights violations".[1120]

It is accordingly reasonable to expect a person's right to natural resources to contribute to dealing with the transboundary environmental concern of climate change.[1121]

4 3 3 Application of the right to economic self-determination

4 3 3 1 Introduction

Having established that the right to dispose freely of natural resources exists for all people as a constituent of the right of self-determination, it remains to be discussed how this right can be applied. Even though international instruments allow the right to dispose freely of natural resources to be considered as an important human right and there are many instances where state decisions about national natural resources have impacted negatively on this right, there has been very little legal action taken by people directly on the basis of this right. Legal action has mostly been limited to a couple of contexts, discussed in the succeeding sections. The first context is that of indigenous people, which will be discussed in Section 4.3.3.2. The next context is that of international investment law, which will be discussed in Section 4.3.3.3. It is possible that in these contexts the right to dispose freely of natural resources would be a useful remedy in situations where states do not act in the interest and well-being of their people.[1122]

1120 African Union resolution 153 of 25 November 2009, entitled "Resolution on Climate Change and Human Rights and the Need to Study its Impact in Africa".

1121 S McInerney-Lankford "Climate Change and Human Rights: An Introduction to Legal Issues" (2009) 33 *Harvard Environmental Law Review* 431-437; E Blanco & J Razzaque *Globalisation and Natural Resources Law; Challenges, Key Issues and Perspectives* (2011) 133.

1122 J Gilbert "The right to freely dispose of natural resources: Utopia or forgotten right?" (2013) 31(2) Netherlands Quarterly of Human Rights 314 332.

It has been submitted by Gilbert that the right of people to dispose freely of natural resources is perhaps an idealistic idea and not practically useful;[1123] however, this is not the view forwarded within the research project. This does not mean that the exercise of people's right to economic self-determination will be free of obstacles.[1124] The next section of the research project will aim to show that the right of people to dispose freely of natural resources is not an idealistic idea, but a viable future application of the principle of PSNR that will not only serve the well-being of people, but will also promote state compliance with climate commitments made in terms of the Kyoto Protocol and the Paris Agreement.

4 3 3 2 Application of economic self-determination by indigenous people

PSNR and the rights of indigenous people form a substantial research topic in its own right, and a complete exposition would fall outside the scope of the research project. This research project is concerned only with discussing the right of indigenous people to dispose freely of natural resources as it pertains to a relatively common instance of enforcement of the right of economic self-determination of people. As was stated in Section 1.5 and 2.5.3, the term 'people' is utilised within the research project, also in the context of indigenous people, and includes the whole of the state population.

The right of people to dispose freely of their natural resources is often applied in the context of indigenous people, as it is a principal right of indigenous people.[1125] One of the most important rights of indigenous peo-

1123 J Gilbert "The right to freely dispose of natural resources: Utopia or forgotten right?" (2013) 31(2) Netherlands Quarterly of Human Rights 314 326.

1124 As articulated in the legal study of the Secretariat of the International Commission of Jurists in the 1970s: "[a] problem which often confronts us and to which as yet no acceptable answer has been found in the provisions of the Charter, is the conflict between the principles of the integrity of sovereign States and the assertion of the right to self-determination […]. Here again, as in the case of human rights, a dangerous deadlock can paralyse the ability of the U.N. to help those involved." [International Commission of Jurists "The Events in East Pakistan, 1971" (1972) (page numbers unavailable) <http://icj.wpengine.netdna-cdn.com/wp-content/uploads/1972/06/Bangladesh-events-East-Pakistan-1971-thematic-report-1972-eng.pdf> (accessed 24-02-2015).].

1125 J Gilbert "The right to freely dispose of natural resources: Utopia or forgotten right?" (2013) 31(2) Netherlands Quarterly of Human Rights 314 327.

ple is the right to ancestral land and territory – this right includes a right to dispose freely of their natural resources in this territory or area.[1126] When the term 'sovereignty' is used in the context of indigenous people, it refers to "the right to manage, govern, or regulate the use of the resources by the indigenous people itself, by individuals, or by others."[1127] The number of indigenous people around the world is estimated to be between 100 and 200 million people, and more than 40 states across the world are home to indigenous people.[1128] This application of the right to dispose freely of natural resources is therefore not to be overlooked. The Final Report of the Special Rapporteur, on the Prevention of Discrimination and Protection of Indigenous Peoples, states that the right to self-determination applies to indigenous people, especially if they suffer from unfair, unequal and oppressive arrangements or agreements concluded with other parties, such as states.[1129] Indigenous people further have a right to development and the right to participate in the realisation of this right.[1130] In order to give effect to these rights, the implementation of sovereignty over natural resources for indigenous people is a prerequisite.[1131] The right to natural resources

1126 J Gilbert "The right to freely dispose of natural resources: Utopia or forgotten right?" (2013) 31(2) Netherlands Quarterly of Human Rights 314 327.

1127 EIA Daes "Prevention of discrimination and protection of indigenous peoples – Indigenous peoples' permanent sovereignty over natural resources" (2004) (page numbers unavailable) ECOSOC <http://www.un.org/esa/socdev/unpfii/documents/Report%20by%20Erica%20Irene%20A.%20Daes.pdf> 15 (accessed 26-02-2015).

1128 P Malanczuk Akehurst's Modern Introduction to International Law 7ed (1997) 106.

1129 EIA Daes "Prevention of discrimination and protection of indigenous peoples – Indigenous peoples' permanent sovereignty over natural resources" (2004) (age numbers unavailable) ECOSOC <http://www.un.org/esa/socdev/unpfii/documents/Report%20by%20Erica%20Irene%20A.%20Daes.pdf> (accessed 26-02-2015).

1130 EIA Daes "Prevention of discrimination and protection of indigenous peoples – Indigenous peoples' permanent sovereignty over natural resources" (2004) (page numbers unavailable) ECOSOC <http://www.un.org/esa/socdev/unpfii/documents/Report%20by%20Erica%20Irene%20A.%20Daes.pdf> (accessed 26-02-2015).

1131 EIA Daes "Prevention of discrimination and protection of indigenous peoples – Indigenous peoples' permanent sovereignty over natural resources" (2004) (page numbers unavailable) ECOSOC <http://www.un.org/esa/socdev/unpfii/documents/Report%20by%20Erica%20Irene%20A.%20Daes.pdf> (accessed 26-02-2015).

and the right to development are therefore closely connected in the context of indigenous people.

The Declaration on the Granting of Independence to Colonial Countries and Peoples, contained in UNGA resolution 1514 (XV)[1132] affirms that "peoples may, for their own ends, dispose freely of their natural wealth and resources". Furthermore, the Declaration also states that there is a need to create stability and well-being, which is based on respect for self-determination of people. The relatively recent UNGA 61/295,[1133] entitled "United Nations Declaration on the Rights of Indigenous Peoples" of 2007 (UNDRIP), recognises the right of indigenous people to resources and self-determination.[1134]

It is clear that the right of indigenous people to natural resources can be applied in two ways. Firstly, when indigenous people claim their right to ancestral land, they also claim the natural resources contained in the territory, and with it the right freely to dispose of their natural resources.[1135] In other words, it can be stated that indigenous people have a right to use and enjoy the natural resources that lie within their ancestral land and territories.[1136] Secondly, indigenous people's right to dispose freely of natural resources includes the right to free, prior and informed consent – a right mentioned in article 32 of UNDRIP.[1137] In other words, indigenous people

---

1132 UNGA resolution 1514 (XV) 14 December 1960.

1133 UNGA resolution 61/295 of 13 September 2007.

1134 UNGA resolution 61/295 of 13 September 2007.
The UNDRIP recognizes the right to resources, and reads as follows: "the urgent need to respect and promote the inherent rights of indigenous peoples which derive from their political, economic and social structures and from their cultures, spiritual traditions, histories and philosophies, especially their rights to their ands, territories and resources."
Article 3 of the UNDRIP recognizes the right to self-determination and reads as follows: "Indigenous peoples have the right to self-determination. By virtue of that right they freely determine their political status and freely pursue their economic, social and cultural development.".

1135 J Gilbert "The right to freely dispose of natural resources: Utopia or forgotten right?" (2013) 31(2) Netherlands Quarterly of Human Rights 314 327.

1136 J Gilbert "The right to freely dispose of natural resources: Utopia or forgotten right?" (2013) 31(2) Netherlands Quarterly of Human Rights 314 329.

1137 J Gilbert "The right to freely dispose of natural resources: Utopia or forgotten right?" (2013) 31(2) Netherlands Quarterly of Human Rights 314 328.
UNGA resolution 61/295 of 13 September 2007 contains UNDRIP. Article 32(2) of UNDRIP reads as follows: "States shall consult and cooperate in good faith with the indigenous peoples concerned through their own representative in-

have a right to give or withhold their consent to any exploitation, exploration and extraction of natural resources in their territories.[1138] To exercise the right to free, prior and informed consent would be an application of the right to dispose freely of natural resources, albeit not as directly as would ideally be the case.[1139]

The Saramaka case[1140] concerned the granting of logging and mining concessions by the state (Suriname) on the territory of the Saramaka people without the full and effective participation of the Saramaka people.[1141] The court found that it is the responsibility of the state to ensure that the Saramaka people are allowed effective participation and reasonable benefit from the exploration and extraction of natural resources within their territory.[1142] The court found that Suriname had not implemented measures that would allow effective participation and reasonable benefit for the Saramaka people, and accordingly the court found that Suriname had violated article 21 of the American Convention on Human Rights 1969 (American Convention on Human Rights) and article 1 of the International Covenants on Human Rights.[1143] The Inter-American Court of Human

stitutions in order to obtain their free and informed consent prior to the approval of any project affecting their lands or territories and other resources, particularly in connection with the development, utilization or exploitation of mineral, water or other resources.".

1138 J Gilbert "The right to freely dispose of natural resources: Utopia or forgotten right?" (2013) 31(2) Netherlands Quarterly of Human Rights 314 329.

1139 J Gilbert "The right to freely dispose of natural resources: Utopia or forgotten right?" (2013) 31(2) Netherlands Quarterly of Human Rights 314 329.

1140 Inter-American Court of Human Rights: Case of the Saramaka People v Suriname (Preliminary Objections, Merits, Reparations, and Costs) 28 November 2007 Inter-American Court of Human Rights Series C No 172 para 122.

1141 MA Orellana "Saramaka People V. Suriname" (2008) 102(4) The American Journal of International Law 841 844.

1142 N Schrijver "Self-determination of peoples and sovereignty over natural wealth and resources" in United Nations Human Rights Office of the High Commissioner (ed) *Realizing the Right to Development* (2013) 95 99.

1143 N Schrijver "Self-determination of peoples and sovereignty over natural wealth and resources" in United Nations Human Rights Office of the High Commissioner (ed) *Realizing the Right to Development* (2013) 95 99.
Article 21 of the American Convention on Human Rights is entitled "Right to Property" and reads as follows: "(1) Every person has the right to use and enjoyment of his property. The law may subordinate such use and enjoyment to the interest of society (2) No one shall be deprived of his property except upon payment of just compensation, for reasons of public utility or social interest, and in

Rights set out certain measures for the state to adhere to in order not to repeat the violation of the rights of indigenous people in the future.[1144] Some of these measures include: implementing effective consultation in accordance with traditions and customs of the indigenous people, allowing indigenous people to withhold their free, informed and prior consent in regard to projects that affect their indigenous territory, and ensuring that the indigenous people obtain a reasonable share in the benefits of a project in their indigenous territory.[1145] The Inter-American Commission on Human Rights (IACHR) found, in the Maya Indigenous Communities case, that the right to dispose freely of natural resources means that the community has the right to prior effective consultation and must be granted an effective opportunity to participate.[1146] Furthermore, in the Saramaka case, the Inter-American Court of Human Rights considered the right of indigenous people to use and enjoy their ancestral land and territory in accordance with their traditions and customs meaningless if not connected to the right to enjoy natural resources that lie on and within the land.[1147]

---

the cases and according to the forms established by law. (3) Usury and any other form of exploitation of man by man shall be prohibited by law.".

1144 Inter-American Court of Human Rights: Case of the Saramaka People v Suriname (Preliminary Objections, Merits, Reparations, and Costs) 28 November 2007 Inter-American Court of Human Rights Series C No 172 para 194.
Note that the Inter-American Human Rights System does not have an equivalent provision to article 21 of the Right to Natural resources in the Banjul Charter. The Inter-American Human Rights System turns to article 21 of the American Convention on Human Rights in this context and reads the right to natural resources into the right to property contained in article 21. [ACHPR: Centre for Minority Rights Development (Kenya) and Minority Rights Group International (on behalf of Endorois Welfare Council) v Kenya 25 November 2009 ACHPR 276/03 par 256.].

1145 Inter-American Court of Human Rights: Case of the Saramaka People v Suriname (Preliminary Objections, Merits, Reparations, and Costs) 28 November 2007 Inter-American Court of Human Rights Series C No 172 para 194; IACHR "Maya Indigenous Communities of the Toledo District v Belize, Merits Report 40/04, Case 12.053" (12 October 2004) (par 5, 117, 142-143) <http://www.cidh.org/annualrep/2004eng/belize.12053eng.htm> (accessed 08-09-2016).

1146 IACHR "Maya Indigenous Communities of the Toledo District v Belize, Merits Report 40/04, Case 12.053" (12 October 2004) (par 5, 117, 142-143) <http://www.cidh.org/annualrep/2004eng/belize.12053eng.htm> (accessed 08-09-2016).

1147 Inter-American Court of Human Rights: Case of the Saramaka People v Suriname (Preliminary Objections, Merits, Reparations, and Costs) 28 November 2007 Inter-American Court of Human Rights Series C No 172 par 118.

In the SERAC case,[1148] it was alleged that the dealings of the Nigerian government with oil consortiums led to oil exploitation and environmental degradation in Ogoniland.[1149] It was found that the Nigerian government did not involve the indigenous Ogoni people in these dealings and related decisions, and the Ogoni people received no benefits from the government's dealings.[1150] It was found that the actions of the Nigerian government constitute a violation of article 21 of the Banjul Charter, which deals with the right of people to dispose freely of natural resources.[1151] However, the decision is mostly considered for its significant contribution to promoting the human right to a satisfactory environment,[1152] and not insofar as it reflects on the right of people to dispose of natural resources in terms of PSNR.

The Endorois case[1153] involved the displacement of the indigenous Endorois community from their ancestral lands in Kenya, which resulted in various violations of the Banjul Charter. The ACHPR had to decide whether article 21 of the Charter, regarding the right to dispose freely of natural resources, was also violated. The Endorois community's displacement from their ancestral lands meant that they were not allowed access to vital resources such as traditional medicine, salt licks for cattle, and fertile

---

1148 ACHPR: Social and Economic Rights Action Center (SERAC) and the Center for Economic and Social Rights v Nigeria 27 October 2001 ACHPR 155/96.

1149 ACHPR: Social and Economic Rights Action Center (SERAC) and the Center for Economic and Social Rights v Nigeria 27 October 2001 *ACHPR 155/96* par 55; N Schrijver "Self-determination of peoples and sovereignty over natural wealth and resources" in United Nations Human Rights Office of the High Commissioner (ed) *Realizing the Right to Development* (2013) 95 99.

1150 ACHPR: Social and Economic Rights Action Center (SERAC) and the Center for Economic and Social Rights v Nigeria 27 October 2001 ACHPR 155/96 par 55.

1151 ACHPR: Social and Economic Rights Action Center (SERAC) and the Center for Economic and Social Rights v Nigeria 27 October 2001 ACHPR 155/96 par 55.

1152 OC Ruppel "Climate Change Law Policy Positions in the African Union and Related Developments in Selected African Countries" in OC Ruppel et al (eds) Climate Change: International Law and Global Governance, Volume II: Policy, Diplomacy and Governance in a Changing Environment (2013) 411 418.

1153 ACHPR: Centre for Minority Rights Development (Kenya) and Minority Rights Group International (on behalf of Endorois Welfare Council) v Kenya 25 November 2009 ACHPR 276/03.

soil.[1154] The ACHPR found that the inaccessibility to natural resources constituted a violation of article 21 of the Banjul Charter.[1155] In addition, mining concessions were granted in regard to the land without giving the indigenous community a share in these resources.[1156] The ACHPR found that article 21 protects the right of the community to potential wealth from their land, and a violation of article 21 occurred when the community did not benefit from their land's potential wealth.[1157] The right of development, as contained in article 22 of the Banjul Charter, pertaining to the right of development, was also found to be violated by the state.[1158] The court found that violating either the procedural or substantive element of the right to development constitutes a violation of the right to development.[1159] In other words, the court found that fulfilling only one element of the right to development (either the procedural element or the substantive element) cannot lead to the fulfilment of the right.[1160] In order for the right to development to be fulfilled, the fulfilment of the following criteria is a useful indicator: equitability, non-discrimination, participation, accountability and transparency.[1161] Furthermore, participation must include

1154 ACHPR: Centre for Minority Rights Development (Kenya) and Minority Rights Group International (on behalf of Endorois Welfare Council) v Kenya 25 November 2009 ACHPR 276/03 par 120.

1155 ACHPR: Centre for Minority Rights Development (Kenya) and Minority Rights Group International (on behalf of Endorois Welfare Council) v Kenya 25 November 2009 ACHPR 276/03 par 120.

1156 ACHPR: Centre for Minority Rights Development (Kenya) and Minority Rights Group International (on behalf of Endorois Welfare Council) v Kenya 25 November 2009 ACHPR 276/03 par 120.

1157 ACHPR: Centre for Minority Rights Development (Kenya) and Minority Rights Group International (on behalf of Endorois Welfare Council) v Kenya 25 November 2009 ACHPR 276/03 par 124.

1158 ACHPR: Centre for Minority Rights Development (Kenya) and Minority Rights Group International (on behalf of Endorois Welfare Council) v Kenya 25 November 2009 ACHPR 276/03 par 124.

1159 ACHPR: Centre for Minority Rights Development (Kenya) and Minority Rights Group International (on behalf of Endorois Welfare Council) v Kenya 25 November 2009 ACHPR 276/03 par 277.

1160 ACHPR: Centre for Minority Rights Development (Kenya) and Minority Rights Group International (on behalf of Endorois Welfare Council) v Kenya 25 November 2009 ACHPR 276/03 par 277.

1161 ACHPR: Centre for Minority Rights Development (Kenya) and Minority Rights Group International (on behalf of Endorois Welfare Council) v Kenya 25 November 2009 ACHPR 276/03 par 277.

free, prior and informed consent.[1162] The approach to the right to dispose freely of natural resources and the right to development is closely connected, and it can be submitted that the approach, suggested by the court above in regard to the right of development, should also be implemented in regard to the right to dispose freely of natural resources.

Even though the right to free, prior and informed consent has been applied in the context of the right freely to dispose of natural resources, it remains a limited and indirect application of the right to dispose freely of natural resources. The limitation lies in the fact that people are giving consent to states to exercise their right to dispose freely of natural resources on their behalf. Such consent occurs after states have made the decision how to dispose of natural resources, which means that this decision has not been made by people and therefore a limit in application exists.[1163] Nonetheless, the fact that the right to dispose freely of natural resources (and in effect the right to economic self-determination) can be applied, even in an indirect manner as seen above (as the right to free, prior and informed consent), is a notable advancement for the right of all people to dispose freely of natural resources.[1164]

As established, the people's right to self-determination (and economic self-determination) also means that people may not be deprived of their means of subsistence.[1165] There is some overlap observed between cases involving indigenous people's right to dispose freely of natural resources and subsistence rights cases of indigenous people. In regard to cases involving subsistence rights the General Comment of the United Nations Committee on Economic, Social and Cultural Rights (CESCR) provides that: "a people may not be deprived of its means of subsistence" and that states should ensure in this regard that there is "adequate access to water for subsistence farming and for securing the livelihoods of indigenous

---

1162 ACHPR: Centre for Minority Rights Development (Kenya) and Minority Rights Group International (on behalf of Endorois Welfare Council) v Kenya 25 November 2009 ACHPR 276/03 par 281-283.

1163 J Gilbert "The right to freely dispose of natural resources: Utopia or forgotten right?" (2013) 31(2) Netherlands Quarterly of Human Rights 314 329.

1164 J Gilbert "The right to freely dispose of natural resources: Utopia or forgotten right?" (2013) 31(2) Netherlands Quarterly of Human Rights 314 329.

1165 A Kiss "The Peoples' Right to Self-Determination" (1986) 7 *Human Rights Law Journal* 165 171.

peoples."[1166] The right of people to dispose freely of natural resources is therefore considered to promote other stand-alone human rights, since it can provide the grounding upon which people can claim subsistence rights, such as, for example, the right to water or the right to food.[1167]

From the above application of the right to economic self-determination, it is evident that in order for the right of people to dispose freely of natural resources to be fully implemented, certain related rights must also be realised. For instance, the full implementation of the right of people to dispose freely of natural resources is also dependent on the right of people to participate in environmental decisions[1168] and the right of people to have access to environmental information in order to review procedures and challenge public environmental decisions. Especially in the case of indigenous people, it is accepted that the right to dispose freely of natural resources goes hand in hand with the right to give free, prior and informed consent.[1169] It is abundantly evident from the case law mentioned above that indigenous people have a right to free, prior and informed consent to any disposal of natural resources in their territories.[1170]

1166 CESCR "General Comment No. 15: The Right to Water (arts. 11 and 12 of the International Covenant on Economic, Social and Cultural Rights)" (20 January 2003) (4) <http://www2.ohchr.org/english/issues/water/docs/CESCR_GC_15.pdf> (accessed 08-09-2016).

1167 J Gilbert "The right to freely dispose of natural resources: Utopia or forgotten right?" (2013) 31(2) Netherlands Quarterly of Human Rights 314 332.

1168 A Farmer "Towards a meaningful rebirth of economic self-determination: Human rights realization in resource-rich countries" 2006 (39) International Law and Politics 417 453.

1169 R Pereira & O Gough "Permanent sovereignty over natural resources in the 21st century: Natural resource governance and the right to self-determination of indigenous peoples under international law" (2013) 14 *Melbourne Journal of International Law* 26; E Blanco & J Razzaque *Globalisation and Natural Resources Law; Challenges, Key Issues and Perspectives* 2011 134; J Gilbert "The right to freely dispose of natural resources: Utopia or forgotten right?" (2013) 31(2) *Netherlands Quarterly of Human Rights* 314 328-329.

1170 The Saramaka case, illustartes this particularly well. In the Saramaka case, the Court made an order, that inter alia provided for the Samaraka people to be "effectively consulted, in accordance with their traditions and customs" and given "the right to give or withhold their free, informed and prior consent, with regards to development or investment projects that may affect their territory" and also determined that the Samaraka people had to share reasonably in the benefits of their territory. [Inter-American Court of Human Rights: Case of the Saramaka People v Suriname (Preliminary Objections, Merits, Reparations, and Costs)

The right to free, prior and informed consent is also well ingrained in international instruments, which apply the right in a broader context than indigenous people. The right to free, prior and informed consent is mentioned in article 32 of UNDRIP[1171] and article 15(2) of the Convention Concerning Indigenous and Tribal Peoples in Independent Countries 1989 (Convention Concerning Indigenous and Tribal Peoples in Independent Countries).[1172] Furthermore, the United Nations Economic Commission for Europe (UNECE) *Convention on Access to Information, Public Participation in Decision-Making and Access to Justice in Environmental Matters* 1998 (Aarhus Convention) is an example of an instrument promoting the participation of all people and not only indigenous people in the environmental decision-making process. The preamble of the Aarhus Convention recognises that "every person has the right to live in an environment adequate to his or her health and well-being, and the duty, both individually and in association with others, to protect and improve the environment for the benefit of present and future generations".[1173] The Aarhus Convention clearly establishes people's rights (and duties) in terms of their environment and it is very relevant in determining the rights of peoples in the context of natural resources and PSNR. The Aarhus Convention, as a regional instrument, establishes environmental rights of the public (on the

---

28 November 2007 Inter-American Court of Human Rights Series C No 172 para 214(8).].

1171 UNGA resolution 61/295 of 13 September 2007 contains UNDRIP. Article 32(2) of UNDRIP reads as follows: "States shall consult and cooperate in good faith with the indigenous peoples concerned through their own representative institutions in order to obtain their free and informed consent prior to the approval of any project affecting their lands or territories and other resources, particularly in connection with the development, utilization or exploitation of mineral, water or other resources.".

1172 Article 15(2) of the Convention concerning Indigenous and Tribal Peoples in Independent Countries reads as follows: "In cases in which the State retains the ownership of mineral or sub-surface resources or rights to other resources pertaining to lands, governments shall establish or maintain procedures through which they shall consult these peoples, with a view to ascertaining whether and to what degree their interests would be prejudiced, before undertaking or permitting any programmes for the exploration or exploitation of such resources pertaining to their lands. The peoples concerned shall wherever possible participate in the benefits of such activities, and shall receive fair compensation for any damages which they may sustain as a result of such activities.".

1173 Preamble of the UNECE Convention on Access to Information, Public Participation in Decision-Making and Access to Justice in Environmental Matters.

level pertaining to individuals and associations).[1174] In order for the rights established by the Aarhus Convention, parties to the Convention are required to enable appropriate national, regional or local authorities to give effect to these rights.[1175] The main rights afforded by the Convention to people (including the associations of individuals) pertain to three areas, namely access to information, participation in decision-making and the ability to challenge decisions made. It follows that the right of people to dispose freely of natural resources in terms of PSNR also gives effect to three more procedural rights of people in this regard, namely access to information, participation in decision-making and the ability to challenge decisions made.

It should be reiterated that the applications mentioned above are discussed in the context of indigenous people because they often lay claim to ancestral or indigenous territory and, because of their marginalised existence, they have a relatively well-recognised right to free, prior and informed consent.[1176] However, the author contends that the right to dispose freely of natural resources is a right applicable to all people of the state, regardless of whether they can be classified as indigenous people or not.[1177] It is submitted that this connection between self-determination and consent can therefore be utilised by all people, and not only indigenous people, and that in no way should any group of people be impeded from claiming this right.[1178] Furthermore, the research project considers people's right to dispose of natural resources as a manifestation of economic

---

1174 EC "The Aarhus Convention" (07-05-2015) (page numbers unavailable) *EC* <http://ec.europa.eu/environment/aarhus/> (accessed 03-09-2015).

1175 EC "The Aarhus Convention" (07-05-2015) (page numbers unavailable) *EC* <http://ec.europa.eu/environment/aarhus/> (accessed 03-09-2015).

1176 J Gilbert "The right to freely dispose of natural resources: Utopia or forgotten right?" (2013) 31(2) Netherlands Quarterly of Human Rights 314 330.

1177 See, for instance: TT Onifade "Peoples-Based Permanent Sovereignty over Natural Resources: Toward Functional Distributive Justice?" (2015) 16 Human Rights Review 343 344; J Gilbert "The right to freely dispose of natural resources: Utopia or forgotten right?" (2013) 31(2) Netherlands Quarterly of Human Rights 314 330.

1178 E Duruigbo "Permanent Sovereignty and Peoples' Ownership of Natural Resources in International Law" (2006) 38 The George Washington International Law Review 33 55; R Pereira & O Gough "Permanent sovereignty over natural resources in the 21st century: natural resource governance and the right to Self-determination of indigenous peoples under international law" (2013) 14 Melbourne Journal of International Law 1 4; ME Salomon & A Sengupta The Right

self-determination, which is a constituent of self-determination. Since self-determination is generally organised as a third generation human right and accordingly a group right,[1179] the research project contends that the right to dispose freely of natural resources is a group right and not an individual right.[1180]

4 3 3 3 Growing possibilities of the application of economic self-determination in the context of foreign investment

Foreign investment is, as provided in Section 4.3.3.1, another context in which legal action is most likely to occur should state decisions about national natural resources negatively affect the right of people to dispose freely of natural resources, and there is some development indicating the gaining of momentum of the right of people to dispose freely natural resources within international foreign investment-related arbitration. As was submitted in Section 4.2.2.6.1, a full exposition of PSNR within the context of foreign investment falls outside the scope of the research project, and accordingly this growing possibility of the application of economic self-determination in the context of foreign investment will be briefly discussed below.

As evident during the drafting stages of the International Covenants on Human Rights, international investment has always presented somewhat of an obstacle to the people's right to dispose freely of natural re-

---

to Development: Obligations of States and the Rights of Minorities and Indigenous Peoples (2003) 35.

1179 In this regard it is noted that there is some debate regarding the classification of human rights according to first, second and third generations, but it falls outside the scope of the research project to include an examination of this system of classification of human rights. For more information on the first, second and third generation human rights, see Section 4.3.2.3.

1180 The understanding of the research project that people's rights in terms of PSNR are group rights and not individual rights is in line with the understanding supported by Miranda [LA Miranda "The role of international law in intrastate natural resource allocation: Sovereignty, human rights, and peoples-based development" (2012) 45 Vanderbilt Journal of Transnational Law 785 785-840] and Cambou and Smis [D Cambou D & S Smis "Permanent sovereignty over natural resources from a human right perspective: Natural resources exploitation and indigenous peoples' rights in the artic" (2013) 22(1) Michigan State International Law Review 347 347–376].

sources.[1181] This is because international investment agreements attempt to balance rights of the investor (which involve ensuring security and confidence on the side of the investor) with rights of the host state (which involve ensuring a measure of control over the natural resources being exploited).[1182] Mostly left out of the equation in international investment law agreements or treaties (at least until recently) is, *inter alia*, the human right to natural resources.[1183] However, with the more recent international trend of privatisation of essential services to international investors, more references to health and safety and the environment (and accordingly human rights) have found their way into bilateral investment treaties – which has resulted in an increase in foreign investment arbitration related to human rights.[1184] Thus far, a number of arbitrations have already occurred in the context of the privatisation of water and sewage systems, electricity generation and supply, telecommunication and public transport services.[1185] Most notably, the right to water has been a concern in international arbitration cases, such as the case of Aguas de Aconquija, which concerns the privatisation of water services in Argentina.[1186]

The increasing role of human rights and specifically how the right of people to dispose freely of natural resources can become more integrated

1181 UNGA "Annotations on the text of the draft International Covenants on Human Rights" (1955) (15) UNGA <http://www2.ohchr.org/english/issues/opinion/articles1920_iccpr/docs/A-2929.pdf> (accessed 12-02-2015); J Gilbert "The right to freely dispose of natural resources: Utopia or forgotten right?" (2013) 31(2) Netherlands Quarterly of Human Rights 314 334.

1182 J Gilbert "The right to freely dispose of natural resources: Utopia or forgotten right?" (2013) 31(2) Netherlands Quarterly of Human Rights 314 334.

1183 J Gilbert "The right to freely dispose of natural resources: Utopia or forgotten right?" (2013) 31(2) Netherlands Quarterly of Human Rights 314 336.

1184 See in this regard the relatively newly emerged topic of 'privatisation of human rights', as explained by U Kriebaum "Privatizing Human Rights, The Interface between International Investment Protection and Human Rights" in A Reinisch & U Kriebaum (eds) The Law of International Relations (2007) 165 165-189.

1185 F Marella "On the Changing Structure of International Investment Law: The Human Right to Water and ICSID Arbitration " (2010) 12 International Community Law Review 335 348; U Kriebaum "Privatizing Human Rights, The Interface between International Investment Protection and Human Rights" in A Reinisch & U Kriebaum (eds) The Law of International Relations (2007) 165 166.

1186 ICSID: Case of Compañía de Aguas del Aconquija S.A. & Vivendi Universal S.A. v Argentine Republic (Case of Aguas de Aconquija v. Argentina) 21 November 2000 *International Legal Materials* (2000) 40(2) 462-453.

in investment disputes in the future is illustrated in the Piero Foresti case.[1187] In this foreign investment dispute, foreign investors filed a claim against the host state (South Africa) for violating provisions under two bilateral investment treaties.[1188] The foreign investors submitted *inter alia* that the South African government had unfairly discriminated against them by adopting new legislation,[1189] that provided that all existing mineral rights had to be reverted to the government unless companies and corporations having mineral rights converted their "old order" exploration and mining rights into "new" rights under terms specified in the new legislation.[1190] The right of people to natural resources is indirectly relevant to this case, because the new legislation that the South African government introduced was *inter alia* aimed at advancing previously disadvantaged persons of South Africa by ensuring that the benefits of natural resource exploitation in the country are shared equally between the people of the country.[1191] Therefore, as put by Gilbert, this case grappled with "the issue

---

1187 ICSID: In the Matter of an Arbitration between Piero Foresti, Laura de Carli and others v Republic of South Africa ICSID Case No ARB(AF)/07/01 8 January 2007.

1188 J Gilbert "The right to freely dispose of natural resources: Utopia or forgotten right?" (2013) 31(2) Netherlands Quarterly of Human Rights 314 336.

1189 MPRDA, adopted by the South African Parliament in 2002.
For more information on the MPRDA, see: FT Cawood "The Mineral and Petroleum Resources Development Act of 2002: A paradigm shift in mineral policy in South Africa" (2004) January/February The Journal of The South African Institute of Mining and Metallurgy 53-64.

1190 J Gilbert "The right to freely dispose of natural resources: Utopia or forgotten right?" (2013) 31(2) Netherlands Quarterly of Human Rights 314 336; ICSID: In the Matter of an Arbitration between Piero Foresti, Laura de Carli and others v Republic of South Africa ICSID Case No ARB(AF)/07/01 8 January 2007 par 54, 55.

1191 J Gilbert "The right to freely dispose of natural resources: Utopia or forgotten right?" (2013) 31(2) Netherlands Quarterly of Human Rights 314 336; ICSID: In the Matter of an Arbitration between Piero Foresti, Laura de Carli and others v Republic of South Africa ICSID Case No ARB(AF)/07/01 8 January 2007 par 56.
Section 24 of the South African Constitution 1996 provides that "Everyone has the right (a) to an environment that is not harmful to their health or wellbeing; and
(b) to have the environment protected, for the bene t of present and future generations, through reasonable legislative and other measures that (i) prevent pollution and ecological degradation;
(ii) promote conservation; and

of defining whether a country could introduce legislation aiming at ensuring a fairer disposal of the country's natural resources in the name of its people" and this issue is close to a direct application of the right of people to dispose freely of their natural resources.[1192]

Accordingly, the possibility that the right of people to dispose of natural resources may become a more common issue on which to decide on in the context of international investment is increasing.

4 4 Expanding the means to combat climate change through people's rights in terms of the principle of sovereignty over natural resources

4 4 1 Introduction and initial steps for people-based action in holding states accountable for climate commitments

There are a number of platforms on which to base climate change litigation. This research project is concerned with climate change litigation based specifically on the human right to dispose freely of natural resources. It is argued that the possibility of litigation based on the human right to dispose freely of natural resources (a right of people in terms of PSNR) can constitute a negative incentive for state compliance with climate commitments. In this way it is possible to expand the means by which climate change can be combated.

'Climate commitments' is the term used in this research project to refer to the legally binding emission targets in terms of the Kyoto Protocol and the NDCs in terms of the Paris Agreement. As mentioned in Chapter 3, climate change moved from being a scientific issue to being an intergovernmental issue in 1988. Since 1988, the remedies to curb climate change have been overwhelmingly state-oriented, and state sovereignty has made it difficult to hold states accountable on an international level in terms of the international legal climate change regime. Since it is difficult for an international organisation or body to hold states accountable on an international level, a possible solution to this problem is for people of a state to hold states accountable on a national or regional level. In the particular

---

(iii) secure ecologically sustainable development and use of natural resources while promoting justifiable economic and social development.".

1192 J Gilbert "The right to freely dispose of natural resources: Utopia or forgotten right?" (2013) 31(2) Netherlands Quarterly of Human Rights 314 336.

context of climate change, allowing people to initiate climate change action on a national or regional level can constitute a negative incentive for states to abide by their climate commitments,[1193] and can accordingly present a method of holding states accountable to the international legal climate change regime. It is suggested that the people's right to dispose freely of their natural resources is "largely forgotten"[1194] as a way in which to influence state behaviour. Within the international climate change regime, the role of people is highly neglected and the value that they can add to the regime by increasing state compliance to climate commitments constitutes an untapped resource within climate change law. Recently, cases brought against states by people of the state have started to unlock this people-based potential and reveal new possibilities for ensuring the future compliance of states with climate commitments. If the suggested approach proposed in this research project is promoted and fostered, this type of action by people will be able to contribute greatly to increasing the effectiveness of the international legal climate change regime.

The success of this approach is contingent on a few initial steps, introduced as initial steps or assumptions in Section 1.4. Firstly, any state of which compliance is to be enforced should actually have made a climate commitment in terms of the Kyoto Protocol or the Paris Agreement whereby the people of the state could hold the state accountable. There has to be some legal obligation to which the state has committed itself, but failed to deliver upon. However, as will be shown in the Urgenda case[1195] discussed in Section 4.4.3, the people actually held the state accountable for not doing more than it committed to since it was possible to do so. This leads to the conclusion that this initial step is not entirely set in stone.

The second step is that within a state where people wish to ensure state compliance, the state must function in a way that would be congruent with the people-based action of this kind. A significant hurdle in this respect would be if a state has not incorporated rights (such as those found in the

1193 S Barrett & R Stavins "Increasing Participation and Compliance in International Climate Change Agreements" (2002) 94 *Nota di Lavoro* 1 18-19.

1194 J Gilbert "The right to freely dispose of natural resources: Utopia or forgotten right?" (2013) 31(2) *Netherlands Quarterly of Human Rights* 314 336.

1195 The Hague District Court: Urgenda Foundation v The Kingdom of the Netherlands (Ministry of Infrastructure and the Environment) 24 June 2015 *Rechtbank Den Haag C/09/456689 / HA ZA 13-1396 (English translation).*

Aarhus Convention) into its domestic legal system.[1196] In other words, the state should implement a national legal system whereby it is possible for people of the state to hold their state accountable. In this regard, state recognition and the functioning of a state also become important. Both of these elements are discussed in Section 2.5. Although there may appear to be a stumbling block here, in the way of the practical implementation of the idea that people are able to hold their state accountable for climate commitments, there should be no serious hindrances in reality. When one considers the states which might not allow these types of mechanisms to be available to their citizens, it can be suggested – although generally – that these might tend to be less developed states and, and as shown in Chapter 3, less developed states often contribute least to the greenhouse gas emissions responsible for climate change. On the other hand, developed states which contribute massively to greenhouse gas emissions would – again generally – tend to have the developments enshrined in their legal systems that would enable people to challenge state action in a court of law of the state. Should a state have the appropriate legal system by which to pursue this people-based action, further stumbling blocks in this context could still include the efficacy and affordability of the legal system.[1197]

There is also another possible limitation or initial step that comes to light in this context, which is that only the nationals of a specific state may be able to hold their state accountable for its climate commitments. It is possible to argue that human right claims are generally directed at the state by the state's citizens (whether on a person's own behalf, by means of a representative, or as a group action).[1198] However, here it is suggested, especially in regard to transboundary environmental concerns such as climate change, that there is "no reason to limit the application of international human rights law only to violations by a state against its own citi-

---

1196 E Blanco & J Razzaque *Globalisation and Natural Resources Law; Challenges, Key Issues and Perspectives* (2011) 133.

1197 D Shelton "Human Rights and the Environment: Problems and Possibilities" (2008) 38(1) *Environmental Policy and Law* 42 as quoted by E Blanco & J Razzaque *Globalisation and Natural Resources Law; Challenges, Key Issues and Perspectives* (2011) 132-133.

1198 M Doelle "Climate Change and Human Rights: The Role of the International Human Rights in Motivating States to Take Climate Change Seriously" (2004) 1(2) *Macquarie Journal of International and Comparative Environmental Law* 179 (page numbers unavailable in electronic version).

zens especially if there are no avenues for the citizen's own state to protect its citizens from the harm".[1199] For the purposes of this research project, it is not necessary to solve this debate. The author of the research project takes the view that people of the state, encompassing the whole of the state population or specific groups of people (including indigenous people), would hold the state accountable in the suggested people-based action, but the author accepts that it might be possible for people other than state nationals to also hold the state accountable. Should it be convincingly argued that any group of state nationals can hold any state (and not necessarily their own state) accountable, the main research objectives of the research project remain unaffected.

4 4 2 People-based action against the state in terms of the Oslo Principles on Global Climate Change

Under the Kyoto Protocol states have made legally binding commitments in regard to their emission levels, and under the Paris Agreement states have submitted nationally determined contributions. In both instances it is possible for the people of the state to know what a state has pledged to do in regard to climate change, and then institute action that serves to hold the state accountable should it not adhere to these commitments or even agree to commit itself. However, the Oslo Principles support the idea that states ought to be held accountable for their actions or lack thereof in the context of climate change, even if the state has not made a climate commitment as described above.[1200] In other words, the Oslo Principles ask what can legally be expected from states to combat climate change in the

---

1199 M Doelle "Climate Change and Human Rights: The Role of the International Human Rights in Motivating States to Take Climate Change Seriously" (2004) 1(2) *Macquarie Journal of International and Comparative Environmental Law* 179 (page numbers unavailable in electronic version).

1200 The commentary on the Oslo Principles provides that "[s]tates are *legally obliged* to reduce their GHG- emissions, even if they do not conclude (further) international agreements or conventions." [Oslo Principles on Global Climate Change Obligations "Oslo Principles Commentary" (06-04-2015) (8) *Oslo Principles on Global Climate Change Obligations* <http://www.osloprinciples.org/macmillan/globaljustice/Oslo%20Principles%20Commentary.pdf> (accessed 14-08-2016)]. This idea finds practical application in the Urgenda case discussed in Section 4.4.4.3.

absence of any specific agreement or treaty.[1201] The Oslo Principles are an endeavour to set out "as concretely as possible" what the legal obligations of states and enterprises are in regard to climate change.[1202] The Oslo Principles try to remove the uncertainty of what must be done by each state to curb climate change, since there cannot be any effective action if it is unclear what the action should be.[1203] The Oslo Principles can be categorised as soft law. Soft law is succinctly explained by Hinz as follows:

> "The category of soft law was developed in order to have a characterisation of norms which had not yet received the quality of law, but which enjoyed wide support by all who were dealing with a specific matter with authority. [...] States are not obliged to implement existing soft law, but they may be confronted with critical questions as to why they have ignored it. The fact that something has the quality of soft law changes the burden of proof to the deviating authority to give good reasons why certain rules equipped with high authority have been ignored."[1204]

Therefore, the Oslo Principles are a valuable contribution to addressing the issue of climate change on an international level, but it has not attained the status of law. The Oslo Principles are considered to be a positive contribution to solve the political deadlock within the international climate change negotiations at the time during which the Principles were formulated, in which the attitude of states can be described as follows:

1201 Global Justice Program "The Oslo Principles on Global Climate Change Obligations Launched by Expert Group Including GJP Director Thomas Pogge" (date of publication unavailable) (page numbers unavailable) *Yale Global Justice Group* <http://www.yale.org/macmillan/globaljustice/news.html> (accessed 08-09-2016).

1202 Oslo Principles on Global Climate Change Obligations "Oslo Principles Commentary" (06-04-2015) (8) *Oslo Principles on Global Climate Change Obligations* <http://www.osloprinciples.org/macmillan/globaljustice/Oslo%20Principles%20Commentary.pdf> (accessed 14-08-2016).

1203 Global Justice Program "The Oslo Principles Commentary" (date unavailable) (4) *Yale Global Justice Program* <http://www.yale.edu/macmillan/globaljustice/OsloPrinciplesCommentary.pdf> (accessed 11-05-2015).

1204 MO Hinz "Agenda 21 and Climate Protection: The Development of Global and Local Governance for Environment and Development - Obeservations from Research in Namibia" in OC Ruppel et al (eds) Climate change: International law and global governance. Vol. 1 Policy, diplomacy and governance in a changing environment. (2013) 621 649.

> "We all agree that something needs to be done, but we cannot agree on who has to do what and how much. In the absence of any such agreement, we have no obligation to do anything."[1205]

The Oslo Principles were launched on 30 March 2015 and originated from the research of a group of experts from across the globe within the fields of international environmental law,[1206] human rights law,[1207] and the law of delict.[1208] The Oslo Principles hold that, under existing human rights law, international environmental law and law of delict, it is incorrect for governments to assert that they have no obligation to do anything to combat climate change.[1209]

1205 J Powles & T Khan "Climate change: at last a breakthrough to our catastrophic political impasse?" (30-03-2015) (page numbers unavailable) *The Guardian* <http://www.theguardian.com/commentisfree/2015/mar/30/climate-change-paris-talks-oslo-principles-legal-obligations> (accessed 21-04-2015).

1206 In this regard the customary international law principle to avoid significant transboundary-harm is cited as one of the Oslo Principles by the committee. [Global Justice Program "The Oslo Principles Commentary" (date unavailable) (15) *Yale Global Justice Program* <http://www.yale.edu/macmillan/globaljustice/OsloPrinciplesCommentary.pdf> (accessed 11-05-2015).].

1207 In this regard the principle of human dignity, the right to life, the right to property (especially as it pertains to dignity), the right to health, the right to food, the right to water, and the right to a clean and healthy or satisfactory environment is particularly important as a legal underpinnings of the Oslo Principles. [The Global Justice Program "The Oslo Principles Commentary" (date unavailable) (16-30) *Yale Global Justice Program* <http://www.yale.edu/macmillan/globaljustice/OsloPrinciplesCommentary.pdf> (accessed 11-05-2015).]
Furthermore, the following human rights were called upon in the case of *Urgenda Foundation v Kingdom of the Netherlands* (as discussed in section 4.4.4.3) in terms of the ECHR: the right to health (Article 2) and the right to respect for private and family life (Article 8).

1208 In this regard, the point of departure is that an act or omission is deemed unlawful if it "subjects the life, well-being or property of others to a risk of damage if the risk is considerable, the potential damage is colossal, and if the risk can be avoided without undue detriment to the party/parties causing that risk." [The Global Justice Program "The Oslo Principles Commentary" (date unavailable) (31) *Yale Global Justice Program* <http://www.yale.edu/macmillan/globaljustice/OsloPrinciplesCommentary.pdf> (accessed 11-05-2015).].
The application of tort law in this regard will differ according to the jurisdiction.

1209 Also see the complementary draft principles of the ILA: ILA "The Legal Principles relating to Climate Change" (date of publication unavailable) (page numbers unavailable) *ILA* <http://www.ila-hq.org/en/committees/index.cfm/cid/1029> (accessed 11-05-2015). The commentary on the Oslo Principles admit that there is some overlap between the draft principles of the ILA and the Oslo Prin-

According to the preamble of the Oslo Principles, these experts set out to "identify and articulate a set of Principles that comprise the essential obligations States and enterprises have to avert the critical level of global warming".[1210] The formulation of the Oslo Principles in themselves represent people-based action in the context of climate change in some form, but, more pertinently, the Oslo Principles offer support for people-based action in the absence of international government agreements.[1211]

The Oslo Principles are important because they can establish the *opinio juris*,which can be described as the subjective belief that a particular state activity is legally required,[1212] that is needed in addition to state action, to create customary international law. These Principles can therefore help judges to decide if specific states are complying with their legal obligations to combat climate change.[1213] The importance of the Oslo Principles in this context has been showcased in the Urgenda case,[1214] as the principles were considered to support the plaintiffs' arguments.

The Oslo Principles focus on preventing climate change from causing a temperature increase above 2 degrees Celsius from pre-industrial

---

ciples, but this does not affect the purpose of the discussion in Section 4.4.4.2. [Global Justice Program "The Oslo Principles Commentary" (date unavailable) (10) *Yale Global Justice Program* <http://www.yale.edu/macmillan/globaljustice/OsloPrinciplesCommentary.pdf> (accessed 11-05-2015).].

1210 Preamble, Oslo Principles.

1211 The commentary on the Oslo Principles provides: "In our view, States are *legally obliged* to reduce their GHG- emissions, even if they do not conclude (further) international agreements or conventions." [Oslo Principles on Global Climate Change Obligations "Oslo Principles Commentary" (06-04-2015) (8) *Oslo Principles on Global Climate Change Obligations* <http://www.osloprinciples.org/macmillan/globaljustice/Oslo%20Principles%20Commentary.pdf> (accessed 14-08-2016)]; Urgenda "Climate case" (date of publication unavailable) (page numbers unavailable) *Urgenda* <http://www.urgenda.nl/en/climate-case/> (accessed 21-04-2015).

1212 RMM Wallace *International Law* 5ed (2005) 16; MN Shaw *International Law* 6ed (2008) 84.

1213 Global Justice Program "The Oslo Principles on Global Climate Change Obligations Launched by Expert Group Including GJP Director Thomas Pogge" (date of publication unavailable) (page numbers unavailable) *Global Justice Group* <http://www.yale.edu/macmillan/globaljustice/news.html> (accessed 11-05-2015).

1214 The Hague District Court: Urgenda Foundation v The Kingdom of the Netherlands (Ministry of Infrastructure and the Environment) 24 June 2015 *Rechtbank Den Haag C/09/456689 / HA ZA 13-1396 (English translation).*

times.[1215] Commentary on the Oslo Principles set out the three obstacles to achieving this goal and the committee responsible for the inception of the Principles admits that any enforceable legal obligation must also deal with these three obstacles.[1216] Firstly, the majority of states and enterprises that are required to act now are minimally responsible for causing the problem thus far.[1217] The second obstacle concerns the political issue of the deference from judges to politicians.[1218] The third obstacle is considered to be the lack of adequate obligations and targets set by international and national law.[1219] To some extent the Paris Agreement addresses some of these obstacles. For instance, it could be argued that NDCs in terms of the Paris agreement are addressing the hurdle of setting adequate obligations and targets in international and national law. However, it remains to be seen what the exact contribution of the Paris Agreement will be to the effectiveness of the international legal climate change regime.

1215 The Oslo Principles do not discuss adaptation, damages or climate refugees. [Oslo Principles on Global Climate Change Obligations "Oslo Principles Commentary" (06-04-2015) (5) *Oslo Principles on Global Climate Change Obligations* <http://www.osloprinciples.org/macmillan/globaljustice/Oslo%20Principles%20Commentary.pdf> (accessed 14-08-2016)].

1216 Oslo Principles on Global Climate Change Obligations "Oslo Principles Commentary" (06-04-2015) (6) *Oslo Principles on Global Climate Change Obligations* <http://www.osloprinciples.org/macmillan/globaljustice/Oslo%20Principles%20Commentary.pdf> (accessed 14-08-2016).

1217 Oslo Principles on Global Climate Change Obligations "Oslo Principles Commentary" (06-04-2015) (5) *Oslo Principles on Global Climate Change Obligations* <http://www.osloprinciples.org/macmillan/globaljustice/Oslo%20Principles%20Commentary.pdf> (accessed 14-08-2016).

1218 Oslo Principles on Global Climate Change Obligations "Oslo Principles Commentary" (06-04-2015) (5) *Oslo Principles on Global Climate Change Obligations* <http://www.osloprinciples.org/macmillan/globaljustice/Oslo%20Principles%20Commentary.pdf> (accessed 14-08-2016).

1219 Oslo Principles on Global Climate Change Obligations "Oslo Principles Commentary" (06-04-2015) (5-6) *Oslo Principles on Global Climate Change Obligations* <http://www.osloprinciples.org/macmillan/globaljustice/Oslo%20Principles%20Commentary.pdf> (accessed 14-08-2016).

4 4 3 Urgenda case and general movement toward people-based climate change litigation

The importance of the Oslo Principles, as explained above, has been showcased in the Urgenda case.[1220] In this case, the principles were considered to support the plaintiffs' (Urgenda and co-plaintiffs) argument. The Urgenda case moved to the forefront of international news on 14 April 2015, when the first public hearings took place. The summons was filed in the District Court in The Hague on 20 November 2013.[1221] The case was decided on 24 June 2015, in favour of Urgenda and its co-plaintiffs.

This is the first case in Europe where citizens attempted to hold a government responsible for inaction, or more specifically "the failure of the state to take sufficient actions to prevent dangerous climate change".[1222] This is a landmark case, as it is the first case in the world where citizens of a country use existing human rights law, as well as law of delict, to protect themselves against climate change. It is the first case in the world, where a court ordered a state to limit greenhouse gas emissions for reasons other than statutory mandated reasons.[1223] It is particularly significant that people used human rights law as a ground upon which to hold states accountable for their actions in terms of climate change. Since the right of people to dispose freely of natural resources is an accepted human right, it is ar-

---

1220 The Hague District Court: Urgenda Foundation v The Kingdom of the Netherlands (Ministry of Infrastructure and the Environment) 24 June 2015 *Rechtbank Den Haag C/09/456689 / HA ZA 13-1396 (English translation).*
Note that Section 4.4.4.3 contains all developments up until June 2016.

1221 Preceding legal action, the Urgenda Foundation wrote a letter to the Dutch government on 12 November 2012, requesting that the Dutch government take all necessary measures to reduce greenhouse gas emissions of the Netherlands by 40% relative to levels in 1990. [Urgenda "Letter to the Government" (12-11-2012) (page numbers unavailable) *Urgenda* <http://www.urgenda.nl/documents/Staat-der-NL-Engels1.pdf> (accessed 21-04-2015)].

1222 The Hague District Court: Urgenda Foundation v The Kingdom of the Netherlands (Ministry of Infrastructure and the Environment) 24 June 2015 *Rechtbank Den Haag C/09/456689 / HA ZA 13-1396 (English translation of summons, 25 June 2015 version).*

1223 J Klein "July 2015 Update to Climate Litigation Charts" (07-07-2015) (page numbers unavailable) *Columbia Law School* <http://blogs.law.columbia.edu/climatechange/2015/07/07/july-2015-update-to-climate-litigation-charts/> (accessed 11-09-2016).

gued by the author that this case opens the door for using this particular right as a ground for a similar claim. The author contends that climate change impacts PSNR by *inter alia* warranting a reinterpretation and reconfiguration of people's rights in terms of PSNR that repositions the principle as a tool within the international legal climate change regime. This contention is based on the fact that human rights was used as a basis for action in the Urgenda case.

In this case, the Dutch Urgenda Foundation and a group of 886 Dutch citizens brought a case against the Dutch government (any reference hereafter to Urgenda includes the 886 Dutch citizens as well) in the context of climate change. Issues which lay at the very core of this case were the lack of action by the government (considered a violation of the rights of individuals who would suffer because of the lack of action) and the failure of the government to act regardless of whether or not it was successful at negotiating international solutions to climate change.[1224]

The following specific three requests were made to the court by Urgenda: Firstly, Urgenda sought a declaration from the Court of The Hague that the Dutch government was acting unlawfully (contributing to climate change and posing a threat to human rights) by failing to reduce emissions by 40% or the minimum of 25% by 2020 relative to the 1990 level of emissions (as these are the levels of emissions reduction required from industrialised countries by the UNFCCC).[1225] It is no secret that the Netherlands only plans to reduce emissions by 16% relative to its 1990 emission levels.[1226] Secondly, Urgenda asked that the Court of The Hague compel the Dutch government to act in accordance with the European Convention on Human Rights (ECHR) and by means of a mandatory order (or a mandatory injunction) compel the Dutch government to take whatever actions are needed to achieve the reduction limits between 25% to 40% as stipulated in the UNFCCC by 2020.[1227]

1224 K Rigg "Landmark Dutch Lawsuit Puts Governments Around the World on Notice" (04-12-2015) *Huff Post Green* (page numbers unavailable) <http://www.huffingtonpost.com/kelly-rigg/landmark-dutch-lawsuit-pu_b_7025126.html> (accessed 21-04-2015).

1225 RHJ Cox "The Liability of European States for Climate Change" (2014) 30(78) *Utrecht Journal of International and European Law* 125 126.

1226 RHJ Cox "The Liability of European States for Climate Change" (2014) 30(78) *Utrecht Journal of International and European Law* 125 128.

1227 RHJ Cox "The Liability of European States for Climate Change" (2014) 30(78) *Utrecht Journal of International and European Law* 125 126.

As mentioned in Chapter 3, an important outcome of the Cancun COP was the commitment of parties to the 2 degrees Celsius threshold, which means not allowing the maximum global average temperature to increase by more than 2 degrees Celsius above pre-industrial levels.[1228] Urgenda argued that the Dutch government (and other European governments), as it adopted the Cancun Agreements, was knowingly in breach of its duty of care by not acting in a way that would prevent it from breaching the 2 degrees Celsius threshold.[1229] Therefore, the third request of Urgenda was that the Dutch government be compelled to reduce emissions so that it would not breach the 2 degrees Celsius threshold and accordingly reduce emissions between 25–40% by 2020 compared to 1990 levels of emissions.

Furthermore, Urgenda contended that the Dutch government had no excuse not to protect human rights, desist from negligence and ensure socially proper conduct (by reducing emissions as stipulated above), since areas of energy efficiency and renewable energy are well developed within the state and strategies involving their implementation would be economically achievable.[1230] Urgenda also set out the science to substantiate their claims, mostly supported by the IPCC, the findings of which are recognised by all parties to the UNFCCC, including the Netherlands.[1231]

The court decision of 24 June 2015 granted the plaintiff's claims and compelled the Dutch government to implement stricter and more effective environmental (particularly climate-related) policies that will contribute to lowering Dutch carbon emissions.[1232] The court found that the then plans of the Dutch government to curb emissions by 14 to 17% were inadequate,

1228 UNFCCC "Meetings" (date of publication unavailable) (page numbers unavailable) UNFCCC <http://unfccc.int/meetings/cancun_nov_2010/meeting/6266.php> (accessed 13-05-2013).

1229 RHJ Cox "The Liability of European States for Climate Change" (2014) 30(78) *Utrecht Journal of International and European Law* 125 126.

1230 RHJ Cox "The Liability of European States for Climate Change" (2014) 30(78) *Utrecht Journal of International and European Law* 125 127.

1231 RHJ Cox "The Liability of European States for Climate Change" (2014) 30(78) *Utrecht Journal of International and European Law* 125 127.

1232 Urgenda "Urgenda wins the case for better Dutch climate policies" (date of publication unavailable) (page numbers unavailable) *Urgenda* <http://www.urgenda.nl/en/climate-case/> (accessed 26-06-2015).

and ordered the government to cut emissions by 25% within five years.[1233] The court decision was based *inter alia* on the article 21 of the Dutch Constitution,[1234] the emission targets of the EU in terms of the Kyoto Protocol, the principles of the ECHR, and principles of international law embodied in the UNFCCC (such as the precautionary principle and the principle of sustainable development).[1235] The court found that the Dutch government was in breach of its obligation to protect the climate, was acting negligently, and was not meeting its duty of care.[1236] In order for the government to meet its duty of care, the government had to reduce greenhouse gas emissions. The court found that due to the "severity of the consequences of climate change and the great risk of climate change occurring", the state has a duty to reduce emissions.[1237]

The Dutch case, although set in a European context, may signal the beginning of a new era of climate change litigation around the world. The case is directly applicable to other people-based action of the same kind in Europe, since the Dutch law system applies the principle of duty of care,

---

1233 M Minnesma "Hague climate change verdict: 'Not just a legal process but a process of hope'" (25-06-2015) (page numbers unavailable) *The Guardian* <http://www.theguardian.com/global-development-professionals-network/2015/jun/25/hague-climate-change-verdict-marjan-minnesma> (accessed 30-06-2015); A Neslen "Dutch government ordered to cut carbon emissions in landmark ruling" (24-06-2015) (page numbers unavailable) *The Guardian* <http://www.theguardian.com/environment/2015/jun/24/dutch-government-ordered-cut-carbon-emissions-landmark-ruling> (accessed 30-06-2015).

1234 Article 21 of the Dutch Constitution reads as follows: "It shall be the concern of the authorities to keep the country habitable and to protect and improve the environment.".

1235 J Klein "July 2015 Update to Climate Litigation Charts" (07-07-2015) (page numbers unavailable) *Columbia Law School* <http://blogs.law.columbia.edu/climatechange/2015/07/07/july-2015-update-to-climate-litigation-charts/> (accessed 11-09-2016).

1236 K Boom et al "Climate Justice: The international momentum towards climate litigation" (2016) (27) *Climate Justice Programme* <http://climatejustice.org.au/wp-content/uploads/2016/06/Report-Climate-Justice-2016.pdf> (accessed 08-09-2016).

1237 The Hague District Court: Urgenda Foundation v The Kingdom of the Netherlands (Ministry of Infrastructure and the Environment) 24 June 2015 *Rechtbank Den Haag C/09/456689 / HA ZA 13-1396* (English translation of decision) 28; K Boom et al "Climate Justice: The international momentum towards climate litigation" (2016) (28) *Climate Justice Programme* <http://climatejustice.org.au/wp-content/uploads/2016/06/Report-Climate-Justice-2016.pdf> (accessed 08-09-2016).

as well as the principle of several or proportionate liability, in the same way as it is applied in many other European legal systems.[1238] In addition, the human rights claims in the Dutch case were made under the ECHR, which also binds other European states.[1239] The decision is considered to be able to provide persuasive force to common law jurisdictions.[1240] As already discussed, the initiation of many similar cases around the world would depend on country-specific legal systems, specifically including how broad government environmental responsibilities can be interpreted within environmental laws and the available legal provision for public initiation of climate cases against the state. The verdict is capable at the very least of constituting a warning (and incentive) to all developed and financially able countries capable of doing more to mitigate climate change to do so.

The case also brings to light a question, possibly with great impact on future climate change litigation, namely: Can emission reductions be claimed through judicial action or is this strictly an action confined to the political realm by means of political policy formulation and implementation?[1241] The Urgenda case answers this question the Court admits that the question before the Dutch court may be a question of law with political implications, but it is nevertheless a question of law and should not be re-

1238 RHJ Cox "The Liability of European States for Climate Change" (2014) 30(78) *Utrecht Journal of International and European Law* 125 127.
For the arguments in terms of Dutch tort law of negligence, Dutch constitutional law and European human rights law, refer to Cox: RHJ Cox "The Liability of European States for Climate Change" (2014) 30(78) *Utrecht Journal of International and European Law* 125 128-133.

1239 RHJ Cox "The Liability of European States for Climate Change" (2014) 30(78) *Utrecht Journal of International and European Law* 125 127.

1240 Boom K et al "Climate Justice: The international momentum towards climate litigation" (2016) (30) Climate Justice Programme <http://climatejustice.org.au/wp-content/uploads/2016/06/Report-Climate-Justice-2016.pdf> (accessed 08-09-2016).

1241 RHJ Cox "The Liability of European States for Climate Change" (2014) 30(78) *Utrecht Journal of International and European Law* 125 133.

duced to exist only in the political realm.[1242] In the Oregon case,[1243] the court's approach differed in that it found climate change to be overachingly limited to the political realm. The Oregon case is relevant within the context of the atmosphere as a natural resource, as the court finds that the atmosphere can fall within a broad definition of the term 'resource'.[1244] However, the court finds that the atmosphere is not included in Oregon's public trust doctrine and decides the case on Oregon's public trust doctrine,[1245] therefore placing the decision outside the scope of the research project.

1242 RHJ Cox "The Liability of European States for Climate Change" (2014) 30(78) *Utrecht Journal of International and European Law* 125 134; Boom K et al "Climate Justice: The international momentum towards climate litigation" (2016) (30) *Climate Justice Programme* <http://climatejustice.org.au/wp-content/uploads/2016/06/Report-Climate-Justice-2016.pdf> (accessed 08-09-2016).

1243 Circuit Court of the State of Oregon for Lane County: Olivia Chernaik, Lisa Cherniak, Kelsey CR Juliana & Cathy Juliana v Kate Brown & the State of Oregon (USA) 11 May 2015 *Circuit Court of the State of Oregon for Lane County Case No. 16-11-09273.*

1244 Circuit Court of the State of Oregon for Lane County: Olivia Chernaik, Lisa Cherniak, Kelsey CR Juliana & Cathy Juliana v Kate Brown & the State of Oregon (USA) 11 May 2015 *Circuit Court of the State of Oregon for Lane County Case No. 16-11-09273* par 11.

1245 Circuit Court of the State of Oregon for Lane County: Olivia Chernaik, Lisa Cherniak, Kelsey CR Juliana & Cathy Juliana v Kate Brown & the State of Oregon (USA) 11 May 2015 *Circuit Court of the State of Oregon for Lane County Case No. 16-11-09273* pg 10-11.
This case was filed against the government in 2011, and initiated on the basis of the government's failure to adhere to the public trust doctrine, by failing to protect the atmosphere from the impacts of climate change. The case is argued on the basis that the atmosphere forms part of the public trust and must therefore be protected for future generations. The case was dismissed *inter alia* based on the court's lack of authority to prescribe state official action regarding the reduction of carbon emissions in the atmosphere. In other words, the court decided that climate change is a political issue and not a legal issue. Furthermore, the Court submits that the public trust doctrine of each state of the USA, is developed in accordance with that state's common law and general principles of law, and that in Oregon, the atmosphere is not included in the state's public trust doctrine. [Circuit Court of the State of Oregon for Lane County: Olivia Chernaik, Lisa Cherniak, Kelsey CR Juliana & Cathy Juliana v Kate Brown & the State of Oregon (USA) 11 May 2015 *Circuit Court of the State of Oregon for Lane County Case No. 16-11-09273* pg 10.] The public trust doctrine is not employed in all state jurisdictions, and can be explained as "the essential part of the common

The non-profit organisation, Klimaatzaak Belgium which consists of a number of concerned citizens of Belgium, has already followed Urgenda's example and filed a case against the governments of Flanders, Wallonia and Brussels, as well as the Belgian federal government.[1246] The Belgian case was initiated in December 2014 and the case was filed in April 2015. The hearing will take place after written submissions by the plaintiffs to the court have been submitted.[1247] The case focuses more strongly on the human rights aspect of the case than the Urgenda case did, and the court's decision will therefore be significant within the context of the research project.[1248] The case was brought against the government by the people, in an effort to force the government to do more to fight climate change. The plaintiffs argue *inter alia* that the federal and regional governments are vi-

---

heritage of mankind". [K Baslar *The Concept of the Common Heritage of Mankind in International Law* (1998) 68]. In terms of the Public Trust Doctrine, the state is the trustee, which is responsible to manage the property in the interest of the public. [L Bento "Searching for Intergenerational Green Solutions: The Relevance of the Public Trust Doctrine to Environmental Preservation" (date unavailable) (page numbers unavailable) *Common Law Review* <http://www.commonlawreview.cz/searching-for-intergenerational-green-solutions-the-relevance-of-the-public-trust-doctrine-to-environmental-preservation> (accessed 25-08-2016).].

1246 Belgium, Brussels, Court of first instance: VWZ Klimaatzaak v Kingdom of Belgium et al 2015; Klimaatzaak "Press Release" *Klimaatzaak* <http://www.klimaatzaak.eu/en/wp-content/uploads/2014/11/Klimaatzaak_PERS_EN.pdf> (accessed 21-04-2015); J Ayre "Belgian Citizen Coalition Pursuing Legal Case Against Government Over Inaction on Climate Change" (04-05-2015) (page numbers unavailable) *Clean Technica* <https://cleantechnica.com/2015/05/04/belgian-citizens-coalition-pursuing-legal-case-government-inaction-climate-change/> (accessed 08-09-2016); A Hope "Klimaatzaak takes governments to court over climate policies" (28-04-2015) (page numbers unavailable) *Flanders Today* <http://www.flanderstoday.eu/politics/klimaatzaak-takes-governments-court-over-climate-policies> (accessed 08-09-2016).

1247 J Klein "July 2015 Update to Climate Litigation Charts" (07-07-2015) (page numbers unavailable) *Columbia Law School* <http://blogs.law.columbia.edu/climatechange/2015/07/07/july-2015-update-to-climate-litigation-charts/> (accessed 11-09-2016).
This section contains all developments until June 2016.

1248 Urgenda "Landmark Climate Change Lawsuit" (date of publication not available) (page numbers unavailable) *Urgenda* <http://www.urgenda.nl/documents/Press_release_Urgenda_climate_case_hearing_14_april_2015_FIN.pdf> (accessed 21-04-2015); A Cooper "The Oslo Principles and climate change" (2015) 20(1) *Coventry Law Journal* 45 48.

olating human rights law by not reducing greenhouse gas emissions.[1249] The outcome of the case will be relevant to the EU jurisdiction in particular, but also to other jurisdictions based on both civil law and common law respectively.[1250]

In Pakistan, a case was brought against the government by a Pakistani farmer, based on human rights.[1251] The objective of the case was to order the government to implement the government's existing climate change policy.[1252] The court found that the government was not implementing its climate change policy and accordingly in breach of fundamental rights of Pakistani citizens relating to food, water and energy security.[1253] The court ordered different lines of action in order to remedy this breach and orient the country towards a climate-resilient future, which included the creation of new bodies and the appointment of responsible persons to oversee the implementation of the country's climate policy.[1254] The case is considered to be a judicial acceptance of climate justice, and is heralded specifically for this reason.[1255] The decision is deemed applicable to jurisdictions

---

1249 J Klein "July 2015 Update to Climate Litigation Charts" (07-07-2015) (page numbers unavailable) *Columbia Law School* <http://blogs.law.columbia.edu/climatechange/2015/07/07/july-2015-update-to-climate-litigation-charts/> (accessed 11-09-2016).

1250 Boom K et al "Climate Justice: The international momentum towards climate litigation" (2016) (1-67) *Climate Justice Programme* <http://climatejustice.org.au/wp-content/uploads/2016/06/Report-Climate-Justice-2016.pdf> (accessed 08-09-2016).

1251 Pakistan Lahore High Court, Green Bench: Ashgar Leghari v Federation of Pakistan *W.P No 25501/2015* 2015.

1252 Boom K et al "Climate Justice: The international momentum towards climate litigation" (2016) (31) *Climate Justice Programme* <http://climatejustice.org.au/wp-content/uploads/2016/06/Report-Climate-Justice-2016.pdf> (accessed 08-09-2016).

1253 Boom K et al "Climate Justice: The international momentum towards climate litigation" (2016) (31) *Climate Justice Programme* <http://climatejustice.org.au/wp-content/uploads/2016/06/Report-Climate-Justice-2016.pdf> (accessed 08-09-2016).

1254 Boom K et al "Climate Justice: The international momentum towards climate litigation" (2016) (31) *Climate Justice Programme* <http://climatejustice.org.au/wp-content/uploads/2016/06/Report-Climate-Justice-2016.pdf> (accessed 08-09-2016).

1255 Boom K et al "Climate Justice: The international momentum towards climate litigation" (2016) (31) *Climate Justice Programme* <http://climatejustice.org.au/wp-content/uploads/2016/06/Report-Climate-Justice-2016.pdf> (accessed 08-09-2016).

where the right to life could be interpreted by the judiciary as including the right to a healthy environment.[1256] The case is therefore applicable to general human rights based climate litigation, and not necessarily specifically concerned with the human right of people to dispose freely of natural resources.

There are a myriad of other cases around the world indicating the momentum towards people-based climate litigation against states or companies in the broad context of climate change.[1257] However, these cases fall outside the scope of the research project as they do not pertain specifically to the right of all people to dispose freely of natural resources in terms of PSNR (as a basis upon which to fight climate change).[1258] When

---

1256 Boom K et al "Climate Justice: The international momentum towards climate litigation" (2016) (31) *Climate Justice Programme* <http://climatejustice.org.au/wp-content/uploads/2016/06/Report-Climate-Justice-2016.pdf> (accessed 08-09-2016).

1257 In addition to the legal action against states, people are also increasingly bringing action against companies. Legal action coming to light in the Philippines and Pacific Island nations includes the formation of the People's Declaration for Climate Justice, which was signed by participants of the Philippines, Tuvalu, Kiribati, Fiji, Vanuatu and the Solomon Islands in which they vow the following in the People's Declaration for Climate Justice 2015: "We commit to bring a case that would investigate the human rights implications of climate change and hold the big carbon polluters accountable to appropriate international bodies or processes." [Greenpeace International "People from the Philippines and Pacific Island nations sign declaration to hold big polluters accountable" (09-06-2015) (page numbers unavailable) *Greenpeace International* <http://www.greenpeace.org/international/Global/international/briefings/climate/2015/People's%20Declaration%20for%20Climate%20Justice.pdf> (accessed 30-06-2015).].

1258 Examples of case law in this context, falling outside the scope of the research project, include case law in the USA concerning constitutional and public trust litigation against the federal and state governments. For example, the Oregon case briefly mentioned in this current section of the research project. [Circuit Court of the State of Oregon for Lane County: Olivia Chernaik, Lisa Cherniak, Kelsey CR Juliana & Cathy Juliana v Kate Brown & the State of Oregon (USA) 11 May 2015 *Circuit Court of the State of Oregon for Lane County Case No. 16-11-09273.*] For more examples of climate litigation in general, falling outside the scope of the research project, see: K Boom et al "Climate Justice: The international momentum towards climate litigation" (2016) (1-67) *Climate Justice Programme* <http://climatejustice.org.au/wp-content/uploads/2016/06/Report-Climate-Justice-2016.pdf> (accessed 08-09-2016); J Spier & U Magnus (eds) *Climate Change Remedies, Injunctive Relief and Criminal Law Responses* (2014).

it comes to the judicial consideration of the specific human right of people to dispose freely of natural resources as the basis for climate litigation against states, the views of the courts across the globe remains to be seen. Dispite the lack of case law yet available in this particular context, the author sets out to merely highlight the possibility that such people-based action on the right to dispose of natural resources in terms of PSNR may be instituted in the future as a means by which to hold states accountable to climate commitments. Should this line of action be received favourably by courts (Urgenda representing this possibility), this reinterpreted and reconfigured application of PSNR can constitute a significant addition to the international legal climate change regime and the effort to improve compliance to the regime.

4 5 Conclusion

Chapter 4 explored the consequences of the effects of climate change on the development of PSNR. These consequences can be grouped into two subjects of discussion. The first subject of discussion involves the rights and duties of states in terms of PSNR, and that these are shifting owing to increased global environmental concerns, in particular because of the concern related to the effects of climate change. The result of the effects of climate change on PSNR is that the state rights in terms of PSNR are being restricted while state duties (especially the state duties to its state nationals and in the interest of their well-being) are gaining momentum and import. This discussion confirms PSNR as a generally state-centric principle, and the observed shift in importance attached to state duties also points to the principle of PSNR as an evolutionary concept. In this regard, the first subject of discussion in Chapter 4 completes the examination of PSNR, commenced in Chapter 2, and introduces the first round of impacts that the effects of climate change have on the development of PSNR.

For more information on the increase in climate litigation generally, see: K Boom et al "Climate Justice: The international momentum towards climate litigation" (2016) (1-67) *Climate Justice Programme* <http://climatejustice.org.au/wp-content/uploads/2016/06/Report-Climate-Justice-2016.pdf> (accessed 08-09-2016); J Spier & U Magnus (eds) *Climate Change Remedies, Injunctive Relief and Criminal Law Responses* (2014).

Building upon the developments noted above in regard to state rights and duties in terms of PSNR, a development in regard to people's rights in terms of PSNR is also noticed along with growing global environmental concerns, such as the concern related to the effects of climate change. The second subject of discussion therefore involves the fact that the effects of climate change are so severe that a reinterpretation and reconfiguration of PSNR is warranted. The reinterpretation and reconfiguration should focus on the people's rights in terms of PSNR, as such rights can form the basis upon which people are able to institute actions against states for not adhering to their climate commitments, and therefore present a negative incentive for state compliance to climate commitments. In order to determine the aforementioned, an investigation is conducted, firstly, into what people's rights in terms of PSNR are; and, secondly, what the prospects are concerning litigation (implemented by people in the context of instituting climate change litigation against states), based on infringements of people's rights in terms of PSNR. In so doing, it is determined that people have a human right to self-determination; or, more specifically relevant to the context of the research project and in terms of PSNR, a right to economic self-determination, which gives effect to the right to dispose freely of natural resources. This right to natural resources can be enforced directly (as mostly seen in the cases related to indigenous people, even though this right belongs to all people of the state and not only to indigenous people). The right to dispose freely of natural resources can also be enforced indirectly, as is evident in cases related to indigenous people and subsistence rights. Having stipulated that the author considers the right to dispose freely of natural resources to belong to all people and seeing that human rights form part of the basis upon which peoples are able to hold states accountable for climate commitments (as seen in the Urgenda case), it can be deducted that the human right to dispose freely of natural resources (as a human right) can also constitute part of the basis upon which people will be able to hold states accountable for climate commitments in the future. Accordingly, it is possible to contend that people's right to dispose freely of natural resources in terms of PSNR is in the infant stage of providing the basis upon which people are able to hold states accountable for their climate commitments. If developed in future, this type of and threat of litigation could prove a powerful negative incentive for states to adhere to their climate commitments and do what can be reasonably expected to combat the effects of climate change.

It can be submitted that the main objective of the research project, which is to determine the impact of climate change law on the principle of sovereignty over natural resources, has been addressed. Chapter 4 provides that the gravity of climate change impacts PSNR, as it warrants a reinterpretation and reconfiguration of PSNR, whereby state rights may be restricted, state duties may gain momentum, and people's rights (especially the human right to dispose freely of natural resources) are repositioned as part of the international legal climate change regime. This reinterpretation and reconfiguration of PSNR can constitute a significant basis from which people can hold states accountable to their climate commitments in future.

CHAPTER 5: Conclusion

5 1 Remarks on the original research objectives and hypotheses

This research project set out to determine the answer to the primary research question, which asks if climate change has an impact on the principle of sovereignty over natural resources. It is determined that climate change does have an impact on the principle. More precisely, it is determined that the gravity of climate change warrants the reinterpretation and reconfiguration of PSNR as it is currently applied in the sphere of international law, to apply the principle also within the international legal climate change regime.[1259] It was determined that the effects of climate change are so grave that a reinterpretation or reconfiguration of any principle of international law is warranted, if this principle is capable of contributing to increasing the effectiveness of the international legal climate change regime. The proposed reinterpretation or reconfiguration of PSNR would mean that state duties and particularly people's rights constitute the renewed focus of the principle, instead of state rights, which have, till now, generally been the focus in application of the principle. The research project therefore calls for the repositioning of PSNR as a principle focused on people with the people's right to dispose freely of natural resources constituting the potential basis upon which people are able to hold states accountable for their climate commitments in terms of the international legal climate change regime. The research project therefore goes beyond the familiar application of the principle (in the context of indigenous people), and calls for all people to utilise this right in terms of PSNR in order to hold states accountable to their climate commitments. The right of people to dispose freely of natural resources in terms of PSNR has the potential to present a

1259 The phrase 'reinterpretation and reconfiguration' should be clarified: firstly, by 'reinterpreting' the principle, the principle is interpreted or rather construed in a different light, meaning in this case that the principle is construed in light of the grave effects of climate change and in the context of climate change; and, secondly, by 'reconfiguring' the principle, constitutive elements of the principle are organised in a different form, meaning in this case that people's rights as a constitutive element of PSNR are organised in such a way as to promote or focus on people's rights, rather than state rights.

valuable addition to measures ensuring state compliance to climate commitments made by states in terms of the international legal climate change regime. More generally, the research project aims to designate a method by which an established principle of international law, namely PSNR, can be deployed to address a source of ineffectiveness within the international legal climate change regime.

The overarching objective of the research project is to constitute a positive contribution in the global efforts to address the problem of climate change through the repositioning of a international law principle, namely PSNR. Note that the research project does not set out to examine the extensive practical implications of applying the principle in this manner within the international legal climate change regime. Since the principle has not yet been applied in this manner by people and accordingly considered in this light by courts, it becomes difficult to predict how this practical application of PSNR will be received by courts. However, the author considers the proposition of this possibility as significant, and feasible, especially considered in light of the outcome of the Urgenda case discussed in Section 4.4.4.3. In this case, people use *inter alia* existing human rights law to protect themselves against climate change and to address government inaction in this regard. It is therefore not inconceivable that the specific human right of people to dispose freely of natural resources in terms of PSNR should be a basis for people upon which to do the same in the future. As mentioned in Section 1.5, a point of future research can consider the precise practical constraints of implementing such people-based action within different jurisdictions (as the Urgenda case is an example within the European Union). This research point fell outside the scope of the research project, which is determined by the primary research question of the research project, as mentioned in Section 1.2.

The starting point for the abovementioned research objectives was the formulation of the hypotheses discussed below. It was hypothesised that PSNR is a principle that is capable of evolving and capable of being affected by occurrences with political and legal implications, such as climate change. It was suspected that the scale and gravity of the environmental effects of climate change and the magnitude of the legal response required to address climate change, warrant the reinterpretation and reconfiguration of PSNR as this principle already contains the constituents (namely the right of people to dispose freely of natural resources), which, if implemented, can contribute to increasing state compliance within the international legal climate change regime. After examination, this hypothesis is upheld.

Furthermore, it was hypothesised that people's rights and state duties gain momentum, while state rights are restricted as PSNR evolves by being reinterpreted and reconfigured in light of global environmental concerns, in particular climate change. Furthermore, it was hypothesised that the people's rights in terms of PSNR might be able to form a basis upon which people can institute climate litigation against states, and thereby increase state compliance to climate commitments in terms of the international legal climate change regime. After examination, this hypothesis is also upheld.

5 2 Areas of future research

All the possible areas of future research emanating from the research project are too extensive to be listed in detail. For the author of this research project, some areas of future research seem particularly prominent and are accordingly mentioned below.

It is worthwhile to consider the role that people can play in securing a well-functioning international legal regime. Besides a direct human rights approach to issues concerning international environmental law, the human aspect of international law and, in particular, the well-being of people are subjects rarely included in the discussion in the context of international environmental law. The research project considers the issue of the well-being of people, only where it is appropriate to illustrate the gravity of the environmental, social and economic impacts of climate change on people and, as necessary, to substantiate the argument for the interpretation of PSNR as a principle focused on people. This research project focuses on the rights of people in terms of PSNR and how the principle can contribute to the international legal climate change regime. It can also be meaningful to consider whether it is possible for PSNR to promote the well-being of people in a particular state by being interpreted as a principle focused on people and implementing the rights of people in terms of PSNR. This is not a research question that the research project endeavours to answer and can accordingly be considered as an avenue for future research.

Furthermore, as is evident in the research project, for PSNR to support the international legal climate change regime, the interpretation of PSNR focuses on examining the rights of people in terms of the principle. To limit the scope of the research project, the duties of people are not discussed in this study. A possible area of future research that springs from the research project is therefore the (rather theoretical) examination of the duties of people in terms of PSNR, and the possibility of the practically useful manifestation of such duties.

As is evident throughout the research project, the author makes the deliberate decision to refer to 'people' (as opposed to 'peoples') throughout the research project. As discussed in Section 2.5.3, the term 'people' is understood by the author as encompassing the whole of the state population, but also being able to refer to a specific group of people forming a part of the state population such as indigenous communities or people.[1260] There is an ongoing debate in regard to the meaning of the term 'peoples' within international law (and specifically within human rights law).[1261] To include a thorough examination of this extensive debate falls outside the scope of the research project, and accordingly the debate will merely be introduced in Section 2.5.3, when determining the meaning of the term 'people' as a subject of PSNR and as employed throughout the research project. The term 'people' is therefore understood by the author as a term that includes 'peoples', regardless of the differing interpretations of the term 'peoples'. A future point of research can therefore include a determination of the exact meaning ascribable to the term 'peoples' within the context of rights granted in terms of PSNR.

As mentioned in Section 1.4 and Section 4.4.1, initial steps or assumptions to show that people's rights in terms of PSNR, when enforced, can present a valuable addition to ensuring state compliance with climate commitments made by states in terms of the international legal climate change regime. These initial steps or assumptions include *inter alia* that the state implements a national legal system whereby people are able to hold the state accountable, and that it is possible that only the nationals of a specif-

---

1260 This is a possibility according: RM Kiwanuka "The meaning of 'People' in the African Charter on Human and Peoples' Rights" (1988) 82 American Journal of International Law 80. The definition of the term 'people' is discussed further in Section 2.5.3.

1261 For an introduction to the debate, see: B Algan "Rethinking 'Third Generation' Human Rights" (2004) 1(1) Ankara Law Review 121 131; TT Onifade "Peoples-Based Permanent Sovereignty over Natural Resources: Toward Functional Distributive Justice?" (2015) 16 Human Rights Review 343 344; D Makinson "On Attributing Rights to all Peoples: Some Logical Questions" (1989) 8(1) Law and Philosophy 53 62-63.

ic state might be able to hold their state accountable. An area of future research might include setting out the exact legal environment of a jurisdiction that presents the ideal environment for people to hold states accountable for climate commitments. In other words, future research would set out to ascertain the specific legal system(s) that would allow this type of people-based action (based in particular on the human right to dispose freely of natural resources) to flourish.

5 3 Chapter summaries and outline of research outcomes

5 3 1 Chapter 2: The principle of sovereignty over natural resources

The primary research question of this study asks whether climate change has an impact on the principle of sovereignty over natural resources. This question means to enquire, more precisely, whether the effects of the phenomenon of climate change affect the interpretation and construction of PSNR. It is determined that this question can be answered in the affirmative. In order to make this determination, the content and characteristics of PSNR must be ascertained. Chapter 2 serves this objective by determining the content and characteristics of PSNR. Accordingly, the origin, legal nature, and the terminology of the principle are examined in Chapter 2. It is established that the principle finds its legal roots in UNGA resolutions and is cemented today as a fundamental principle of international law, a right of international law, and a part of treaty and customary international law. The object of PSNR is identified as natural resources and an examination of the scope of the term 'natural resources' reveals that the atmosphere and the climate are both 'natural resources' in terms of PSNR.

It is further determined that the foundational pillars of PSNR are composed of the economic development of underdeveloped states and the self-determination of people. The main subjects of the principle, and accordingly right-bearers of the principle, are identified as states and people. In determining the subjects of PSNR as *inter alia* states, the meaning of the term 'state' is examined as well as the concept of and role of state sovereignty as a component of PSNR, which is found also to contribute to establishing the evolutionary characteristic of PSNR. It is also evident that the principle is customarily more state-oriented than people-oriented. The examination of the subjects of PSNR also reveals the shifting focus between states and people as subjects of the principle, which serves to showcase *inter alia* the evolutionary character of the principle alongside major shifts in the political and legal landscape. Having established the evolutionary nature of PSNR, it is possible to determine in Chapter 2 that PSNR is a principle that is capable of evolving alongside major shifts in the political, environmental and legal landscape, and that climate change, as a significant environmental, legal and political phenomenon, has an impact on the evolution of PSNR. This ultimately supports the idea that climate change has an impact on the principle of sovereignty over natural resources and climate change affects the contemporary interpretation and construction of the principle. This idea is examined further in Chapter 3. Exactly how climate change affects the interpretation and construction of PSNR is explored in Chapter 3 and Chapter 4.

5 3 2 Chapter 3: The international legal climate change regime and the question of whether the gravity of climate change warrants a reinterpretation and reconfiguration of the principle of sovereignty over natural resources

Chapter 3 establishes the nexus between climate change and PSNR as a rationale for why the impact of climate change is studied particularly in respect of PSNR, as opposed to any other principle of international law. Furthermore, this chapter examines the international legal climate change regime as a core constituent of the research project, by considering the terminology, tracing the developmental milestones of the regime, and examining the most important constituents of the regime. The key constituents of the regime are identified as the UNFCCC, the Kyoto Protocol and the Paris Agreement. The principles on which these legal instruments of the regime are centred include equity and other equity principles such as the principle of common but differentiated responsibility; the precautionary principle; the polluter-pays principle; and the principle of sustainable development. These principles are also examined as the foundation of the international legal climate change regime.

Chapter 3 serves to address the primary research objective of the research project, which is to determine whether climate change has an impact on the principle of sovereignty over natural resources. As submitted above, it is accepted that the impact of climate change on PSNR might not be automatic. In fact, it is submitted that it is the gravity of the environmental, social and economic effects of climate change that warrants the reinterpretation and reconfiguration of any principle of international environmental law that is able to contribute to making the international legal climate change regime more effective. In this context, Chapter 3 sets out the scientific basis of climate change and outlines the environmental, social and economic effects of climate change, and the resulting effects on the well-being of people, all in an effort to illustrate the urgency, scale and gravity of the problem of climate change. This means that the grave effects of climate change warrant the reinterpretation and reconfiguration of any principle of international environmental law that can contribute to the international legal climate change regime, in this case PSNR, owing to the nexus established between PSNR and climate change. Accordingly, it is determined that climate change might not have a direct impact on the principle of PSNR, but that the effects of climate change are so severe that they warrant an indirect impact on this principle, since the principle can contribute to increasing the effectiveness of the international legal climate change regime.

Chapter 3 also addresses a shortcoming of the international legal climate change regime of particular relevance to the research project, which is the lack of state compliance to climate commitments undertaken in terms of the international legal climate change regime. Chapter 3 introduces the idea that in order for PSNR to offer a possible solution to this shortcoming, people's rights in terms of PSNR should be promoted, as the people's right to dispose freely of natural resources can constitute a basis upon which people can institute legal action against states for not adhering to their climate commitments. Chapter 3 therefore addresses possible approaches to increasing the effectiveness of the international legal climate change regime, and elucidates the people-based action described above as a negative incentive which may improve state compliance with climate commitments. In this regard PSNR, which encapsulates the right of people to dispose freely of natural resources, can contribute to the international legal climate change regime by constituting a basis upon which people can hold states accountable to their climate commitments. In this manner Chapter 3 also addresses the question of how climate change affects PSNR.

5 3 3 Chapter 4: The reinterpretation and reconfiguration of the principle of sovereignty over natural resources and the expansion of the means to combat climate change

After establishing the ability of PSNR to evolve in Chapter 2 and demonstrating the existing nexus or relationship between climate change and PSNR in Chapter 3, it is concluded in Chapter 4 that the severe effects of climate change have an impact on the evolution and interpretation of PSNR. Chapter 4 specifically explores the consequences of the effects of climate change on the development and interpretation of the rights and duties of states and the rights of people in terms of PSNR. In short, Chapter 4 concludes that the effects of climate change restrict state rights and promote state duties, and further warrant the repositioning of the principle as an aid to the international legal climate change regime which entails a refocus of the rights of people in terms of PSNR.

More specifically, this chapter explores the rights and duties of states as a component of PSNR, thus completing the exposition of PSNR commenced in Chapter 2. The discussion reveals the customary state-centric approach of the principle and further illustrates that the effects of climate change have had an impact on the evolution of PSNR by illustrating how some of the more relevant rights and duties of states have been affected by climate change as part of the broader noticeable international movement towards environmental conservation and sustainable development. The result of the effect of climate change on PSNR is that states' rights in terms of PSNR are being restricted, while state duties (especially the state duties to its state nationals and in the interest of their well-being) are gaining momentum and import.

Chapter 4 also discusses the further consequences of the gravity of climate change on PSNR as they pertain to people's rights. It is concluded in Chapter 3 that the effects of climate change are so severe that they warrant the repositioning of PSNR, whereby people's rights are brought to the fore and contribute to the international legal climate change regime. Chapter 4 explores the repositioning of PSNR that focuses on the rights of people in terms of the principle. An examination is conducted into how people's rights in terms of PSNR can form the basis upon which people are able to institute actions against states for not adhering to their climate commitments in terms of the international legal climate change regime. This ability of people to institute actions against states can constitute a negative incentive for states to adhere to their climate commitments. In order to arrive at the aforementioned conclusion, it is firstly determined what rights people have in terms of PSNR. It is determined that all people have a third generation human right to economic self-determination in terms of PSNR that translates into a right to dispose freely of natural resources. This right to dispose freely of natural resources can be enforced directly (as mostly seen in the cases related to indigenous people even though this right belongs to all people and not only to indigenous people), and indirectly (as mostly evidenced in cases involving indigenous people and their claims to subsistence rights). Secondly, consideration is given to the real prospects of success when people institute climate litigation against states on the basis of their right to dispose freely of natural resources, by examining cases around the world relevant to this context. It is deduced that there is potential for the human right to dispose freely of natural resources to constitute part of the basis upon which people are able to hold states accountable for climate commitments in the future. Accordingly, it is possible to contend that the people's right to dispose freely of natural resources in terms of PSNR is in the infant stage of constituting an accepted basis upon which people are able to hold states accountable for their climate commitments, and that this right can in future constitute a significant basis from which people can hold states accountable to their climate commitments. If developed in future, this threat of litigation could prove to be a powerful negative incentive for states to adhere to their climate commitments and could, in effect, play a significant role in increasing the effectiveness of the international legal climate change regime. Finally, it is evident that a mutually beneficial relationship can exist between PSNR and the international legal climate change regime, whereby PSNR is invigorated by the possibility of acting as the basis for people-based action against states in climate litigation, while the international legal climate change regime adds to its arsenal to increase the effectiveness of the regime by increasing state compliance. In this way Chapter 4 has shown how climate change impacts PSNR.

5 4 Recommendations

Schrijver's prediction of the future of PSNR was included in Chapter 1, and reads as follows:

> "It would be incorrect to assert that, as a result of the trend towards globalization, fragmentation and privatization, the principle of permanent sovereignty

> is dead or no longer serves any function in international law and in international relations. […] The challenge in the next two or three decades will be how to balance permanent sovereignty over natural resources with other basic principles and emerging norms of international law – including the duty to observe international agreements, grant fair treatment of foreign investors, pursue sustainable development at national and international levels and to respect human […] rights – and in this way to serve the best interests of present and future generations."[1262]

This research project considered climate change is the emerging area of international environmental law, as described above by Schrijver, and ascertained the impact of climate change on PSNR. The conclusions reached in this research project can also be framed as recommendations to influence future legal development that will promote the continued usefulness of PSNR and to improve compliance of states with the international legal climate change regime.

It is recommended that PSNR, as an existing principle of international law, be considered as a possible aid in addressing a specific shortcoming of the international legal climate change regime, namely the lack of state compliance. This construction means that PSNR should be repositioned as a people-oriented, rather than a state-oriented principle, that places more emphasis on the rights of people and the duties of states than it places on the rights of states. The right to dispose freely of natural resources, which is a right of people in terms of PSNR, can constitute a basis for the litigation instituted by people against states, as seen in the recent Urgenda case. This people-based potential for climate litigation against states can constitute a negative incentive for states to abide by their climate commitments, which can contribute to addressing a shortcoming of the international legal climate change regime, namely the lack of state compliance. This research project therefore recommends a repositioning of PSNR, so that the principle finds a renewed application within the international legal climate change regime, where it can constitute a basis for people to hold states accountable to their climate commitments. Accordingly, as was mentioned in Chapter 1, the research conducted in this research project is partly reform-oriented, since the research project examines the current understanding of PSNR and (owing to the global environmental, social, economic

1262 N Schrijver Sovereignty Over Natural Resources, Balancing Rights and Duties (1997) 379-380.

and legal concern of climate change) the adequacy or suitability of PSNR as a tool to combat climate change.

Viewed broadly, there are two further recommendations that flow from the research project. Firstly, it would be a worthwhile pursuit to endeavour to examine if and how existing formulated principles of international law (in this case, PSNR in particular) can be repositioned as aids to the international legal climate change regime. Secondly, the approach of the research project to PSNR and the international legal climate change regime shows that it is possible for constituents of international environmental law to operate in a mutually beneficial manner. In the context of the research project, it becomes apparent that the principle of sovereignty over natural resources and the international legal climate change regime can function in a mutually beneficial manner. It is suggested that these two constituents of international law can function in a mutually beneficial manner, since application of PSNR within the international legal climate change regime can offer renewed relevance and applicability to the principle within international law. Furthermore, the incorporation by the international legal climate change regime of PSNR can benefit the regime by increasing state compliance to the regime when people's rights in terms of PSNR constitutes a basis upon which people can hold states accountable to the regime.

The research project represents but the tip of the iceberg for the renewed applicability of PSNR within international law, and in particular within the international legal climate change regime as a method by which to hold states accountable to their climate commitments and increase the effectiveness of the regime.

Bibliography

Adger WN, S Huq, K Brown, D Conway & M Hulme "Adaptation to climate change in the developing world" (2003) 3(3) Progress in Development Studies 179.

Aldy JE et al "Beyond Kyoto, Advancing the international effort against climate change" (2003) Pew Center on Global Climate Change 1.

Algan B "Rethinking 'Third Generation' Human Rights" (2004) 1(1) Ankara Law Review 121.

Amechi EP The Millennium Development Goals (MGD's) and National and International Policy Reform: Realising the Right to a Healthy Environment in Africa PhD Dissertation, University of the Witwatersrand (2009).

Andresen S & S Agrawala "Leaders, pushers and laggards in the making of the climate change regime" (2012) 12 Global Environmental Change 41.

Anghie A Imperialism, Sovereignty and the Making of International Law (2005).

Araujo R "Sovereignty, Human Rights, and Self-Determination: The Meaning of International Law" (2000) 24(5) Fordham International Law Journal 1476.

Armstrong AJ & J Hamrin "The Renewable Energy Policy Manual" (date of publication unavailable) (page numbers unavailable) United States Export Council for Renewable Energy <http://www.oas.org/dsd/publications/unit/oea79e/ch05.htm> (accessed 02-03-2015).

Asche H & M Schüller "China's Engagement in Africa – Opportunities and Risks for Development" (2008) Deutsche Gesellschaft für Technische Zusammenarbeit (GTZ) GmbH 1.

Ashton J & X Wang "Equity and climate, In principle and practice" in JE Aldy et al (eds) Beyond Kyoto, Advancing the international effort against climate change (2003) 61.

Auty RM & RF Mikesell Sustainable Development in Mineral Economies (1998).

Badrinarayana D "Global Warming: A Second Coming for International Law?" (2010) 85 Washington Law Review 253.

Baer P "Equity, Greenhouse Gas Emissions, and Global Common Resources" in SH Schneider, A Rosencranz & JO Niles (eds) Climate Change Policy, A Survey (2002) 393.

Baer P et al "Climate Change: Equity and Greenhouse Gas Responsibility" (2000) 289(5488) Science 2287.

Barnard CE "Key drivers in environmental legislation towards good governance" (2007) 9 Water Policy 31.

Barrera-Hernández L "Sovereignty over Natural Resources under Examination: The Inter-American System for Human Rights and Natural Resource Allocation" (2006) 12 Annual Survey of International & Comparative Law 43.

Barrett S & R Stavins "Increasing Participation and Compliance in International Climate Change Agreements" (2003) 3 International Environmental Agreements: Politics, Law and Economics 349.

Baslar K The Concept of the Common Heritage of Mankind in International Law (1998).

Bastida E, T Wälde & J Warden-Fernández (eds) International and Comparative Mineral Law and Policy, Trends and Prospects (2005).

Bedjaoui M (ed) International Law: Achievements and Prospects (1991).

Behera S "National and International Environmental Principles – Changing Dimensions" (2008) (April) Orissa Review 34.

Benjamin A et al "Oslo Principles Commentary" (15-7-2015) (11) Yale Global Justice Program <http://globaljustice.macmillan.yale.edu/sites/default/files/files/Oslo%20Principles%20Commentary.pdf> (accessed 11-05-2015).

Bento L "Searching for Intergenerational Green Solutions: The Relevance of the Public Trust Doctrine to Environmental Preservation" (date unavailable) (page numbers unavailable) Common Law Review <http://www.commonlawreview.cz/searching-for-intergenerational-green-solutions-the-relevance-of-the-public-trust-doctrine-to-environmental-preservation> (accessed 25-08-2016).

Biersteker TJ & C Weber (eds) State Sovereignty as Social Construct (1996).

Birnie P & A Boyle International Law & the Environment 2ed (2002).

Birnie P, A Boyle & C Redgwell International Law & The Environment 2 ed, 3ed (2002, 2009).

Blake J & M Bonilla (eds) Regional Economic Outlook: Sub-Saharan Africa (2011).

Blanco E & J Razzaque Globalisation and Natural Resources Law: Challenges, Key Issues and Perspectives (2011).

Bodansky D "The History of the Global Climate Change Regime" in U Luterbacher & DF Sprinz (eds) International Relations and Global Climate Change (2001) 23.

Bodansky D "The Legitimacy of International Governance: A Coming Challenge for International Environmental Law?" (1999) 93 American Journal of International Law 596.

Bodansky D & L Rajamani "The Evolution and Governance Architecture of the Climate Change Regime" (2013) (2) SSRN <http://papers.ssrn.com/sol3/papers.cfm?abstract_id=2168859> (accessed 12-05-2015).

Bodansky D The art and craft of international environmental law" (2010).

Bodansky D, SD O'Connor & E Diringer "Issues for a 2015 climate agreement" Center for Climate and Energy Solutions (May 2015) (2) <http://www.c2es.org/docUploads/issues-for-a-2015-climate-agreement.pdf> (accessed 04-05-2015).

Bodle R, L Donat & M Duwe "The Paris Agreement: Analysis, Assessment and Outlook" (28-01-2016) (19) Ecologic <http://ecologic.eu/sites/files/event/2016/ecologic_institute_2016_paris_agreement_assessment_0.pdf> (accessed 09-07-2016).

Bond TC et al "Bounding the role of black carbon in the climate system: A scientific assessment" (2013) 18 Journal of Geophysical Research 5380.

Boom K, J Richards & S Leonard "Climate Justice: The international momentum towards climate litigation" (2016) (1-67) Climate Justice Programme <http://climatejustice.org.au/wp-content/uploads/2016/06/Report-Climate-Justice-2016.pdf> (accessed 08-09-2016).

Borah R "BRICS: The New Great Game" (26-06-2011) (page numbers unavailable) Geopolitical Monitor <http://www.geopoliticalmonitor.com/brics-the-new-great-game-4398> (accessed 16-07-2011).

Bösl A & J Diescho (eds) Human Rights in Africa, Legal Perspectives on their Protection and Promotion (2009).

Bosshard P "China's Environmental Footprint in Africa" (2008) 3 SAIIA China in Africa Policy Briefing 1.

Botin JA (ed) Sustainable Management of Mining Operations (2009).

Boyle A & M Anderson (eds) Human Rights Approaches to Environmental Protection (1996).

Bragdon SH "National Sovereignty and Global Environmental Responsibility: Can the Tension Be Reconciled for the Conservation of Biological Diversity?" (1992) 33(2) Harvard International Law Journal 381.

Brandl E & H Bungert "Constitutional Entrenchment of Environmental Protection: A Comparative Analysis of Experiences Abroad" (1992) 16 Harvard Environmental Law Review 1.

Brennan G "Discounting the future, yet again" (2007) 6 Politics Philosophy Economics 259-284.

Brown O, A Hammill & R Mcleman "Climate change as the 'new' security threat: implications for Africa" (2007) 83 International Affairs 1141.

Brownlie I Principles of Public International Law 5 ed, 7 ed (1998, 2008).

Bruch C, W Coker & C VanArsdale "Breathing Life into Fundamental Principles: Implementing Constitutional Environmental Protections in Africa" (2001) (page numbers unavailable) World Resources Institute <http://pdf.wri.org/eaa_bruch.pdf> (accessed 13-06-2011).

Bungenberg M & S Hobe (eds) Permanent Sovereignty over Natural Resources (2015).

Burdon P "Wild Law, The Philosophy of Earth Jurisprudence" (2010) 35 Alternative Law Journal 58.

Busby J "After Paris: Good Enough Climate Governance" (2016) (January) Current History 1.

Caldwell LK "Concepts in Development of International Environmental Policies" (1973) (April) International Environmental Policies 190.

Cambou D & S Smis "Permanent sovereignty over natural resources from a human rights perspective: Natural resources exploitation and indigenous peoples' rights in the arctic" (2013) 22(1) Michigan State International Law Review 347.

Cameron J & J Abouchar "The Precautionary Principle: A Fundamental Principle of Law and Policy for the Protection of the Global Environment" (1991) XIV(1) Boston College International & Comparative Law Review 2.

Cameron J & W Wade-Cery "Addressing Uncertainty: Law, Policy and the Development of the Precautionary Principle" (1992) 43 CSERG Working Paper GEC 1.

Camilleri JA & J Falk The End of Sovereignty? The Politics of a Shrinking and Fragmenting World" (1992).

Campbell B (ed) "Regulating Mining in Africa. For whose benefit?" (2004).

Campbell KM et al "The Age of Consequences: The Foreign Policy and National Security Implications of Global Climate Change" (2007) Centre for Strategic & International Studies 1.

Cano GJ A legal and institutional framework for natural resources management 2ed (1983).

Carzola M & M Toman "International Equity and Climate Change Policy" (2000) 27 Climate Issue Brief 1.

Cassese A "The Martens Clause: Half a Loaf or Simply Pie in the Sky?" (2000) 11(1) EJIL 187.

Cassese A International Law 2ed (2005).

Cawood FT "The Mineral and Petroleum Resources Development Act of 2002: A paradigm shift in mineral policy in South Africa" (2004) January/February The Journal of The South African Institute of Mining and Metallurgy 53.

Chaytor B & KR Gray International environmental law and policy in Africa (2003).

Chowdhury SR, EMG Denters & PJIM de Waart The Right to Development in International Law (1992).

Clark LS "International Law and Natural Resources" 1976-1977 (4) Syracuse Journal of International Law and Commerce 377.

Clémençon R
"The Two Sides of the Paris Climate Agreement: Dismal Failure or Historic Breakthrough" (2016) 25(1) Journal of Environment & Development 3.

Clémençon R
"What future for the Global Environment Facility?" (2006) 15 The Journal of Environmental Development 50.

Cohen JL Globalization and Sovereignty - Rethinking Legality, Legitimacy, and Constitutionalism (2012).

Conca K "Rethinking the Ecology-Sovereignty Debate" (1994) 23 Millennium Journal of International Studies 701.

Cooper A "The Oslo Principles and climate change" (2015) 20(1) Coventry Law Journal 45.

Corden WM & JP Neary "Booming Sector and De-Industrialisation in a Small Open Economy" (1982) 92 The Economic Journal 825.

Corntassel JJ, TH Primeau "Indigenous 'Sovereignty' and International Law: Revised Strategies for Pursuing Self-Determination" (2006) 2 Hawaiian Journal of Law & Politics 52.

Corvalan C et al "Ecosystems and Human Well-Being" (2005) (2) World Health Organization <http://www.who.int/globalchange/ecosystems/ecosys.pdf> (accessed 19-07-2016).

Costanza R et al "Quality of life: An approach integrating opportunities, human needs, and subjective well-being" (2007) 61 Ecological Economics 268.

Cottier T, O Nartova & SZ Bigdeli (eds) International Trade Regulation and the Mitigation of Climate Change (2010).

Cox RHJ "The Liability of European States for Climate Change" (2014) 30(78) Utrecht Journal of International and European Law 125.

Crawford E "Arrhenius' 1896 Model of the Greenhouse Effect in Context" (1997) 26(1) Ambio 6.

Cross FB "Paradoxical Perils of the Precautionary Principle" (1996) 53(3) Washington and Lee Law Review 851.

D'Arge RC "Climate and economic activity" (1979) Proceedings of the World Climate Conference, A Conference of Experts on Climate and Mankind 652.

Daes EIA "Prevention of discrimination and protection of indigenous peoples – Indigenous peoples' permanent sovereignty over natural resources" (2004) (page numbers unavailable) ECOSOC <http://www.un.org/esa/socdev/unpfii/documents/Report%20by%20Erica%20Irene%20A.%20Daes.pdf> (accessed 26-02-2015).

Danner RA & MLH Bernal Introduction to Foreign Legal Systems (1994).

Darmstadter J (ed) Global Development and the Environment, Perspectives on Sustainability (1992).

David R & JEC Brierley Major Legal Systems in the World Today, An Introduction to the Comparative Study of Law 2ed (1978).

Davis GA "The minerals sector, sectoral analysis, and economic development" (1998) 24 Resources Policy 217.

De Aréchaga JE "International Law in the Past Third of a Century" (1979) Recueil des Cours 159.

De Waart P, P Peters & E Denters (eds) International Law and Development (1998).

Dean JM, ME Lovely H Wang "Are foreign investors attracted to weak environmental regulations? Evaluating the evidence from China" (2009) 90 Journal of Development Economics 1.

Dellapenna JW "The customary international law of transboundary fresh waters" (2001) 1(3/4) International Journal of Global Environmental Issues 264.

Dellink RB & A Ruijs (eds) Economics of Poverty, Environment and Natural-Resource Use (2008).

Denters E & N Schrijver (eds) Reflections on International Law form the Low Countries, in Honour of Paul de Waart (1998).

Derret JDM An Introduction to Legal Systems (1968).

Devenish GE "Some thoughts on socio-economic rights" (2003) (6) Potchefstroom Electronic Law Journal 44-45 < http://dspace.nwu.ac.za/handle/10394/1960> (accessed 23-06-2011).

Devlin RA & Q Grafton Economic Rights and Environmental Wrongs, Property Rights for the Common Good (1998).

Doelle M “Climate Change and Human Rights: The Role of the International Human Rights in Motivating States to Take Climate Change Seriously” (2004) 1(2) Macquarie Journal of International and Comparative Environmental Law 179.

Dolzer R & M Stevens Bilateral Investment Treaties (1995).

Dowling AC “'Un-Locke-Ing' a 'Just Right' Environmental Regime: Overcoming the Three Bears of International Environmentalism – Sovereignty, Locke, and Compensation” (2002) 26(3) William & Mary Environmental Law and Policy Review 891.

Dröge S “The Paris Agreement 2015, Turning Point for the International Climate Regime” (2016) (February) German Institute for International and Security Affairs 1.

Du Plessis A “Public Participation, Good Environmental Governance, and Fulfilment of Environmental Rights” (2008) 11 Potchefstroom Electronic Law Journal 170-252 <http://dspace.nwu.ac.za/bitstream/10394/1885/3/2008x2x_6a_Du_Plessis_art.pdf> (accessed 23-06-2011).

Du Plessis AA & C Rautenbach “Legal Perspectives on the Role of Culture in Sustainable Development” (2010) (13) Potchefstroom Electronic Law Journal 27-71 <http://dspace.nwu.ac.za/bitstream/10394/4041/1/Du_Plessis_and_Rautenbach_APotgieter.pdf> (accessed 23-06-2011).

Du Plooy P “Rethink Investment in South Africa” (2008) WWF 1.

Dugard J International Law, A South African Perspective 3ed, 4ed (2005, 2011).

Dupuy PM “Soft law and the international law of the environment” (1991) 12 Michigan Journal of International Law 420.

Duruigbo E “Permanent Sovereignty and Peoples' Ownership of Natural Resources in International Law” (2006) 38 The George Washington International Law Review 33.

Ebbesson J & P Okowa (eds) Environmental Law and Justice in Context (2009).

Eckstein G “Water Scarcity, Conflict, and Security in a Climate Change World: Challenges and Opportunities for International Law and Policy” (2009) 27(3) Wisconsin International Law Journal 409.

Eckstein G & Y Eckstein “A Hydrogeological Approach to Transboundary Ground Water Resources and International Law” (2003) 19(2) American University International Law Review 201.

Economic Analysis Division of the International Energy Agency “World Energy Outlook 2007, China and India Insights” (2007) OECD, IEA 1.

Egede E Africa and the Deep Seabed Regime: Politics and International Law of the Common Heritage of Mankind (2011).

Elian G The Principle of Sovereignty over Natural Resources (1979).

Engberg-Pedersen L “Climate Change negotiations and their implications for international development cooperation” (2011) (1-50) Danish Institute for International Studies <https://www.ciaonet.org/attachments/18066/uploads> (accessed 10-09-2016).

Falk R Law, Morality and War in the Contemporary World (1963).

Farmer A "Towards a meaningful rebirth of economic self-determination: Human rights realization in resource-rich countries" (2006) 39 International Law and Politics 417.

Faure M & N Niessen (eds) Environmental Law in Development, Lessons from the Indonesian Experience (2006).

Fawcett JES & A Parry Law and International Resource Conflicts (1981).

Fawcett JES & A Parry Law and International Resource Conflicts (1981).

Feris L "The socio-economic nature of section 24(b) of the Constitution - some thoughts on HTF Developers (Pty) Ltd v Minister of Environmental Affairs and Tourism (HTF)" (2008) 23 SA Publiekreg/SA Public Law 194.

Feris LA "The Role of Good Environmental Governance in the Sustainable Development of South Africa" (2010) 13 Potchefstroom Electronic Law Journal 73.

Feris LA The Conceptualisation of Environmental Justice within the context of the South African Constitution LLD Dissertation, Stellenbosch University (2000).

Ferreira-Snyman MP The Erosion of State Sovereignty in Public International Law: Towards a World Order? LLD Dissertation, University of Johannesburg (2009).

Fischer-Kowalski M et al "Decoupling Natural Resource Use and Environmental Impacts from Economic Growth, A Report of the Working Group on Decoupling to the International Resource Panel" (2011) UNEP.

Fisher-Vanden K "Debate: A Case Study of the United States" (1997) (6) ENRP Discussion Paper 1.

Fitzmaurice M Contemporary Issues in International Environmental Law (2009).

Franklin JH (editor, translator) Jean Bodin On Sovereignty Four chapters from The Six Books of the Commonwealth (1992).

Freestone D & C Streck (eds) Legal Aspects of Implementing the Kyoto Protocol Mechanisms, Making Kyoto Work (2005).

French D "'From Seoul with Love' – The continuing relevance of the 1986 Seoul ILA Declaration on progressive development of principles of public international law relating to a new international economic order" (2008) Netherlands International Law Review 3.

French DA "A reappraisal of sovereignty in the light of global environmental concerns" (2001) 21(3) Legal Studies 376.

Gardner RC "Respecting Sovereignty" 1996 (VIII) Fordham Environmental Law Journal 133.

Gess KN "Permanent sovereignty over natural resources: An analytical review of the United Nations declaration and its genesis" (1964) 13 International and Comparative Law Quarterly 398.

Getu M "Accommodating the Interests of Developing Countries in the Climate Change Regime: Lessons from the Ozone Layer Regime" (2012) 6(1) Mizan Law Review 1.

Gilbert J "The right to freely dispose of natural resources: Utopia or forgotten right?" (2013) 31(2) Netherlands Quarterly of Human Rights 314.

Ginsburg N "Natural Resources and Economic Development" (1957) 47 Annals of the Association of American Geographers 197.

Glazewski J "National sovereignty v the global common good" (22-04-2010) (page numbers unavailable) UCT <https://www.uct.ac.za/downloads/uct.ac.za/news/media/oped/Jan_Glazewsk_22_04_10.pdf> (accessed 07-08-2016).

Glazewski J "The Environment, Human Rights and a new South African Constitution" (1991) 167 South African Journal on Human Rights 167.

Glazewski J & S Posnik "Compliance with international environmental standards and expectations: review of international developments" (2000) July/August The Journal of The South African Institute of Mining and Metallurgy 211.

Glazewski J Environmental Law in South Africa 2ed (2005).

Godden L "Death, Desire, Modernity and Redemption: Climate Change and Public International Environmental Law" (2009) 10 Melbourne Journal of International Law (page numbers unavailable on electronic copy).

Grant TD The Recognition of States, Law and Practice in Debate and Evolution (1999).

Grimm S, J Humphrey, E Lundsgaarde & SJ de Sousa "European Development Cooperation to 2020: Challenges by New Actors in International Development" (2009) (page numbers unavailable) Institute of Development Studies <http://www.edc2020.eu/fileadmin/publications/EDC_2020_Working_Paper_No_8_-_November_2010.pdf> (accessed 23-06-2011).

Guillaumont P "Measuring Structural Vulnerability to Allocate Development Assistance and Adaptation Resources" (05-03-2012) (1-22) Fondation Pour Les Études Et Recherches Sur Le Développement International (page numbers unavailable in electronic copy).

Gümplova P "Restraining permanent sovereignty over natural resources" (2014) 53 Quaderns de Filosofia 93.

Guzman AT "Why LDCs Sign Treaties That Hurt Them: Explaining the Popularity of Bilateral Investment Treaties" (1997) 38 Virginia Journal of International Law 639.

Gylfason T & G Zoega "Natural Resources and Economic Growth: The Role of Investment" (2002) Central Bank of Chile 1.

Haas PM, RO Keohane & MA Levy Institutions for the Earth, Sources of Effective International Environmental Protection (1994).

Handl G "Territorial Sovereignty and the Problem of Transnational Pollution" (1975) 69(1) The American Journal of International Law 50.

Hansjürgens B, R Antes & Strunz M Permit Trading in Different Applications (2011).

Hansohm D, W Breytenbach, T Hartzenberg & C McCarthy (eds) Monitoring Regional Integration in Southern Africa Yearbook Vol. 5 (2005).

Harris PG "Collective Action on Climate Change: The Logic of Regime Failure" (2007) 47 Natural Resources Journal 195.

Haugen HM "Peoples' right to self-determination and self-governance over natural resources: Possible and desirable" (2014) 8(1) Nordic Journal of Applied Ethics 3.

Hayward T "Constitutional Environmental Rights: a Case for Political Analysis" (2000) 48 Political Studies 558.

Heinzerling L "Climate Change, Human Health, and the Post-Cautionary Principle" (2008) 96 The Georgetown Law Journal 445.

Held D "The Changing Structure of International Law: Sovereignty Transformed?" in D Held, A McGrew (eds) The Global Transformations Reader: an Introduction to the Globalization Debate (2003) 162.

Helm D Climate-change Policy (2005).

Henderson C "Fossil Fuel-Fired Power Generation, Case Studies of Recently Constructed Coal- and Gas-Fired Power Plants" (2007) OECD, IEA.

Henkin L "Human Rights and State 'Sovereignty'" (1995/1996) 25 Georgia Journal of International and Comparative Law 31.

Henkin L (ed) The International Bill of Human Rights: The Covenant on Civil and Political Rights (1981).

Hickey JE "Some legal Impacts of the Emerging International Climate Change Regime on Energy Prices" (2013) 4 Global Business Law Review 1.

Hickey JE & VR Walker "Refining the precautionary principle in international environmental law" (1994-1995) 14 Virginia Environmental Law Journal 423.

Hilson G & B Murck "Sustainable development in the mining industry: clarifying the corporate perspective" (2000) 26 Resources Policy 227.

Hinsley FH Sovereignty 2ed (1986).

Hinz MO "Agenda 21 and Climate Protection: The Development of Global and Local Governance for Environment and Development - Obeservations from Research in Namibia" in OC Ruppel, C Roschmann & K Ruppel-Schlichting (eds.) Climate change: International law and global governance. Vol. 1 Policy, diplomacy and governance in a changing environment. (2013) 621.

Hirji R, H Mackay, PS Maro, SADC Defining and mainstreaming environmental sustainability in water resources management in Southern Africa: a summary (2002).

Hofbauer JA The Principle of Permanent Sovereignty over Natural Resources and Its Modern Implications LLM Dissertation, University of Iceland (2009).

Holder J & M Lee Environmental Protection, Law and Policy 2ed (2007).

Hossain K & SR Chowdhury (eds) Permanent Sovereignty Over Natural Resources in International Law, Principle and Practice (1984).

Huang PCC "The Theoretical and Practical Implications of China's Development Experience: The Role of Informal Economic Practices" (2011) 37 Modern China 3.

Hunter D, J Salzman & D Zaelke International Environmental Law and Policy (2007)

Hurrel A & B Kingsbury (eds) The International Politics of the Environment (1992).

Hurrell A and S Sengupta "Emerging powers, North-South relations and global climate politics" (2012) 88(3) International Affairs 463.

Huse MD & SL Muyakwa "China in Africa: lending, policy space and governance" (2008) Norwegian Campaign for Debt Cancellation, Norwegian Council for Africa 1.

Hutchinson T Researching and Writing in In Law 3ed (2010).

Hyde JN "Permanent Sovereignty over Natural Wealth and Resources" (1956) 50(4) The American Journal of International Law 854.

IPCC [Arent DJ et al (eds)] Key Economic Sectors and Service in Climate Change 2014: Impacts, Adaptation, and Vulnerability. Part A: Global and Sectoral Aspects (2014) 662.

IPCC [Edenhofer OR et al (eds)] Summary for Policymakers in Climate Change 2014: Mitigation of Climate Change (2014) 1.

IPCC [Field CB et al (eds)] Summary for Policymakers in Climate Change 2014: Impacts, Adaptation, and Vulnerability, Part A: Global and Sectoral Aspects (2014) 1.

IPCC [Mach KJ et al (eds)] Annex II: Glossary. In: Climate Change 2014: Synthesis Report (2014) 117.

IPCC [Palutikof JP et al (eds)] Summary for Policymakers. In: Climate Change 2007: Impacts, Adaptation and Vulnerability (2007) 1.

IPCC [Parry ML et al (eds)] Summary for Policymakers. In: Climate Change 2007: Impacts, Adaptation and Vulnerability (2007) 1.

IPCC [Solomon S et al (eds)] Climate Change 2007: The Physical Science Basis (2007) 1.

IPCC [Stocker TF et al (eds)] Summary for Policymakers in Climate Change 2013: The Physical Science Basis (2013) 1.

Jacobsen T, C Sampford & R Thakur (eds) Re-envisioning sovereignty, the end of Westphalia? (2008).

Jaenicke G "Zur Frage des Internationalen Ordre Public" (1967) 7 Berichte der Deutschen Gesellschaft für Völkerrecht 77.

Jamieson D "Climate Change, Consequentialism, and the Road Ahead" (2013) 13(2) Chicago Journal of International Law 439.

Jiang W "Fuelling the Dragon: China's rise and its energy and resource extraction in Africa" (2009) 199 The China Quarterly 585.

Johnson AK "The Influence of Institutional Culture on the Formation of Pre-regime Climate Change Policies in Sweden, Japan and the United States" (1998) 7(2) Environmental Values 223.

Joyner CC International Law in the 21st Century, Rules for Global Governance (2005).

Judkins RR & W Fulkerson "The Dilemma of Fossil Fuel Use and Global Climate Change" (1993) 7 Energy & Fuels 14.

Karousakis K "Joint Implementation: Current Issues and Emerging Challenges" (2006) 9 OECD 1.

Kasimbazi EB "Policy and legal dimensions of CDM projects in the forestry sector: implications for climate change mitigation and adaptation in Uganda" in BJ Richardson, Y Le Bouthillier, H McLeod-Kilmurray & S Wood Climate Law and Developing Countries, Legal and Policy Challenges for the World Economy (2009) 285.

Kates RW, JH Ausubel & M Berberian (eds) Climate Impact Assessment: Studies of the Interaction of Climate and Society (1985).

Keeling CD "The concentration and isotopic abundances of carbon dioxide in the atmosphere" (1960) 12(2) Tellus 200.

Keohane RO & K Raustiala "Toward a Post-Kyoto Climate Change Architecture: a Political Analysis" in JA Aldy & RN Stavins (eds) Post-Kyoto International Climate Policy (2010) 372.

Kerr TM "Legal Aspects of Storing CO2, Update and Recommendations" (2007) OECD, IEA 1.

Khan MR "Polluter-Pays-Principle: The Cardinal Instrument for Addressing Climate Change" (2015) (4) Laws 638.

Kidd M "Greening the Judiciary" (2006) 3 Potchefstroom Electronic Law Journal 1.

Kidd M Environmental Law (2008).

Kirgis FL "Standing to Challenge Human Endeavors That Could Change the Climate" (1990) 84 The American Journal of International Law 525.

Kirk EA The Changing Shape of Sovereignty in International Environmental Law LLM Dissertation, The University of British Columbia (1994).

Kirton JJ & MJ Trebilcock (eds) Hard Choices, Soft Law: Voluntary Standards in Global Trade, Environment and Social Governance (2004).

Kiss A "The Peoples' Right to Self-Determination" (1986) Human Rights Law Journal 165.

Kiss A & D Shelton International Environmental Law 3ed (2004).

Kiwanuka RN "The meaning of "People" in the African Charter on Human and Peoples' Rights" (1988) 82 American Journal of International Law. 80.

Kliot N "Institutions for management of transboundary water resources: their nature, characteristics and shortcomings" (2001) 3 Water Policy 229.

Koskenniemi M "What Use for Sovereignty Today?" (2011) 1 Asian Journal of International Law 61.

Kotzé LJ "Improving Unsustainable Environmental Governance in South Africa: the Case for Holistic Governance" (2006) 9 Potchefstoom Electronic Law Journal 75.

Kotzé LJ & AR Paterson (eds) The Role of the Judiciary in Environmental Governance: Comparative Perspectives (2009).

Krasner S (ed) International Regimes (1983).

Krasner SD "The Case For Shared Sovereignty" (2005) 16(1) Journal of Democracy 69.

Kriebaum U "Privatizing Human Rights, The Interface between International Investment Protection and Human Rights" in A Reinisch & U Kriebaum (eds) The Law of International Relations (2007) 165.

Kubasek NK & GS Silverman Environmental Law 2ed (1997).

Landsberg H "Climate as a Natural Resource" (1946) 3(4) The Scientific Monthly 293.

Lang W "Is the Ozone Depletion Regime a Model for an Emerging Regime on Global Warming?" (1991) 9(2) UCLA Journal of Environmental Law & Policy 16.1

Lang W "UN-Principles and International Environmental Law" (1999) 3 Max Planck United Nations Yearbook 157.

Langford M, W Vandenhole, M Scheinin & W van Genugten Global Justice, State Duties – The Extraterritorial Scope of Economic, Social and Cultural Rights in International Law (2013).

Lapalme L "The Social Dimension of Sustainable Development and the Mining Industry, A Background Paper" (2003) Natural Resources Canada, Minerals and Mining Sector 1.

Lazarus RJ "Changing Conceptions of Property and Sovereignty in Natural Resources Law: Questioning the Public Trust Doctrine" (1986) 71 Iowa Law Review 631.

Lederman D & W Maloney "Open Questions About the Link Between Natural Resources and Economic Growth: Sachs and Warner Revisited" (2002) Central Bank of Chile 1.

Lefèvre N "Energy Security and Climate Policy, Assessing Interactions" (2007) OECD, IEA 1.

Lenzertini F "Sovereignty Revisited: International Law and Parallel Sovereignty of Indigenous Peoples" (2006) 42 Texas International Law Journal 155.

Liftin KT "Sovereignty in Ecopolitcs" (1997) (41) Mershon International Studies Review 167.

Ma X & L Ortolano Environmental Regulation in China: Institutions, Enforcement, and Compliance (2000).

Makinson D "On Attributing Rights to all Peoples: Some Logical Questions" (1989) 8(1) Law and Philosophy 53.

Malanczuk P Akehurst's Modern Introduction to International Law 7ed (1997).

Malcolm R A Guidebook to Environmental Law (1994).

Malzbender D & A Earle "Water Resources of the SADC: Demands, Dependencies and Governance Responses" (2007) African Centre for Water Research 1.

Mancilla A "Shared Sovereignty over Migratory Natural Resources" (2016) 22 Res Publica 21.

Marafa LM "Africa's Business and Development Relationship with China: Seeking moral and capital values of the last economic frontier" (2009) Nordiska Afrikainstitutet 1.

Maranga K "The Changing Role of State Responsibility: Comparative Approach" (2011) 1(1) Journal of Global Affairs and Public Policy 1.

Marella F "On the Changing Structure of International Investment Law: The Human Right to Water and ICSID Arbitration " (2010) 12 International Community Law Review 335.

May JR "Constituting Fundamental Environmental Rights Worldwide" (2005-2006) 23 Pace Environmental Law Review 113.

Mayeda G "Where Should Johannesburg Take Us? Ethical and Legal Approaches to Sustainable Development in the Context of International Environmental Law" (2004) 15 Colorado Journal of International Environmental Law and Policy 29.

Mayur R "Environmental Problems of Developing Countries" (1979) 444 Annals of the American Academy of Political and Social Science 89.

McClure W World Prosperity as sought through the economic work of the League of Nations (1933).

McCorquodale R "Beyond State Sovereignty: The International Legal System and Non-State Participants" (2006) 8 International Law: Revista Colombiana de Derecho Internacional 103.

McElwee CR Environmental Law in China, Mitigating Risk and Ensuring Compliance (2011).

McInerney-Lankford S "Climate Change and Human Rights: An Introduction to Legal Issues" (2009) 33 Harvard Environmental Law Review 431.

Meijer KS Human Well-Being Values of Environmental Flows, Enhancing Social Equity in Integrated Water Resources Management PhD Dissertation, Technische Universiteit Delft (2007).

Metcalf GE & DA Weisbach "The Design of a Carbon Tax" (2009) Public Law and Legal Theory Working Papers 1.

Metz B "International equity in climate change policy" (2000) 1 Integrated Assessment 111.

Mgbeoji I "Beyond Rhetoric: State Sovereignty, Common Concern, and the Inapplicability of the Common Heritage Concept to Plant Genetic Resources" (2003) 16(4) Leiden Journal of International Law 821.

Mikesell RF "Explaining the resource curse, with special reference to mineral-exporting countries" (1997) 23 Resources Policy 191.

Miranda LA "The Role of International Law in Intrastate Natural Resource Allocation: Sovereignty, Human Rights, and Peoples-Based Development" (2012) 45 Vanderbilt Journal of Transnational Law 785.

Mitchell T "Carbon Democracy" (2009) 38(3) Economy and Society 399.

Morris CW An Essay on the Modern State (1998).

Mosner SC "Communicating climate change: history, challenges, process and future directions" (2010) 1 WIREs Climate Change 31.

Mubangizi JC "Towards a new approach to the classification of human rights with specific reference to the African context" (2004) African Human Rights Law Journal (2004) 4(1) 93.

Muldavin J "The Paradoxes of Environmental Policy and Resource Management in Reform-Era China" (2000) 76 Economic Geography 244.

Mushkat R International Environmental Law and Asian Values: Legal Norms and Cultural Influences (2004).

Mwiturubani DA & J van Wyk "Climate Change and Natural Resources Conflicts in Africa" (date unavailable) (page numbers unavailable) Institute for Security Studies <https://www.issafrica.org/uploads/Mono170.pdf> (accessed 07-09-2016).

Nel P & PJ McGowan (eds) Power, Wealth and Global Order, An International Relations Textbook for Africa (1999).

Nelson EJ et al "Climate change's impact on key ecosystem services and the human well-being they support in the US" (2013) 11(9) Frontiers in Ecology and the Environment 483.

Neumayer E "In defence of historical accountability for greenhouse gas emissions" (2000) 33(2) Ecological Economics 185.

Northcott HA "Realisation of the right of indigenous peoples to natural resources under international law through the emerging right to autonomy" (2012) 16(1) The International Journal of Human Rights 73.

Nyst C "Solidarity in a Disaggregated World, Universal Jurisdiction and the Evolution of Sovereignty" (2012) 8 Journal of International Law and International Relations 36.

O'Brien J International Law (2002).

Odendahl K "Conservation and Utilization of Natural Resources and Common Spaces" (2002) (page numbers unavailable) Encyclopaedia of Life Support Systems <http://www.eolss.net/Sample-Chapters/C04/E4-21-06.pdf> (accessed 27-04-2012).

OECD "South-South Trade: Vital for Development" (2006) (page numbers unavailable) Policy Brief <http://www.oecd.org/dataoecd/30/50/37400725.pdf> (accessed 16-05-2011).

Olivastri B "Reflecting on The Paris Climate Agreement Makes Me Glad, Sad and Mad" (13-12-2016) (page numbers unavailable) Friends of the Earth / Les Ami(e)s de la Terre <http://foecanada.org/en/2015/12/reflecting-on-the-paris-climate-agreement-makes-me-glad-sad-and-mad/> (accessed 06-07-2016).

Onifade TT "Peoples-Based Permanent Sovereignty over Natural Resources: Toward Functional Distributive Justice?" (2015) 16 Human Rights Review 343.

Opello WC & SJ Rosow The Nation-State and Global Order: A Historical Introduction to Contemporary Politics 2ed (2004).

Orakhelashvili A Peremptory Norms in International Law (2006).

Orellana MA "Saramaka People V. Suriname" (2008) 102(4) The American Journal of International Law 841.

Orts EW "Environmental Law with Chinese Characteristics" (2003) 11 William & Mary Bill of Rights Journal 545.

Osofsky HM "Learning from Environmental Justice: A New Model for International Environmental Rights" (2005) 24 Stanford Environmental Law Journal 71.

Otto JM & J Cordes (eds) Sustainable Development and the Future of Mineral Investment (2000).

Palmade V & A Anayiotas "FDI Trends: Looking beyond the current gloom in developing countries" (2004) Public Policy for the Private Sector.

Paterson A & LJ Kotzé (eds) Environmental Compliance and Enforcement in South Africa: Legal Perspectives (2009).

Paterson M "Principles of Justice in the Context of Global Climate Change" (2001) (121) Graduate Institute Geneva <http://graduateinstitute.ch/files/live/sites/iheid/files/shared/iheid/800/luterbacher/luterbacher%20chapter%206%20106.pdf> (accessed 09-09-2016).

Pegg S "Can Policy Intervention Beat the Resource Curse? Evidence from the Chad-Cameroon Pipeline Project" (2005) 105 African Affairs 1.

Penrose E, G Joffe & P Stevens "Nationalisation of Foreign-owned Property for a Public Purpose: An Economic Perspective on Appropriate Compensation" (1992) 55(3) The Modern Law Review 351.

Penz P "Environmental Victims and State Sovereignty: A Normative Analysis" (1996) 23(4) Social Justice 41.

Pereira P & O Gough "Permanent sovereignty over natural resources in the 21st century: Natural resource governance and the right to self-determination of indigenous peoples under international law" (2013) 14 Melbourne Journal of International Law 13.

Perrez FX "The Relationship between "Permanent Sovereignty" and the Obligation Not to Cause Transboundary Environmental Damage" 1996 (26) Environmental Law 1189.

Perrez FX Cooperative Sovereignty, From independence to interdependence in the structure of international environmental law (2000).

Peters P & N Schrijver "Latin America and International Regulation of foreign Investment: Changing Perceptions" (1991) Working Paper Series No. 113 1.

Phillips LC & D Seck (eds) Fixing African Economies, Policy Research for Development (2004).

Prieto IC & R Nocedal (eds) Legal Protection of the Environment in Developing Countries (1976).

Qin JY "Reforming WTO Discipline on Export Duties: Sovereignty over Natural Resources, Economic Development and Environmental Protection" 2012 (4) Wayne State University Law School Research Paper <http://papers.ssrn.com/sol3/papers.cfm?abstract_id=2030477> (accessed 16-03-2015).

Rajaram V, S Dutta & K Parameswaran (eds) Sustainable Mining Practices: A global perspective (2005).

Reinold T Sovereignty and the Responsibility to Protect, The power of norms and the norms of the powerful (2013).

Revesz RL, P Sands & RB Stewart (eds) Environmental Law, the Economy, and Sustainable Development (2000).

Richardson BJ, Y Le Bouthillier, H McLeod-Kilmurray & S Wood (eds) Climate Law and Developing Countries, Legal and Policy Challenges for the World Economy (2009).

Ringius L, A Torvanger & A Underdal "Burden Sharing and Fairness principles in International Climate Policy" (2002) 2 International Environmental Agreements: Politics, Law and Economics 1.

Robbins P, J Hintz & SA Moore Environment and Society, A Critical Introduction (2010).

Rolston H "Rights and Responsibilities on the Home Planet" (1993) 18 Yale Journal of International Law 251.

Ruppel OC "Regional economic communities and human rights in East and southern Africa" in A Bösl & J Diescho (eds) Human Rights in Africa: Legal Perspectives on their Protection and Promotion (2009) 275.

Ruppel OC "Third generation human rights and the protection of the environment in Namibia" in N Horn & A Bösl (eds) Human Rights and the Rule of Law in Namibia (2008) 101.

Ruppel OC & K Ruppel-Schlichting (eds) Environmental Law and Policy in Namibia (2011).

Ruppel OC, C Roschmann & K Ruppel-Schlichting Climate Change: International Law and Global Governance, Vol II: Policy, Diplomacy and Governance in a Changing Environment (2013).

Sachs JD & AM Warner "Natural Resources and Economic Development, The curse of natural resources" (2001) 45 European Economic Review 827.

Sachs JD & AM Warner "The big push, natural resource booms and growth" (1999) 59 Journal of Development Economics 43.

Sagoff M "Economic Theory and Environmental Law" (1981) 79 Michigan Law Review 1393.

Salih MAM (ed) Climate Change and Sustainable Development (2009).

Salomon ME & A Sengupta The Right to Development: Obligations of States and the Rights of Minorities and Indigenous Peoples (2003).

Sampson RN & D Hair Natural Resources for the 21st Century (1990).

Sands P Principles of International Environmental Law 2 ed (2003).

Sands P, R Stewart & R Revesz (eds) Environmental Law, The Economy and Sustainable Development (2000).

Savaresi A "The Paris Agreement: a new beginning?" (2016) 34(1) Energy & Natural Resources Law 16.

Schellnhuber HJ, PJ Crutzen, WC Clark & J Hunt "Earth System Analysis for Sustainability" (2005) 47 Environment 10.

Schiermeier Q "The Kyoto Protocol: Hot Air" (28-11-2012) (page numbers unavailable) Nature <http://www.nature.com/news/the-kyoto-protocol-hot-air-1.11882> (accessed 08-07-2016).

Schneider J World public order of the environment: Towards an international ecological law and organization (1979).

Scholtz W "Custodial sovereignty: reconciling sovereignty and global environmental challenges amongst the vestiges of colonialism" (2008) Netherlands International Law Review 323.

Scholtz W "Environmental harmonisation in the SADC region: an Acute Case of Asymmetry" in K Meesen et al (eds) Economic Law as an Economic Good: Its Rule Function and its Tool Function in the Competition of Systems (2009) 385.

Scholtz W "Equity as the basis for a future international climate change agreement: between pragmatic panacea and idealistic impediment. The optimization of the CBDR principle via realism" (2009) 42(2) The Comparative and International Law Journal of South Africa 166.

Schreuer C "Calvo's grandchildren: the return of local remedies in investment arbitration" (2005) 1 The Law and Practice of International Courts and Tribunals 1.

Schrijver N "Self-determination of peoples and sovereignty over natural wealth and resources" in United Nations Human Rights Office of the High Commissioner (ed) Realizing the Right to Development (2013) 95.

Schrijver N & F Weiss (eds) International law and Sustainable Development (2004).

Schrijver N Sovereignty over natural resources, Balancing rights and duties (1997).

Schrijver N The Evolution of Sustainable Development in International Law: Inception, Meaning and Status (2008).

Schrijver NJ Sovereignty over Natural Resources: Balancing Rights and Duties in an Interdependent World LLD Dissertation, Rijksuniversiteit Groningen (1995).

Schwebel SM "The Effect of Resolutions of the U.N. General Assembly on Customary International Law" (1979) 73 Proceedings of the Annual Meeting (American Society of International Law) 302.

Schwebel SM "The Story of the U.N.'s Declaration on Permanent Sovereignty over Natural Resources" (1963) 49 American Bar Association Journal 463.

Segger MC & A Khalfan Sustainable Development Law: Principles, Practices & Prospects (2004).

Sengupta M Environmental Impacts of Mining: Monitoring, Restoration, and Control (1993).

Shaw MN International Law 6ed (2008).

Shelton D "Common Concern of Humanity" (2009) 1 Iustum Aequum Salutare 33.

Shelton D & A Kiss Judicial Handbook on Environmental Law (2005).

Silecchia LA "Ounces of Prevention and Pounds of Cure: Developing Sound Policies for Environmental Compliance Programs" (1996) VII Fordham Environmental Law Journal 583.

Smith BM "Ethics, the Legacy of the Reverend Doctor Martin Luther King, Jr., and the Movement Toward Environmental Justice" (1994) 1 Environmental Law Journal 1.

Sornarajah M The International Law on Foreign Investment 3ed (2010).

Spier J & U Magnus (eds) Climate Change Remedies, Injunctive Relief and Criminal Law Responses (2014).

Spijkers O The United Nations, The Evolution of Global Values and International Law (2011).

Sprinz D & U Luterbacher (eds) International Relations and Global Climate Change (2001).

Stark W "Phoenix from the Ashes – An Analysis of the Paris Agreement to the United Nations Framework Convention on Climate Change" (2016) 41 Wuppertal Institut für Klima, Umwelt, Energie 1.

Starr JR "Water wars" (1991) 82 Foreign Policy 17.

Stern N "Stern Review: The Economics of Climate Change" (2006) (page numbers unavailable) *Centro de Ciência do Sistema Terrestre* <http://mudancasclimaticas.cptec.inpe.br/~rmclima/pdfs/destaques/sternreview_report_complete.pdf> (accessed 30-04-2015).

Sterner T Policy Instruments for Environmental and Natural Resource Management (2003).

Stoehr J "A Question of Sovereignty, Development, and Natural Resources: A New Standard for Binding Third Party Nonsignatory Governments to Arbitration" (2009) 66(3) Washington and Lee Review 1409.

Strydom HA & ND King (eds) Fuggle & Rabie's Environmental Management in South Africa 2 ed (2009).

Talbott WJ Human Rights and Well-Being (2010).

Tamiotti L, R The, V Kulaçoğlu, A Olhoff, B Simmons & H Abaza "Trade and Climate Change" (2009) UNEP & WTO.

Taylor P An ecological Approach to International Law, Responding to challenges of climate change (1998).

Teunissen JJ & A Akkerman (eds) Africa in the World Economy: The National, Regional and International Challenges (2005).

Toman MA "The Difficulty in Defining Sustainability" in J Darmstadter (ed) Global Development and the Environment, Perspectives on Sustainability (1992) 15.

Tsosie R "Indigenous People and Environmental Justice: The impact of climate change" (2007) 78 University of Colorado Law Review 1625.

Turton A & R Henwood (eds) Hydropolitics in the Developing World, A Southern African Perspective (2002).

Tvedt T, O McIntyre & TK Woldetsadik (eds) A History of Water. Series III, Volume 2: Sovereignty and International Water Law (2015).

United Nations "Report of the Office of the United Nations High Commissioner for Human Rights on the relationship between climate change and human rights. Advance Unedited Version, A/HRC/10/61" (15-01-2009) United Nations 1.

United Nations Economic Commission for Africa Managing Mineral Wealth, Training Materials on 'Management of Mineral Wealth and the Role of Mineral Wealth in Socio-economic Development' (2002).

Uppenbrink J "Arrhenius and Global Warming" (1996) 272 Science 1122.

Van der Vyver J "The environment: state sovereignty, human rights, and armed conflict" (2009) 23 Emory International Law Review 85.

Vasak K "A 30-Year Struggle, The sustained efforts to give force of law to the Universal Declaration of Human Rights" (1977) The UNESCO Courier 29.

Verheyen R Climate Change Damage and International Law, Prevention Duties and State Responsibility (2005).

Verschuuren J "Sustainable Development and the Nature of Environmental Legal Principles" (2009) 9 Potchefstroom Electronic Law Journal 209.

Visser F "The principle of permanent sovereignty over natural resources and the nationalisation of foreign interests" (1988) XXI Comparative and International Law Journal of South Africa 76.

Vogler J The Global Commons, Environmental and Technological Governance 2ed (2000).

Von Stein J "The International Law and Politics of Climate Change, Ratification of the United Nations Framework Convention and the Kyoto Protocol" (2008) 52(2) Journal of Conflict Resolution 243.

Walker RBJ & SH Mendlovitz (eds) Contending Sovereignties, Redefining Political Community (1990).

Wallace RMM International Law 5ed (2005).

Watts N et al "Health and climate change: policy responses to protect public health" (23-06-2015) (1) The Lancet < http://www.thelancet.com/pdfs/journals/lancet/PIIS0 140-6736(15)60854-6.pdf> (accessed 30-06-2015).

Weiss EB "In Fairness To Future Generations and Sustainable Development" (1992) 8(1) American University International Law Review 19.

Weiss EB "Intergenerational equity: a legal framework for global environmental change" in E B Weiss (ed) Environmental change and international law: New challenges and dimensions (1992) (page numbers unavailable in electronic copy).

Wheeler T & J von Braun "Climate Change Impacts on Global Food Security" (2013) 341 Science 508.

Winter G (ed) Multilevel Governance of Global Environmental Change, Perspectives from Science, Sociology and the Law (2006).

Woerdman E, A Arcuri & S Clò "Emissions Trading and the Polluter-Pays Principle: Do Polluters Pay Under Grandfathering?" (2008) 14 Rotterdam Institute of Law and Economics Working Paper Series 1.

Wold C, D Hunter & M Powers Climate Change and the Law (2009).

Wolf J "Developmental and Environmental Responsibilities of the 'Rechtsstaat'" (1999) 2 Potchefstroom Electronic Law Journal 23.

World Bank Group "Treasure of Trouble? Mining in Developing Countries" (2002) The World Bank Group, Mining Department 1.

World Commission on Environment and Development Our Common Future (1987).

WWF China's role in global trade, Opportunities and risks in the forestry and mining sector (2011) WWF 1.

Yamin F & J Depledge The International Climate Change Regime, A Guide to Rules Institutions and Procedures (2004).

Zongwe DP "The Legal Justifications for a People-Based Approach to the Control of Mineral Resources in the Democratic Republic of the Congo" (2008) 12 Cornell Law School Inter-University Graduate Student Conference Papers 1.

Zsamboky M, A Fernández-Bilbao, D Smith, J Knight & J Allan Impacts of climate change on disadvantaged UK coastal communities (2011).

Zwick S "The Road From Paris: Green Lights, Speed Bumps, And The Future of Carbon Markets" (01-02-2016) (page numbers unavailable) Ecosystems Marketplace <http://www.ecosystemmarketplace.com/articles/green-lights-and-speed-bumps-on-road-to-markets-under-paris-agreement/> (accessed 08-07-2016).

Case law

ACHPR: Centre for Minority Rights Development (Kenya) and Minority Rights Group International (on behalf of Endorois Welfare Council) v Kenya 25 November 2009 ACHPR 276/03.

ACHPR: Social and Economic Rights Action Center (SERAC) and the Center for Economic and Social Rights v Nigeria 27 October 2001 ACHPR 155/96.

Arbitral Tribunal: "Lake Lanoux Arbitration (Spain v France)" (16 November 1957) (1, 15, 27) Ecolex <http://www.ecolex.org/details/court-decision/lake-lanoux-arbitration-france-v-spain-b09cb956-2cb5-479e-ba3a-bbfd4f7b68fc/> (accessed 07-09-2016).

Arbitral Tribunal: Award in the Dispute between Libyan American Oil Company (Liamco) and the Government of the Libyan Arab Republic relating to Petroleum Concessions 12 April 1977 International Legal Materials (1981) 20(1) 1-87.

Arbitral Tribunal: Award in the Matter of an Arbitration between Kuwait and the American Independent Oil Company (Aminoil) 24 March 1982 International Legal Materials (1982) 21(5) 976-1053.

Arbitral Tribunal: Award on the Merits in Dispute between Texaco Overseas Petroleum Company/California Asiatic Oil Company (Texaco) and the Government of the Libyan Arab Republic 19 January 1977 International Legal Materials 17(1) (1978) 1-37.

Arbitral Tribunal: Awards regarding the Border between Costa Rica and Nicaragua (Costa Rica v Nicaragua) 22 March 1888, 30 September 1897, 20 December 1897, 22 March 1898 & 26 July 1899 Reports of International Arbitral Awards Vol XXVIII 189-236.

Arbitral Tribunal: Island of Palmas Case (Netherlands v United States) 4 April 1928 Reports of International Arbitral Awards Vol II 829-871.

Arbitral Tribunal: Trail Smelter Arbitration (United States v Canada) 16 April 1938 & 11 March 1941 U.N. Repository of International Arbitration Awards 1905-1982.

ASEAN Arbitral Tribunal: Yaung Chi Oo Trading Pte Ltd. V Government of the Union of Myanmar 31 March 2003 International Legal Materials 42(3) (2003) 540-559.

Belgium, Brussels (Court of first instance): VWZ Klimaatzaak v Kingdom of Belgium 2015 (Outcome pending).

Circuit Court of the State of Oregon for Lane County: Olivia Chernaik, Lisa Cherniak, Kelsey CR Juliana & Cathy Juliana v Kate Brown & the State of Oregon (USA) 11 May 2015 Circuit Court of the State of Oregon for Lane County Case No. 16-11-09273.

Hague District Court: Urgenda Foundation v The Kingdom of the Netherlands (Ministry of Infrastructure and the Environment) 24 June 2015 Rechtbank Den Haag C/ 09/456689 / HA ZA 13-1396 (English translation).

IACHR: "Maya Indigenous Communities of the Toledo District v Belize, Merits Report 40/04, Case 12.053" (12 October 2004) (par 5, 117, 142-143) <http://www.cidh.org/annualrep/2004eng/belize.12053eng.htm> (accessed 08-09-2016).

ICJ: Advisory Opinion on the Legality of the Threat or Use of Nuclear Weapons 8 July 1996 ICJ Reports 1996 226.

ICJ: Case Concerning Armed Activities on the Territory of the Congo (Democratic Republic of the Congo v Uganda) 19 December 2005 ICJ Reports 2005.

ICJ: Case Concerning Elettronica Sicula S.P.A. (ELSI) (United States v Italy) 20 July 1989 International Legal Materials (1989) 28(5) 1109-1164.

ICJ: Case Concerning the Gabčíkovo-Nagymaros Project (Hungary v Slovenia) 25 September 1997, ICJ Reports 1997 64-65 & Separate Opinion of Weeramantry on Case Concerning the Gabčíkovo-Nagymaros Project (Hungary v Slovenia) 25 September 1997 ICJ Reports 1997 118.

ICSID: Award in the case of AGIP Company (AGIP) v Popular Republic of the Congo 30 November 1979 International Legal Materials (1982) 21(4) 726-739.

ICSID: Case of Compañía de Aguas del Aconquija S.A. & Vivendi Universal S.A. v Argentine Republic (Case of Aguas de Aconquija v. Argentina) 21 November 2000 International Legal Materials 40(2) 462-453.

ICSID: Final Award in Asian Agricultural Products Ltd. (AAPL) v. Republic of Sri Lanka 21 June 1990 International Legal Materials (1991) 30(3) 624.

ICSID: Final Award in Asian Agricultural Products Ltd. (AAPL) v. Republic of Sri Lanka 21 June 1990 International Legal Materials (1991) 30(3) 624.

ICSID: Generation Ukraine Inc v Ukraine 16 September 2003 International Legal Materials 44 (2005) 404-464.

ICSID: In the Matter of an Arbitration between Piero Foresti, Laura de Carli and others v Republic of South Africa ICSID Case No ARB(AF)/07/01 8 January 2007.

ICSID: Lanco International Inc. v Argentine Republic 8 December 1998 International Legal Materials (2001) 40(2) 457-473.

Inter-American Court of Human Rights: Case of the Saramaka People v Suriname (Preliminary Objections, Merits, Reparations, and Costs) 28 November 2007 Inter-American Court of Human Rights Series C No 172.

Iran-United States Claims Tribunal: Partial Award in Amoco International Finance Corporation (Amoco) v Islamic Republic of Iran 14 July 1987 International Legal Materials (1988) 27(5) 1314-1405.

Pakistan Lahore High Court, Green Bench: Ashgar Leghari v Federation of Pakistan W.P No 25501/2015 2015.

International instruments and legislation

African Charter on Human and Peoples' Rights 1981.

African Convention on the Conservation of Nature and Natural Resources 1968 (revised in 2003).

Agenda 21 1992.

Agreement for the Promotion and Protection of Investments 1987.

Agreement on Promotion, Protection and Guarantee of Investments among Member States of the Organisation of the Islamic Conference 1981.

American Convention on Human Rights 1969.

Antarctic Treaty 1959.

ASEAN Agreement on the Conservation of Nature and Natural Resources 1985.

Atlantic Charter 1941.

Charter of Economic Rights and Duties of States 1974.

Charter of the United Nations 1945.

Constitution of the Kingdom of the Netherlands 2002.

Constitution of the Republic of South Africa 1996.

Constitution of the World Health Organization 1946.

Convention (II) with Respect to the Laws and Customs of War on Land 1899 (and its annex of Regulations Concerning the Laws and Customs of War on Land).

Convention Concerning Indigenous and Tribal Peoples in Independent Countries 1989.

Convention Concerning the Protection of the World Cultural and Natural Heritage 1972.

Convention for the Conservation of Antarctic Marine Living Resources 1980.

Convention for the Protection of the Marine Environment of the North-East Atlantic 1992.

Convention on Access to Information, Public Participation in Decision-Making and Access to Justice in Environmental Matters 1998.

Convention on Long-Range Transboundary Air Pollution 1979.

Convention on the Conservation of European Wildlife and Natural Habitats 1979.

Convention on the Conservation of Migratory Species of Wild Animals 1979.

Convention on the Continental Shelf 1958.

Convention on the Law of Non-navigational Uses of International Watercourses 1997.

Convention on the Settlement of Investment Disputes between States and Nationals of Other States 1965.

Declaration of the United Nations Conference on Environment and Development 1992.

Declaration of the United Nations Conference on the Human Environment 1972.

Declaration on the Establishment of a New International Economic Order 1974.

Declaration on the Granting of Independence to Colonial Countries and Peoples 1960.

Declaration on the Progressive Development of Principles of Public International Law relating to a New International Economic Order 1986.

Declaration on the Right to Development 1986.

Draft OECD Convention on the Protection of Foreign Property 1967.

Earth Charter 2000.

Energy Charter Protocol on Energy Efficiency and Related Environmental Aspects of the Energy Charter Treaty 1994.

Energy Charter Treaty 1994.

European Convention on Human Rights 1950.

European Convention on Human Rights 1950.

Final Act of the United Nations Conference on Trade and Employment 1948.

Helsinki Rules on the Uses of the Waters of International Rivers 1966.

International Convention for the Regulation of Whaling 1946.

International Covenant on Civil and Political Rights 1966.

International Covenant on Economic, Social and Cultural Rights 1966.

International Tropical Timber Agreement 2006.

Johannesburg Declaration on Sustainable Development 2002.

Johannesburg Principles on the Role of Law and Sustainable Development 2002.

Kyoto Protocol 1997.

Lomé IV Convention 1995.

Mineral and Petroleum Resources Development Act (South Africa) 2002.

Montevideo Convention on Rights and Duties of States 1933.

Montreal Protocol on Substances that Deplete the Ozone Layer 1985.

Noordwijk Declaration on Atmospheric Pollution and Climatic Change 1989.

North American Free Trade Agreement 1991.

Oslo Principles on Global Climate Change 2015.

Paris Agreement 2015.

Plan of Implementation of the World Summit on Sustainable Development 2002.

Protocol on Environmental Protection to the Antarctic Treaty 1991.

Tokyo Convention for the High Seas Fisheries of the North Pacific Ocean 1952.

Treaty for Amazonian Cooperation 1978.

Unified Agreement for the Investment of Arab Capital in the Arab States was signed in 1980 .

Unified Agreement for the Investment of Arab Capital in the Arab States 1980.

United Nations Convention on Biological Diversity 1992.

United Nations Convention on the Law of the Sea 1982.

United Nations Convention to Combat Desertification 1994.

United Nations Declaration on the Rights of Indigenous Peoples 2007.

United Nations Framework Convention on Climate Change 1992.

Universal Declaration of Human Rights 1948.

Vienna Convention for the Protection of the Ozone Layer 1985.

Vienna Convention on Succession of States in Respect of State Property, Archives and Debts 1983.

Vienna Convention on Succession of States in Respect of Treaties 1978.

World Charter for Nature 1982.

Seminars, lectures and presentations

Frick C Direct Foreign Investment and the Environment: African Mining Sector paper presented at conference on Foreign direct investment and the environment: Lessons to be learned from the mining sector hosted by the OECD Global Forum on International Investment at the OECD Headquarters, Paris, France (07-02-2002 to 08-02-2002).

Knight J Does economic growth raise happiness in China? seminar at the CGW Shumann Building, Stellenbosch, South Africa (14-03-2011).

Manji F Chinese Investment in Africa: A Blessing or a Curse? (2008) paper presented at conference on China in Africa: Reflections on China's Growing Economic Influence in Africa hosted by the Africa Center for Information and Development, Oslo, Norway (22-10-2008).:

Pring G International Environmental and Human Rights Law Affecting Mining Reform (2008) paper presented at the Mining Law Seminar hosted by the Northern Institute for Environmental and Minority Law Arctic Centre, University of Lapland, Rovaniemi, Finland (25-09-2008 to 26-09-2008).

Tilton JE Is Mineral Depletion a Threat to Sustainable Mining? paper presented at the International Conference on Sustainable Mining hosted at Santiago de Compostela, Spain (17-04-2009).

Yining L Lessons from China's Economic Development (09-06-2011) at the JS Gericke Library Auditorium, Stellenbosch, South Africa.

UNGA resolutions and other resolutions or decisions

African Union resolution 153 (on Climate Change and Human Rights and the Need to Study its Impact in Africa) of 25 November 2009.

OECD "Application of the Polluter-Pays Principle to Accidental Pollution" (adopted 7 July 1989) [c(89)88].

OECD "Guiding Principles concerning International Economic Aspects of Environmental Policies" (adopted 26 May 1972) [C(72)128].

OECD "Implementation of the Polluter-Pays Principle" (adopted 14 November 1974) [C(74)223].

UNGA resolution: 1314 (XIII) of 12 December 1958.

UNGA resolution: 1514 (XV) of 14 December 1960.

UNGA resolution: 1515 (XV) of 15 December 1960.

UNGA resolution: 44/228 of 22 December 1989.

UNGA resolution: 36/6 of 27 October 1981.

UNGA resolution : 2995 (XXVII) of 15 December 1972.

UNGA resolution: 44/207 of 22 December 1989.

UNGA resolution: 46/169 of 19 December 1991.

UNGA resolution*:* 1831 (XVII) of 18 December 1962.
UNGA resolution*:* 1803 (XVII) of 14 December 1962.
UNGA resolution*:* 2145 (XXI) of 27 October 1966.
UNGA resolution*:* 2158 (XXI) of 25 November 1966.
UNGA resolution*:* 2248 (S-V) of 19 May 1967.
UNGA resolution*:* 2372 (XXII) of 12 June 1968.
UNGA resolution*:* 2386 (XXIII) of 19 December 1968.
UNGA resolution*:* 2574 (XXIV) of 15 December 1969.
UNGA resolution*:* 2625 (XXV) of 24 October 1970.
UNGA resolution*:* 2692 (XXV) of 11 December 1970.
UNGA resolution*:* 3005 (XXVII) of 15 December 1972.
UNGA resolution*:* 3016 (XXVII) of 18 December 1972.
UNGA resolution*:* 3021 (S-VI) of 1 May 1974.
UNGA resolution*:* 3037 (XXVII) of 19 December 1972.
UNGA resolution*:* 3082 (XXVIII) of 6 December 1973.
UNGA resolution*:* 3129 (XXVIII) of 13 December 1973.
UNGA resolution*:* 3171 (XXVIII) of 17 December 1973.
UNGA resolution*:* 3175 (XXVIII) of 17 December 1973.
UNGA resolution*:* 31/98 of 15 December 1976.
UNGA resolution*:* 32/161 of 19 December 1977.
UNGA resolution*:* 3201 (S – VI) of 1 May 1974.
UNGA resolution*:* 3281 (XXIX) of 12 December 1974.
UNGA resolution*:* 3326 (XXIX) of 16 December 1974.
UNGA resolution*:* 3295 (XXIX) of 13 December 1974.
UNGA resolution*:* 33/182 A of 21 December 1978.
UNGA resolution*:* 33/182 C of 21 December 1978.
UNGA resolution*:* 33/40 of 13 December 1978.
UNGA resolution*:* 3336 (XXIX) of 17 December 1974.
UNGA resolution*:* 34/136 of 14 December 1979.
UNGA resolution*:* 34/186 of 18 December 1979.
UNGA resolution*:* 3486 (XXX) of 12 December 1975.
UNGA resolution*:* 35/110 of 5 December 1980.
UNGA resolution*:* 3516 (XXX) of 15 December 1975.
UNGA resolution*:* 35/7 of 30 October 1980.
UNGA resolution*:* 36/173 of 17 December 1981.
UNGA resolution*:* 37/135 of 17 December 1982.
UNGA resolution*:* 37/7 of 28 October 1982.
UNGA resolution*:* 38/144 of 19 December 1983.
UNGA resolution*:* 41/128 of 4 December 1986.

UNGA resolution: 43/53 of 6 December 1988.

UNGA resolution: 45/212 of 6 December 1990.

UNGA resolution: 45/94 of 14 December 1990.

UNGA resolution: 523 (VI) of 12 January 1952.

UNGA resolution: 51/229 of 21 May 1997.

UNGA resolution: 53/196 of 22 February 1999.

UNGA resolution: 54/230 of 22 February 2000.

UNGA resolution: 55/209 of 15 February 2001.

UNGA resolution: 56/204 of 21 February 2002.

UNGA resolution: 57/269 of 5 March 2003.

UNGA resolution: 58/229 of 25 February 2004.

UNGA resolution: 59/251 of 7 March 2005.

UNGA resolution: 60/183 of 31 January 2006.

UNGA resolution: 61/184 of 25 January 2007.

UNGA resolution: 61/295 of 13 September 2007.

UNGA resolution: 62/181 of 31 January 2008.

UNGA resolution: 626 (VII) of 21 December 1952.

UNGA resolution: 63/201 of 28 January 2009.

UNGA resolution: 637 (VII) of 16 December 1952.

UNGA resolution: 64/185 of 29 January 2010.

UNGA resolution: 65/179 of 22 March 2011.

UNGA resolution: 66/225 of 25 March 2012.

UNGA resolution: 742 (VIII) of 27 November 1953.

UNGA resolution: 837 (IX) of 14 December 1954.

UNGA resolution: 88 (XII) of 19 October 1973.

Online media

African Development Bank Group "Natural Resources for Sustainable Development in Africa" (2007) (1-201) African Development Report <http://www.afdb.org/fileadmin/uploads/afdb/Documents/Publications/African%20Development%20Report%202007.PDF> (accessed 14-09-2016).

Ayre J "Belgian Citizen Coalition Pursuing Legal Case Against Government Over Inaction on Climate Change" (04-05-2015) (page numbers unavailable) Clean Technica <https://cleantechnica.com/2015/05/04/belgian-citizens-coalition-pursuing-legal-case-government-inaction-climate-change/> (accessed 08-09-2016).

Business & Human Rights Resources Centre "Binding treaty" (2011) (page numbers unavailable) Business & Human Rights Resources Centre <https://business-humanrights.org/en/binding-treaty (accessed 11-08-2016).

Business & Human Rights Resources Centre "UN Guiding Principles" (2011) (page numbers unavailable) Business & Human Rights Resources Centre <https://business-humanrights.org/en/un-guiding-principles> (accessed 11-08-2016).

Carbon Offset Research & Education "Radiative Forcing" (2011) (page numbers unavailable) Carbon Offset Research & Education <http://www.co2offsetresearch.org/aviation/RF.html> (accessed 15-05-2015).

Center for Climate and Energy Solutions "Coal" (date of publication unavailable) (page numbers unavailable) Center for Climate and Energy Solutions <http://www.c2es.org/energy/source/coal> (accessed 10-07-2016).

Center for Climate and Energy Solutions "Outcomes of the U.N. Climate Change Conference in Lima" (date of publication unavailable) (page numbers unavailable) Center for Climate and Energy Solutions <http://www.c2es.org/international/negotiations/cop-19/summary> (accessed 04-05-2015).

Center for Climate and Energy Studies "Submitted Intended Nationally Determined Contributions (INDCS)" (date of publication unavailable) (page numbers unavailable) Center for Climate and Energy Studies <http://www.c2es.org/international/2015-agreement/indcs> (accessed 20-07-2016).

CESCR "General Comment No. 15: The Right to Water (arts. 11 and 12 of the International Covenant on Economic, Social and Cultural Rights)" (20 January 2003) (4) <http://www2.ohchr.org/english/issues/water/docs/CESCR_GC_15.pdf> (accessed 08-09-2016).

Clark D "Has the Kyoto protocol made any difference to carbon emissions?" (26-11-2015) (page numbers unavailable) The Guardian <http://www.theguardian.com/environment/blog/2012/nov/26/kyoto-protocol-carbon-emissions> (accessed 03-06-2015).

Council on Foreign Relations "The Climate Change Regime" (19-06-2013) (page numbers unavailable) Council on Foreign Relations <http://www.cfr.org/climate-change/global-climate-change-regime/p21831> (accessed 17-04-2015).

EC "Development and Relations with African, Caribbean and Pacific States" (date of publication unavailable) EC (page numbers unavailable) <http://ec.europa.eu/development/body/cotonou/lome_history_en.htm> (accessed 07-10-2015).

EC "The Aarhus Convention" (07-05-2015) (page numbers unavailable) EC <http://ec.europa.eu/environment/aarhus/> (accessed 03-09-2015).

Editors of Encyclopaedia Britannica "Biochemical cycle" (04-01-2015) (page numbers unavailable) Encyclopaedia Britannica <https://global.britannica.com/science/biogeochemical-cycle> (accessed 19-07-2016).

EUR-Lex "The precautionary principle" (21-09-2015) (page numbers unavailable) EUR-Lex <http://eur-lex.europa.eu/legal-content/EN/TXT/?uri=URISERV%3Al32042> (accessed 03-09-2016).

Foundation for International Environmental Law and Development "Final Report: Options for EC Emission Trading" (22-02-2000) (11) Foundation for International Environmental Law and Development <http://ec.europa.eu/environment/archives/docum/pdf/0087_field.pdf> (accessed 07-07-2016).

Gallucci M & P Horn "For U.S and China, World's Biggest Climate Polluters, It's Still Business as Usual" (04-02-2013) (page numbers unavailable) Inside Climate News <http://insideclimatenews.org/news/20130129/united-states-china-carbon-greenhouse-gas-emissions-renewable-energy-coal-plants-pollution-global-warming-climate> (accessed 14-04-2015).

Gelis S "Kyoto Protocol, 10 years later: Did deal to combat greenhouse emissions work and what of its future?" (14-02-2015) (page numbers unavailable) National Post <http://news.nationalpost.com/news/world/kyoto-protocol-10-years-later-was-the-deal-to-combat-greenhouse-emissions-successful-and-what-of-its-future> (accessed 08-07-2016).

German Federal Ministry of Education and Research "Rapid Climate Change in the Arctic, Polar Research as a Global Responsibility" (2015) (5) Bundesministerium für Bildung und Forschung <https://www.bmbf.de/pub/Rapid_Climate_Change_in_the_Artic.pdf> (accessed 07-08-2016).

Global Energy Statistical Yearbook 2016 "Coal and Ignite domestic consumption" (2016) (page numbers unavailable) Enerdata <https://yearbook.enerdata.net/#coal-and-lignite-world-consumption.html> (accessed 04-07-2016).

Global Justice Program "The Oslo Principles on Global Climate Change Obligations Launched by Expert Group Including GJP Director Thomas Pogge" (date of publication unavailable) (page numbers unavailable) Yale Global Justice Group <http://www.yale.org/macmillan/globaljustice/news.html> (accessed 08-09-2016).

Gqada I "Mining Rights versus Community Rights" (01-06-2011) (page numbers unavailable) The Times <http://www.saiia.org.za/governance-of-africa-s-resources-opinion/mining-rights-versus-community-rights.html > (accessed 10-06-2011).

Greenpeace International "People from the Philippines and Pacific Island nations sign declaration to hold big polluters accountable" (09-06-2015) (page numbers unavailable) Greenpeace International <http://www.greenpeace.org/international/Global/international/briefings/climate/2015/People's%20Declaration%20for%20Climate%20Justice.pdf> (accessed 30-06-2015).

Greenstone M "Surprisingly, a Voluntary Climate Treaty Could Actually Work" (13-02-2015) (page numbers unavailable) The New York Times <http://www.nytimes.com/2015/02/15/upshot/surprisingly-a-voluntary-climate-treaty-could-actually-work.html?abt=0002&abg=0> (accessed 05-05-2015).

Hope A "Klimaatzaak takes governments to court over climate policies" (28-04-2015) (page numbers unavailable) Flanders Today <http://www.flanderstoday.eu/politics/klimaatzaak-takes-governments-court-over-climate-policies> (accessed 08-09-2016).

Human Rights Committee "Concluding observations of the Human Rights Committee: Canada" (07-04-1999) (page numbers unavailable) CAID <http://caid.ca/UNHRCCan1999.pdf> (accessed 07-09-2016).

ICC "2012 ICC Guidelines for International Investment" (04-05-2012) (7) ICC <http://www.iccwbo.org/Advocacy-Codes-and-Rules/Document-centre/2012/2012-ICC-Guidelines-for-International-Investment/> (accessed 02-10-2015).

ILA "The Legal Principles relating to Climate Change" (date of publication unavailable) (page numbers unavailable) ILA <http://www.ila-hq.org/en/committees/index.cfm/cid/1029> (accessed 11-05-2015).

International Commission of Jurists "The Events in East Pakistan, 1971" (1972) (page numbers unavailable) International Commission of Jurists <http://icj.wpengine.netdna-cdn.com/wp-content/uploads/1972/06/Bangladesh-events-East-Pakistan-1971-thematic-report-1972-eng.pdf> (accessed 24-02-2015).

IPCC "Adaptation and mitigation options" (date of publication unavailable) (page numbers unavailable) IPCC <https://www.ipcc.ch/publications_and_data/ar4/syr/en/spms4.html> (accessed 07-07-2016).

IPCC "Organization" (date of publication unavailable) (page numbers unavailable) IPCC <http://www.ipcc.ch/organization/organization.shtml#.URKQ3KW5clI> (accessed 06-02-2013).

Klein J "July 2015 Update to Climate Litigation Charts" (07-07-2015) (page numbers unavailable) Columbia Law School <http://blogs.law.columbia.edu/climatechange/2015/07/07/july-2015-update-to-climate-litigation-charts/> (accessed 11-09-2016).

Klimaatzaak "Press Release" (date of publication unavailable) (page numbers unavailable) Klimaatzaak <http://www.klimaatzaak.eu/en/wp-content/uploads/2014/11/Klimaatzaak_PERS_EN.pdf> (accessed 21-04-2015).

LDC News Service "Final Paris Agreement: A nail in the coffin for justice for LDC, says LDC Watch" (13-12-2015) (page number unavailable) LDC News Service <http://ldcnews.com/final-paris-agreement-nail-coffin-justice-ldcs-says-ldc-watch/> (accessed 06-07-2016).

Maake M "Mining has become a curse to SADC" (15-06-2011) (page numbers unavailable) Business Live <http://www.businesslive.co.za/Feeds/inet/2011/06/15/mining-has-become-a-curse-to-sadc> (accessed 21-06-2011).

Macaulay C & PR Hensel "Natural Resources and Territorial Conflict" (March 2014) (14) Paul Hensel <http://www.paulhensel.org/Research/isa14.pdf> (accessed 02-03-2015).

Minnesma M "Hague climate change verdict: 'Not just a legal process but a process of hope'" (25-06-2015) (page numbers unavailable) The Guardian <http://www.theguardian.com/global-development-professionals-network/2015/jun/25/hague-climate-change-verdict-marjan-minnesma> (accessed 30-06-2015).

National Snow & Ice Data Center "What is the Cryosphere?" (date of publication unavailable) (page numbers unavailable) National Snow & Ice Data Center <https://nsidc.org/cryosphere/what-is-the-cryosphere.html> (accessed 08-04-2015).

Neslen A "Dutch government ordered to cut carbon emissions in landmark ruling" (24-06-2015) (page numbers unavailable) The Guardian <http://www.theguardian.com/environment/2015/jun/24/dutch-government-ordered-cut-carbon-emissions-landmark-ruling> (accessed 30-06-2015).

OECD "OECD Guidelines for Multinational Enterprises" (date of publication unavailable) (page numbers unavailable) OECD <http://www.oecd.org/corporate/mne/oecdguidelinesformultinationalenterprises.htm> (accessed 02-10-2015).

Office of the Press Secretary "FACT SHEET: U.S.-China Joint Announcement on Climate Change and Clean Energy Cooperation" (11-11-2014) (page numbers unavailable) The Whitehouse <https://www.whitehouse.gov/the-press-office/2014/11/11/fact-sheet-us-china-joint-announcement-climate-change-and-clean-energy-c> (accessed 10-07-2016).

Oslo Principles on Global Climate Change Obligations "Oslo Principles Commentary" (06-04-2015) (8) Oslo Principles on Global Climate Change Obligations <http://www.osloprinciples.org/macmillan/globaljustice/Oslo%20Principles%20Commentary.pdf> (accessed 14-08-2016).

Page S "No, The Paris Agreement Isn't Binding. Here's Why that Doesn't Matter" (14-12-2015) (page numbers unavailable) Climate Progress <http://thinkprogress.org/climate/2015/12/14/3731715/paris-agreement-is-an-actual-agreement/> (accessed 09-07-2016).

Powles J & T Khan "Climate change: at last a breakthrough to our catastrophic political impasse?" (30-03-2015) (page numbers unavailable) The Guardian <http://www.theguardian.com/commentisfree/2015/mar/30/climate-change-paris-talks-oslo-principles-legal-obligations> (accessed 21-04-2015).

Richardson M "China's Green Ambition, US Sees Red" (05-01-2011) (Page numbers unavailable) Yale Global <http://yaleglobal.yale.edu/content/chinas-green-ambition-us-sees-red> (accessed 10-06-2011).

Rigg K "Landmark Dutch Lawsuit Puts Governments Around the World on Notice" (04-12-2015) (page numbers unavailable) Huff Post Green <http://www.huffingtonpost.com/kelly-rigg/landmark-dutch-lawsuit-pu_b_7025126.html> (accessed 21-04-2015).

Rosenthal E "Race is On as Ice Melt Reveals Arctic Treasures" (18-09-2012) (page numbers unavailable) The New York Times <http://www.nytimes.com/2012/09/19/science/earth/arctic-resources-exposed-by-warming-set-off-competition.html?_r=0> (accessed 07-08-2016).

Subsidiary Body for Scientific and Technological Advice "Compilation of economy-wide emission reduction targets to be implemented by Parties included in Annex I to the Convention" (4-15 June 2014) (page numbers unavailable) UNFCCC <http://unfccc.int/resource/docs/2014/sbsta/eng/inf06.pdf> (accessed 06-05-2015).

The Economist "A deal in Durban" (17-12-2011) (page numbers unavailable) The Economist <http://www.economist.com/node/21541806> (accessed 17-04-2015).

Think Global Green "China Overview" (2008) (page numbers unavailable) Think Global Green <http://www.thinkglobalgreen.org/CHINA.html> (accessed 14-04-2015).

U.S Energy Information Administration "International Energy Outlook 2016" (11-05-2016) (page numbers unavailable) U.S Energy Information Administration <http://www.eia.gov/forecasts/ieo/coal.cfm> (accessed: 04-07-2016).

UN "Record: 177 parties signed the Paris Agreement" (date of publication unavailable) (page numbers unavailable) Paris 2015, UN Climate Change Conference, COP 21, CMP 11 <http://www.cop21.gouv.fr/en/a-record-over-160-countries-expected-to-sign-the-paris-agreement-in-new-york-on-22-april-2016/> (accessed 09-07-2016).

UN "Report of the United Nations Water Conference" (14-25 March 1977) (53) UN <http://www.ircwash.org/sites/default/files/71UN77-161.6.pdf> (accessed 06-09-2016).

UN "Report of the United Nations Water Conference" (14-25 March 1977) (53) UN <http://www.ircwash.org/sites/default/files/71UN77-161.6.pdf> (accessed 06-09-2016).

UN "Towards a climate agreement" (date of publication unavailable) (page numbers unavailable) UN <http://www.un.org/climatechange/towards-a-climate-agreement/ > (accessed 02-07-2016).

UNCTAD "Definitions of FDI" (date unavailable) (page numbers unavailable) UNCTAD <http://unctad.org/en/Pages/DIAE/Definitions-of-FDI.aspx> (accessed 08-08-2016).

UNCTAD "Transnational corporations (TNC)" (date unavailable) (page numbers unavailable) UNCTAD <http://unctad.org/en/Pages/DIAE/Transnational-corporations -(TNC).aspx> (accessed 07-10-2015).

UNEP "Climate Change Mitigation" (date of publication unavailable) (page numbers unavailable) UNEP <http://www.unep.org/climatechange/mitigation/> (accessed 07-07-2016).

UNEP "Global Greenhouse Gas Emissions Data" (10-06-2016) (page numbers unavailable) UNEP <https://www3.epa.gov/climatechange/ghgemissions/global.html> (accessed 04-07-2016).

UNEP "Overview of Greenhouse Gases" (26-05-2016) (page numbers unavailable) EPA <https://www3.epa.gov/climatechange/ghgemissions/gases.html> (accessed 30-06-2016).

UNFCCC "Adaptation" (date of publication unavailable) (page numbers unavailable) UNFCCC <http://unfccc.int/adaptation/items/4159.php> (accessed 29-04-2013).

UNFCCC "ANNEX II Countries" (date unavailable) (page numbers unavailable) UNFCCC <http://unfccc.int/cop3/fccc/climate/annex2.htm> (accessed 06-09-2016).

UNFCCC "Clean Development Mechanism" (date of publication unavailable) (page numbers unavailable) <http://cdm.unfccc.int/index.html> (accessed 07-07-2016).

UNFCCC "Conference of the Parties (COP)" (date of publication unavailable) (page numbers unavailable) <http://unfccc.int/bodies/body/6383.php> (accessed 10-07-2016).

UNFCCC "FOCUS: Mitigation – NAMAs, Nationally Appropriate Mitigation Actions" (2014) (page numbers unavailable) UNFCCC <http://unfccc.int/focus/mitigation/items/7172.php> (accessed 06-05-2015).

UNFCCC "FOCUS: Mitigation" (2014) (page numbers unavailable) UNFCCC <http://unfccc.int/focus/mitigation/items/7169.php#intro> (accessed 06-05-2015).

UNFCCC "Green Climate Fund" (date of publication unavailable) (page numbers unavailable) UNFCCC <http://unfccc.int/cooperation_and_support/financial_mechanism/green_climate_fund/items/5869.php> (accessed 03-07-2016).

UNFCCC "International Emission Trading" (date of publication unavailable) (page numbers unavailable) UNFCCC <http://unfccc.int/kyoto_protocol/mechanisms/emissions_trading/items/2731.php> (accessed 07-07-2016).

UNFCCC "Kyoto Protocol, Compliance" (date of publication unavailable) (page numbers unavailable) UNFCCC <http://unfccc.int/kyoto_protocol/compliance/items/3024.php> (accessed: 29-04-2012).

UNFCCC "Kyoto Protocol, Emissions Trading" (date of publication unavailable) (page numbers unavailable) UNFCCC <http://unfccc.int/kyoto_protocol/mechanisms/emissions_trading/items/2731.php> (accessed: 29-04-2012).

UNFCCC "Kyoto Protocol, Reporting" (date of publication unavailable) (page numbers unavailable) UNFCCC <http://unfccc.int/kyoto_protocol/reporting/items/3879.php> (accessed: 29-04-2012).

UNFCCC "Kyoto Protocol" (date of publication unavailable) (page numbers unavailable) UNFCCC <http://unfccc.int/kyoto_protocol/items/2830.php> (accessed 29-04-2013).

UNFCCC "List of Annex I Parties to the Convention" (date unavailable) (page numbers unavailable) UNFCCC <http://unfccc.int/parties_and_observers/parties/annex_i/items/2774.php> (accessed 05-09-2016).

UNFCCC "Meetings" (date of publication unavailable) UNFCCC (page numbers unavailable) <http://unfccc.int/meetings/items/6240.php> (accessed 16-04-2013).

UNFCCC "Paris Agreement – Status of Ratification" (date of publication unavailable) (page numbers unavailable) UNFCCC <http://unfccc.int/paris_agreement/items/9444.php> (accessed 06-07-2016).

UNFCCC "Parties & Observers" (2014) (page numbers unavailable) UNFCCC <http://unfccc.int/parties_and_observers/items/2704.php> (accessed 05-09-2016).

UNFCCC "Project Search" (date of publication unavailable) (page numbers unavailable) UNFCCC <http://cdm.unfccc.int/Projects/projsearch.html> (accessed 07-07-2016).

UNFCCC "Status of Ratification of the Convention" (date of publication unavailable) (page numbers unavailable) UNFCCC <http://unfccc.int/essential_background/convention/status_of_ratification/items/2631.php > (accessed 30-06-2016).

UNFCCC "Status of Ratification of the Kyoto Protocol" (date of publication unavailable) (page numbers unavailable) UNFCCC <http://unfccc.int/kyoto_protocol/status_of_ratification/items/2613.php> (accessed 04-07-2016).

UNFCCC "The Kyoto Protocol Mechanisms" (2010) (7) UNFCCC <https://cdm.unfccc.int/about/cdm_kpm.pdf> (accessed 07-07-2016) 7.

UNFCCC "The Noordwijk Ministerial Declaration on climate change" (01-05-1993) (page numbers unavailable) UNFCCC <http://unfccc.int/resource/ccsites/senegal/fact/fs218.htm> (accessed 02-07-2016).

UNFCCC "The Paris Agreement" (date of publication unavailable) (page numbers unavailable) UNFCCC <http://bigpicture.unfccc.int/#content-the-paris-agreemen> (accessed 06-07-2016).

UNFCCC "The Paris Agreement" (date of publication unavailable) (page numbers unavailable) UNFCCC <http://bigpicture.unfccc.int/#content-the-paris-agreemen> (accessed 06-07-2016).

UNGA "Annotations on the text of the draft International Covenants on Human Rights" (1955) (15) UNGA <http://www2.ohchr.org/english/issues/opinion/articles 1920_iccpr/docs/A-2929.pdf> (accessed 12-02-2015).

United Nations Guiding Principles on Business and Human Rights at: Business & Human Rights Resources Centre "UN Guiding Principles" (2011) (page numbers unavailable) Business & Human Rights Resources Centre <https://business-humanrig hts.org/en/un-guiding-principles> (accessed 11-08-2016).

United Nations Office at Geneva "The League of Nations (1919-1946)" (date of publication unavailable) (page numbers unavailable) Nations Office at Geneva <http://w ww.unog.ch/80256EDD006AC19C/%28httpPages%29/17C8E6BCE10E3F4F8025 6EF30037D733?OpenDocument> (accessed 05-03-2015).

United States Environmental Protection Agency "Global Greenhouse Gas Emissions Data" (10-06-2016) (page numbers unavailable) United States Environmental Protection Agency <https://www3.epa.gov/climatechange/ghgemissions/global.html> (accessed 04-07-2016).

Urgenda "Climate case" (date of publication unavailable) (page numbers unavailable) Urgenda <http://www.urgenda.nl/en/climate-case/> (accessed 21-04-2015).

Urgenda "Landmark Climate Change Lawsuit" (date of publication not available) (page numbers unavailable) Urgenda <http://www.urgenda.nl/documents/Press_rele ase_Urgenda_climate_case_hearing_14_april_2015_FIN.pdf> (accessed 21-04-2015).

Urgenda "Letter to the Government" (12-11-2012) (page numbers unavailable) Urgenda <http://www.urgenda.nl/documents/Staat-der-NL-Engels1.pdf> (accessed 21-04-2015).

Urgenda "Urgenda wins the case for better Dutch climate policies" (date of publication unavailable) (page numbers unavailable) Urgenda <http://www.urgenda.nl/en/clima te-case/> (accessed 26-06-2015).

Vidal J & D Adam "China overtakes US as world's biggest CO2 emitter" (19-07-2007) (page numbers unavailable) The Guardian <http://www.theguardian.c om/environment/2007/jun/19/china.usnews> (accessed 14-09-2015).

WMO "First World Climate Conference" (date of publication unavailable) (page numbers unavailable) WMO <http://www.wmo.int/pages/themes/climate/international_ wcc.php> (accessed 20-05-2013).

World Bank "The economics of adaptation to climate change" (2010) 10 World Bank <http://siteresources.worldbank.org/EXTCC/Resources/EACC_FinalSynthesisRepo rt0803_2010.pdf> (accessed 08-05-2015).

Worldwatch Institute "State of the World 2005 Trends and Facts – Security Redefined" (30-04-2015) (page numbers unavailable) Worldwatch Institute <http://www .worldwatch.org/node/65> (accessed 30-04-2015).